To Fulfill a Vision

Jerusalem Einstein Centennial Symposium on

Gauge Theories and Unification of Physical Forces (1979: Israel Academy of Sciences).

Honorary Sponsors

Académie Royale des Sciences, des Lettres et des Beaux-Arts de Belgique

Accademia Nazionale dei Lincei

Committee on General Relativity and Gravitation

Deutsche Forschungsgemeinschaft

Institut de France—Académie des Sciences

The Institute for Advanced Study (Princeton)

International Council of Scientific Unions (ICSU)

International Union of the History and Philosophy of Science (IUHPS)

International Union of Pure and Applied Physics (IUPAP)

The Japan Academy

National Academy of Sciences (U.S.A.)

The Norwegian Academy of Science and Letters

Royal Danish Academy of Sciences and Letters

Royal Netherlands Academy of Arts and Sciences

Royal Swedish Academy of Sciences

The Royal Society of London

Swiss Academy of Sciences

Association for the Advancement of Science in Israel

Bar-Ilan University

Ben-Gurion University of the Negev

Council for Higher Education in Israel

Haifa University

Technion—Israel Institute of Technology

Tel Aviv University

Weizmann Institute of Science

To Fulfill a Vision

Jerusalem Einstein Centennial Symposium on

Gauge Theories and Unification of Physical Forces

Edited by

Yuval Ne'eman

Tel Aviv University

1981

Addison-Wesley Publishing Company, Inc.
Advanced Book Program
Reading, Massachusetts

London • Amsterdam • Don Mills, Ontario • Sydney • Tokyo

To Fulfill a Vision

Jerusalem Einstein Centennial Symposium on
Gauge Theories and Unification of Physical Forces

Lectures and Discussions on the latest unification theories
presented at The Israel Academy of Sciences and Humanities, Jerusalem,
March 20–23, 1979

Library of Congress Cataloging in Publication Data

Jerusalem Einstein Centennial Symposium on Gauge
  Theories and Unification of Physical Forces, Israel
  Academy of Sciences, 1979.
    To fulfill a vision

  Includes bibliographies and index.
  1. Unified field theories—Congresses. 2. Particles
(Nuclear physics)—Congresses. I. Ne'eman, Yuval.
II. Title.
QC173.68.J47 1979   530.1'42   80-20969

ISBN 0-201-05289-X

Copyright ©1981 by Addison-Wesley Publishing Company, Inc.

Published simultaneously in Canada

All rights reserved. No part of this publication may be reproduced, stored in a retrieval system, or transmitted, in any form or by any means, electronic, mechanical, photocopying, recording, or otherwise, without the prior written permission of the publisher, Addison-Wesley Publishing Company, Inc., Advanced Book Program, Reading, Massachusetts 01867, U.S.A.

Manufactured in the United States of America

ABCDEFGHIJ-HA-8987654321

Contents

	Contributors/Discussants	x
	Introduction	xv
I.	UNIFICATION: AIMS AND PRINCIPLES	1
	1. Chen Ning Yang, Geometry and Physics	3
	Notes	11
	2. Peter G. Bergmann, The Quest for Unity: General Relativity and Unitary Field Theories	12
	Discussion	20
	3. Feza Gürsey, Geometrization of Unified Fields	22
	1. Introduction	22
	2. The Neo-Einsteinian Doctrine	23
	3. Neo-Einsteinian Geometries	27
	4. Description of Unified Interactions by Some Exceptional Structures	29
	5. Concluding Remarks	33
	Notes	35
	Discussion	37
II.	GENERAL RELATIVITY IN THE LARGE	39
	Session Chairman's Remarks	41
	4. Jacob D. Bekenstein, Gravitation, the Quantum, and Statistical Physics	42
	Einstein and the Unity of Physics	42
	The Dawn of the New Developments	42
	Area and Entropy	43
	Black Hole Radiance	45
	Statistical Black Hole Physics	47
	Response of a Black Hole to Incident Radiation	48
	The Probability Distribution for Stimulated Emission	50
	The Einstein Coefficients for a Black Hole	52
	The Fundamental Transition Probability	53
	Black Hole Mass Distribution	54
	The Width Paradox	56
	Notes	58
	Discussion	60

5.	Nathan Rosen, Mach's Principle and the General Relativity Theory	63
	Notes	75
	Discussion	77

III. GAUGE THEORIES OF GRAVITY AND THEIR GENERALIZATIONS 79
Session Chairman's Remarks 81

6.	Daniel Z. Freedman, Unification in Supergravity	82
	Basics of Global Supersymmetry	83
	Irreducible Representations of Supersymmetry	85
	A Global Supersymmetric Field Theory	85
	Superspace	86
	Supergravity	88
	Unified $SO(N)$ Models	90
	Phenomenology	92
	Higher-Spin Gauge Fields	92
	Recent Work on $SO(8)$ Theory	93
	Renormalizability in Quantum Gravity	94
	Conclusion	96
	Notes	96
	Discussion	98
7.	Yuval Ne'eman, Gauged and Affine Quantum Gravity	99
	1. Motivation: The Search for Quantum Gravity	99
	2. The Poincaré and Affine Groups and Gauges	101
	3. The Equations of Motion	104
	4. The Affine Gauge	109
	Notes	112
	Discussion	113

IV. QUANTUM CHROMODYNAMICS 115
Session Chairman's Remarks 117

8.	Yoichiro Nambu, Quark Confinement: The Cases For and Against	118
	Notes	126
	Discussion	127
9.	G. 't Hooft, Beyond Perturbation Expansion	128
	Notes	133
	Discussion	134
10.	Roger Dashen, Toward a Theory of Hadron Structure	136
	Notes	137
	Discussion	138
11.	Harry J. Lipkin, How Can We Measure the Mass, Charge, and Magnetic Moment of Confined Quarks?	139

	1. Introduction	139
	2. Masses and Magnetic Moments	142
	3. Measurement of Colored Quark Charges Below Color Threshold	145
	Notes	153
	Discussion	155
V.	FLAVOR AND ASTHENODYNAMICS	157
	Notes	158
	Session Chairman's Remarks	159
	12. Sheldon L. Glashow, Old and New Directions in Elementary Particle Physics	160
	1. Atomic Synthesis	160
	2. False Synthesis	161
	3. New Synthesis	162
	4. Perverse Directions	164
	5. Groups of Forces; Families of Fermions	165
	6. Unity	166
	Discussion	170
	13. Haim Harari, Quarks and Leptons: The Generation Puzzle	171
	1. Introduction: "Standard Wisdom"	171
	2. The Generation Pattern: Unlikely Alternatives	172
	3. The Electroweak Group: Interesting Extensions of $SU(2) \times U(1)$	172
	4. Simple Unification: An Attractive Idea That Does Not Work	173
	5. The Strong Group: Unlikely Alternatives to $SU(3)$	174
	6. Grand Unification: A Possible Quark-Lepton Connection	175
	7. Calculating $\sin^2\theta_W$: The Only Available Test of Grand Unification	176
	8. What Identifies a Generation?	177
	9. Quark Masses and Cabibbo Angles: The Framework	178
	10. An Interesting Quark Mass-Matrix	179
	11. Some Open Questions and Some Prejudices	182
	Notes	183
	Discussion	184
	14. R. Gatto, The Mass Matrix	185
	1. Symmetry and Quark Masses	185
	2. Naturality Groups	186
	3. The Quark Coupling Matrix	189
	4. A Theorem for $SU(2)_L \times SU(2)_R \times U(1)$	192
	5. Implications of Flavor Conservation in the Extended $SU(2) \times U(1)$ Model	193
	Notes	195
VI.	STRONG-WEAK-ELECTROMAGNETIC UNIFICATION IDEAS	197
	Chairman's Remarks	198
	15. Steven Weinberg, Grand Unification	200
	1. Introduction	200

2. Calculation of M and $\sin^2\theta$		202
3. Stages of Symmetry-Breaking		206
4. Observable Hyperweak Interactions		208
5. Mechanisms for Baryon Nonconservation		210
6. Cosmological Baryon Production		213
7. Conclusion		217
Notes		217
Session Chairman's Remarks		219

16. Jogesh C. Pati, New Physics with Grand Unification 221
    1. Introduction 221
    2. The General Characteristics.. 223
    3. The Important Details 227
       3.1. The Heaviest Gauge Mass, the Grand Unifying Mass Scale, and
            the Weak Angle.. 227
       3.2. The Manner of Violations of B, L, and F.......................... 231
            3.2.1. Practical similarities and differences between the two
                   schemes: Proton decay.................................... 233
            3.2.2. Difference between the two schemes at high energies and
                   high temperatures.. 233
            3.2.3. The possibility of a stable proton? 234
       3.3. The Question of Quark Charges 234

    4. Decaying Integer-Charge Quarks and the Unstable Proton................ 237
       4.1. The Mechanisms for B,L-Violating Decays of Integer-Charge Quarks 237
       4.2. Baryon and Lepton Number-Conserving Quark Decay Modes 240
       4.3. Lifetimes of Liberated Up and Down Quarks..................... 241
       4.4. Quarkosynthesis with Confined or Liberated Quarks: Decay Modes
            and Lifetimes of Cosmological Integer-Charge Quarks 243
            4.4.1. Quarkosynthesis.. 244
            4.4.2. Decay modes and lifetimes of cosmological quarks 245
       4.5. The Problem of Baryon Excess.................................... 246
       4.6. The Unstable Proton 248
            4.6.1. Decay modes... 249

    5. Summary and Questions 251
       5.1. The Decay of the Proton 251
       5.2. The Absolute Universality of Quarks and Leptons and Their Forces . 251
       Notes 252
       Discussion 255

17. Murray Gell-Mann, Summary 257
    Chairman's Concluding Remarks 265

List of Participants 267
Author Index 271
Subject Index 277

Contributors/Discussants	Contribution pp.	Discussion pp.
Y. Achiman *University of Wuppertal*		170, 184
R. Bacher *California Institute of Technology*		159
Jacob D. Bekenstein *Ben-Gurion University*	42–59	61, 62
Peter G. Bergmann *Syracuse University*	12–19	20–21
K. Bleuler *University of Bonn*		138
C. Cooper *Technion–Israel Institute of Technology*		156
J. A. de Azcarraga *University of Valencia and Salamanca*		37
Sybren R. de Groot *Royal Netherlands Academy of Arts and Sciences*		81
Roger Dashen *The Institute for Advanced Study, Princeton*	136–137	138
Daniel Z. Freedman *State University of New York at Stony Brook*	82–97	98
P. G. O. Freund *Enrico Fermi Institute, University of Chicago*		98
R. Gatto *University of Geneva*	185–196	155
J. Geheniau *Université Libre de Bruxelles*		41
Murray Gell-Mann *California Institute of Technology*	257–264	

Contributors/Discussants

Contributors/Discussants	Contribution pp.	Discussion pp.
Sheldon L. Glashow *Harvard University*	160–169	134, 170
M. Goldhaber *Brookhaven National Laboratory*		117, 127, 134, 155
Feza Gürsey *Yale University*	22–36	37–38
Haim Harari *Weizmann Institute*	171–183	134, 184
G. Horwitz *Hebrew University of Jerusalem*		61
Alfred Kastler *Académie des Sciences; Institut de France*		77–78, 198–199, 219–226, 255, 265
S. Lindenbaum *City University of New York*		113, 156
Harry J. Lipkin *Weizmann Institute*	139–154	113, 127, 155, 156, 255
Yuval Ne'eman *Tel Aviv University*	99–112	20, 61, 98, 113–114, 135
H. Pagels *New York Academy of Sciences*		138, 255
A. Pais *Rockefeller University*		20
Jogesh C. Pati *University of Maryland*	221–254	255
Yoichiro Nambu *University of Chicago*	118–126	127
Nathan Rosen *Technion-Israel Institute of Technology*	63–76	77

| **Contributors/Discussants** | Contribution pp. | Discussion pp. |

D. Sciama
Oxford University
60, 61

I. Segal
Massachusetts Institute of Technology
60–61, 113, 134

G. Tauber
Tel-Aviv University
77

Gerald t'Hooft
State University of Utrecht
128–133
134, 135, 184

Steven Weinberg
Harvard University
200–218
155

J. A. Wheeler
University of Texas at Austin
60, 77

Chen Ning Yang
State University of New York at Stony Brook
3–11

Introduction: The Jerusalem Einstein Centennial Symposium

INTRODUCTION: THE JERUSALEM EINSTEIN CENTENNIAL SYMPOSIUM

Yuval Ne'eman

The Jerusalem Einstein Centennial Symposium was held March 14–23, 1979. It was organized by the Israel Academy of Sciences and Humanities, the Hebrew University of Jerusalem (inaugurated by Einstein in 1923), the Van Leer Jerusalem Foundation, the Jerusalem Foundation and the Aspen Institute for Humanistic Studies. The sponsors (see the complete list on an adjacent page) included the International Council of Scientific Unions (ICSU), the International Union of Pure and Applied Physics (IUPAP), the International Union of the History and Philosophy of Science (IUHPS), and the Committee on General Relativity and Gravitation (GRG), an independent professional union of Einstein's most direct spiritual inheritors, affiliated to IUPAP. Professor L. Sosnowski, president of IUPAP, and Professor P. G. Bergmann, president of GRG, honored the meeting with their presence. The list also included twelve national academies of countries connected with Einstein's biography or heritage, the Institute for Advanced Study at Princeton, two Israeli national scientific bodies, and all Israeli universities and institutes of higher learning.

The linkage with the Institute for Advanced Study at Princeton was a particularly important one. Well in advance, the program was coordinated with that of the Princeton Einstein Centennial Symposium. It was thought that although a very large number of symposia would be held to honor Einstein in 1979, the Princeton and Jerusalem celebrations should try to present the most compact message. They should aim both at a retrospective view of Einstein's achievement and at an assessment of present-day physics. The latter would probably correspond to what Einstein himself would have preferred, even though he should have concurred with the first part, having written: "A day of celebration is in the first place dedicated to retrospect, especially to the memory of personages who have gained special distinction for the development of the cultural life. This friendly service for our predecessors must indeed not be neglected, particularly as such a memory of the best of the past is proper to stimulate the well-disposed of today to a courageous effort."[1]

The Jerusalem symposium dealt with two specific aspects of that program. In the historical part, rather than centering on the biographical, or on the evolution of Einstein's scientific ideas, which would be covered by the Princeton meeting, it was decided to study *the impact of Einstein's ideas on the age*. This included the political culture, linguistics, psychoanalysis, philosophy, art, sociology, and international security. A second subject that was considered naturally appropriate

---

Yuval Ne'eman (ed.), To Fulfill A Vision: Jerusalem Einstein Centennial Symposium on Gauge Theories and Unification of Physical Forces
ISBN 0-201-05289-X

Copyright © 1981 by Addison-Wesley Publishing Company, Inc., Advanced Book Program.
All rights reserved. No part of this publication may be reproduced, stored in a retrieval system, or transmitted, in any form or by any means, electronic, mechanical photocopying, recording, or otherwise, without the prior permission of the publisher.

was Einstein's Jewish component, whether spiritual or political. Events in the Middle East developed so that the first formal peace agreement between Israel and an Arab nation (Egypt) was finalized while the symposium was taking place. This lent an added value to the discussion of Einstein's ideas. In this connection Einstein once wrote: "The pursuit of knowledge for its own sake, an almost fanatical love of justice, and the desire for personal independence — these are the features of the Jewish tradition which make me thank my stars that I belong to it."[2] He also said: "I believe that the two great Semitic peoples, each of which in its own way has made some valuable contribution to Western culture, have a great common future, and that, instead of facing each other with barren hostility and mutual distrust, they ought to support one another in their national and cultural aspirations and to seek ways of cooperation in mutual understanding."[3]

In the assessment of present-day physics, it was decided to concentrate on a review of the present efforts and partial successes in unification, whereas the emphasis at Princeton would be more on general relativity and cosmology. To deal with the unification program in its present realization, a highly distinguished group of theoretical physicists gathered in Jerusalem, including many of the most important contributors to the key advances since the early fifties, whose work had made the present progress possible. At the same time, many of the most active researchers of the seventies, directly involved in the present hottest issues, contributed to the vitality of the program. Einstein would have approved of this mix of contributors, most likely, for he once observed: "In science . . . the work of the individual is so bound up with that of his scientific predecessors and contemporaries that it appears almost as an impersonal product of his generation."[4]

Due to various circumstantial developments, it was decided to publish the Jerusalem proceedings in two separate volumes. Hence the physics program, dealing with unification, is published by the Addison-Wesley Advanced Book Program as a twin volume to the Princeton proceedings, thereby providing the interested reader with a scientific coverage going from present-day relativity (Princeton) to today's particle physics (Jerusalem). The humanistic program is published separately.

The present volume is thus a review of the theory of particles and fields as of 1979, attempting a de facto realization of Einstein's unification program. To introduce the subject matter I found it useful to include two of my previously published general papers. The first was originally an invited review of Einstein's achievement in the light of present-day results.[5] It is annotated here so as to relate to the various contributions in this volume and in the Princeton proceedings. The second, another invited review, fills the gap between Einstein's physics and the present in the domain of the fundamental interactions.[6] They are incorporated as separate sections within this introduction. The level of the second is (unavoidably) somewhat technical.

The physics part of the symposium came under the sponsorship of the Commission on Particles and Fields of IUPAP, which was separate from the general IUPAP sponsorship of the Conference. Part I of this volume (from sessions 2 and 11 of the symposium) serves as a general introduction, with P. G. Bergmann describing Einstein's own search for a unified field theory and C. N. Yang discussing the role of geometry in modern physics. F. Gursey points to Einstein's impact on geometry itself.

Part II deals with some aspects of general relativity in the large. Although it may seem to be outside the main thrust of this book, the connections between gravitation and statistical mechanics revealed in the works of Bekenstein and Hawking have been among the most intriguing new developments in general relativity, and Mach's principle is still one of the most elusive features of Einstein's theory and cosmology.

Parts III–VI deal with the modern gauge theories and unification. In Part III is discussed unification between gravity and other interactions, either through supergravity (see the chapter by D. Z. Freedman) or an affine or Poincaré gauge. Part IV discusses the new hypothesis of quantum chromodynamics (QCD) subsequent to 1973 and the discovery of asymptotic freedom, which itself required first the quantization of the Yang-Mills theory, achieved in 1970. It is a gauge theory of the strong interaction. It appears to fit observations at short range, but is also expected to confine quarks and gluons (completely — in QCD, or up to higher energies only — in the Pati-Salam theory described in the chapter by J. Pati). Most of Part IV is involved with the issue of confinement.

Part V discusses "Flavor Dynamics" (generalizations of the original $SU(3)$) and the weak-electromagnetic unification (asthenodynamics). S. L. Glashow's chapter is a general introduction, and Harari and Gatto each discuss actual predictions of the particle spectrum.

Finally, Part VI presents the case for a "grand" unification of strong, weak, and electromagnetic interactions. This is a very bold step and generally presupposes a "desert" (no structure) between 200 GeV and 10^{14} GeV. The topics are highly speculative, but theoretical physicists have been emboldened by the success of quantum asthenodynamics. In my introduction to Part V I suggest an alternative, for which there appears to be much more evidence at the kinematical (and static) level at least.

Coherence, abstraction, and personal involvement: Albert Einstein, physicist and humanist*

STATURE OF THE MAN

I can't help but feel deeply moved whenever I look at Herblock's cartoon in the *Washington Post* published on a certain day of April 1955. The cartoon is a wide-angle view of the universe, showing scores of planets floating by. One of these is special, carrying a sign that reads, "Albert Einstein lived here."

In the procession of minds that have led and shaped the progress of mankind since the beginning of recorded history, Albert Einstein does stand out uniquely. Herblock has replaced the earth's family album of two hundred historical generations, and only one Einstein, by a congruent set of two hundred Einsteinless planets less one. This poetic abstraction manages to convey a moving message, more subtle and more powerful than a direct statement. (Such mechanisms are being studied in the modern theory of linguistics.) In my view, the broader and deeper impact is due to the abstracted message's capability of evoking an abundant and resonant set of thought associations. In what will follow, we shall be able to assess and juxtapose the force of abstraction in the scientific context.

It is worthwhile providing the nonphysicist reader with some indication of the man's achievements in physics, justifying Herblock's cartoon-poem. Indeed, whatever Einstein's impact on the nonphysics world, be it pacifism, Zionism, nuclear preparedness, or nuclear disarmament, all world-important as they may be, his special place in history is due solely to his physics. The rest constituted an important part of his personality — that of an involved scientist — but its effectiveness was due first and foremost to his stature as a physicist. This is also true of his other intellectual pursuits, including philosophy and music. In the case of philosophy, it is true even

*Republished by permission, from Impact of Science on Society **29**, 17-25 (1979).

though that discipline's specific historical role as an ersatz science suffered major collapse at his touch (in the matter of the conception of time and of space-time) similar to that of the Church after its collision with Copernicus and Galileo. The Church — or religion — later wisely retreated into the domain of ethics, where it may be safe (for a while?) as a source of doctrine. (Of course, it has to be attentive to scientific and technological developments — birth control, biological engineering, and so forth — and try hard to find the best ethical answers; otherwise it will lose ethics to some more efficient humanism.) Likewise, post-Einsteinian philosophy has abandoned metaphysics and the direct weighing of natural concepts, leaving them to the relevant sciences and entrenching itself in epistemology and the study of the scientific method. In this field, it is doing well and has enriched our understanding. As to the rest of philosophy, logic has been reshaped after Cantor as a branch of mathematics. In the last century, politics has been replaced by the social sciences. Aesthetics and ethics are still there, in that region where the scientific method has not yet found its foothold.

Brownian Motion and Relativity

Had Einstein's contributions to physics been less outstanding in their originality, depth, and overall importance, they would still have put him in the first row, considering their number and versatility. They straddle most topics and fields that were being dealt with up to the 1930s. In theromodynamics he coinvented the modern statistical mechanics approach independently and simultaneously with W. Gibbs. In quantum theory he elaborated on Planck's idea of the quantum of action and relaunched the corpuscular theory of light in what later became its dual particle-wave nature. In this program, he conceived identical-particle statistics (simultaneously with S. N. Bose, but independently) and explained the photoelectric effect (which we deactivate and reactivate every time we cut the beam at the entrance to a bank or an elevator). These are just some examples of his less epoch-making work. Recently I was talking to a professor of material science about the modern "composite materials" that have become very important in aeronautics, for instance. It turns out that the formalism utilized in the study of these materials is based on the young Einstein's first paper, on Brownian motion (in both cases one is dealing with cores of material A embedded in a background matrix of material B).

Einstein's most important contributions come under the headings of the special and general theories of relativity. I shall discuss the epistemological and metaphysical lessons from these principles in detail in what will follow, but what is relevant to Einstein's personal achievement is the fact that both dealt with the fundamental properties of space and time, and modified them completely. As to the role of Einstein, it has been said that special relativity would have been discovered within a year by either Lorentz or Poincaré if Einstein had not done it. The subject was on the highest priority list of the scientific frontier; there was a negative experimental result to be explained (the failure of Michelson and Morley in their attempt to measure the earth's drift velocity); G. F. Fitzgerald and Lorentz had already calculated the "apparent" contraction of lengths in a moving coordinate frame; and Poincaré was aware of the mathematical group structure required from such transformations. The difficulty lay in the missing time dilations, which were the hardest conceptually. Einstein supplied them together with a change in the interpretation of the entire set of transformations. Rather than think of artificial "apparent" or "effective" changes in matter, due to internal collapse forces, he realized that these were just the specific properties of space-time, the geometrical manifold in which we exist. Instead of a Euclidean or Galilean structure, this manifold preserves the velocity of light (philosophers could never have guessed at this preferred role of the velocity of light, as against that of sound, for instance). Einstein's former teacher H. Minkowski soon gave the system its formal geometric characterization. Everything stems from the requirement that the speed of light be the same for all observers moving at constant velocities with respect to each other. Einstein's 1905 contribution (he was twenty-six in that year in which he launched a series of five articles of the highest-impact model) was thus most important, but it is plausible to assume that by 1908 or 1910 it would have been done by others.[7]

An Intellectual Pursuit

The story of Einstein's theory of gravitation (the general theory of relativity) is very different. He worked on it for ten years and published it in 1916. He had gotten some partial answers several years earlier and published them. He had been helped by Marcel Grossmann, a former fellow student, by that time a mathematician, and by David Hilbert, one of the greatest mathematicians of this century. Still, it was Einstein himself who was obstinately pursuing the goal and building his theory step by step. In the outcome, general relativity is probably the finest achievement ever produced by a scientific mind. Note that the urge to resolve the problem of adapting Newton's gravitation and mechanics to the requirements of special relativity was an entirely intellectual pursuit. There was no pressing experiment to be explained. The topic appeared to most scientists as too far-fetched at that stage.

And when it came to the formulation of the new theory, the 1905–6 conceptual understanding of space-time and of matter residing in it had to be changed again! One clue used by Einstein was a Newtonian "coincidence:" the equality of inertial mass to gravitational mass. It had indeed been checked experimentally with great precision, and thus strengthened Einstein's motivation. His intuitive understanding required the new theory to produce "Mach's principle," a scientifically unsubstantiated but philosophically appealing view, published by Ernst Mach before relativity, in which the inertial forces should themselves result from a gravitational interaction with the distant masses of the universe. (See N. Rosen's contribution in Part II.) With all this, it can be said that another fifty years at least might have passed before the advent of general relativity — had Einstein himself not done it. It might even never have been done at all, though it does resemble some of our current theories (which have, however, general relativity as their model).

From Relativity to Cosmology

Consider that we have nowadays several similar incompatibilities, for example since 1925 between gravitation itself and quantum mechanics. We have not solved the latter problem in fifty-four years, with very many people working on it — although we may be very close now. (See Part III, and contributions by S. W. Hawking, Y. Ne'eman, and by P. van Nieuwenhuizen in the Princeton volume.) I believe that bridging the gap between Newtonian physics and the new kinematics of special relativity (which were obeyed by the electromagnetic theory of Faraday and Maxwell) was a comparable task, and it was done almost singlehandedly within ten years. Moreover, the theory has withstood every experimental test in the last sixty-three years. It has stayed unmodified while everything else has had to be changed or supplemented. Intrinsically, it has a tremendous aesthetic appeal, beyond Chartres or the Taj Mahal in its monumental simplicity, together with an overwhelming power. One very rarely comes across words like "aesthetic" and "beauty," in textbooks of theoretical physics. It is thus noteworthy that the feelings I express here must have been felt by many, if not all, physicists — as witnessed by a line in a famous and excellent textbook by L. D. Landau and E. M. Lifshitz (otherwise written in an extremely dry and unpersonalized style): "The theory of gravitational fields, constructed on the basis of the theory of relativity, is called the general theory of relativity. It was established by Einstein (and finally formulated by him in 1916), and represents probably the *most beautiful* [my emphasis] of all existing physical theories. It is remarkable that it was developed by Einstein in a purely deductive manner and only later was substantiated by astronomical observations." After general relativity, Einstein invented cosmology in 1917. This field has been very lively in the last fifteen years, thanks to new observational material. An alternative approach to cosmology, modifying Einstein's theory, was suggested by a British group in 1948. Though it became very popular at the time, the "steady-state theory," as it was called, has not been vindicated by observations, and Einstein's equations for cosmology still hold. (See contributions by M. J. Rees and D. Sciama in the Princeton volume.)

Einstein's last twenty years were spent in an attempt to unify electromagnetism with gravitation. (See P. G. Bergmann's contribution in Part I.) He did not achieve that aim, but since 1976 the prospects of such a unification have improved, and a look at Einstein's results shows he was on the right track, anticipating some features of work in progress.

THE COHERENCE OF SCIENCE

We have thus seen that Einstein's greatest contribution has consisted in the construction of a more general and more fundamental theory, removing the incompatibility between Newtonian mechanics, based on Galilean kinematics with no preferred velocity (and on action at a distance), and the Einsteinian kinematics of special relativity, discovered by Einstein himself, as well as with those of Faraday–Maxwell electromagnetism. Incidentally, these two clashing theories themselves had been considered, prior to Einstein's achievement, as realizing the final axiomatic-deductive and aesthetic ideals required of a physical theory.

Most developments in theoretical physics deal with the removal of an incompatibility, but on a different scale. Generally, one has a theory and comes across a new observation that appears not to fit. Often the observation itself results from an attempt to test the theory, in the spirit defined by Karl Popper as falsification. The result is generally a modification of the theory, a weakening of one of its postulates. A new free parameter is allowed. In some cases (such as the discovery of time-reversal asymmetry at the fundamental level in 1964), it may require the introduction of an entirely new force. For the Michelson–Morley negative result, it even required a fundamental change in our conception of space-time. But in the case of general relativity, the problem was of another magnitude: to remove the basic incompatibility between two full-fledged and well-grown theories. This is why the answer had to reach that deep. That an answer could be found is in itself an extremely important epistemological clue.

The entire scientific program is based on this success. There is little a priori understanding of why the physical world should obey a mathematical description, even if we believe in the strength of logic and know that mathematics is its most direct realization. The world could still be inextricably complicated and self-interacting, and there would then be no way of guaranteeing the existence of a coherent and complete mathematical system describing it, based upon human observations. Indeed, the gap between disciplines is generally to be found in such regions.

For example, we now have in E. Schroedinger's equation (1925) of quantum mechanics an equation capable in principle of describing correctly any atom (even uranium with 239 protons and neutrons and 92 orbiting electrons) or molecule (even DNA with its millions of atoms). Thus the whole of chemistry and perhaps biochemistry are contained in this one equation. However, the actual calculations become unmanageable beyond two or three electrons or nucleons (protons and neutrons). We have to rest satisfied with the knowledge that the problem is not one of principle but of practical computation, and chemistry continues to evolve at its own composite-structure level. Indeed, G. F. Chew had suggested in the early 1960s that such a situation had arisen for the strong interactions (the nuclear glue). Happily, the 1970s have brought some answers that may resolve those difficulties. (See the following section, "Progress in the Physics of Particles and Fields, 1948-1978," and the chapters dealing with QCD, a gauge hypothesis for the strong interactions.)

Gaps and More Gaps

Conceivably, unbridgeable gaps might exist between various theories dealing with different elements or aspects at the fundamental level. Einstein's achievement encourages us to think that this is not so, and that if we explore deeply enough, we might well achieve a coherent structure. There are nowadays several such gaps that seem to us like real abysses, and it is good to know that one should persevere and not rest satisfied with disconnected theories.

Such a gap existed between quantum mechanics and special relativity. It was effectively bridged in 1948, after twenty years of hard labor. But though this was done by some of the greatest physicists of that period — Dirac, Pauli, Weisskopf, Feynman, Schwinger, Tomonaga, Dyson, to cite just a few names — it was left without the Einsteinian "polish." To this day, the theory lacks a proper mathematical foundation. It is operationally wonderful, yielding the most precise results in any science, yet every such calculation involves a cancellation of infinities on both sides of an equation, a mathematically unsound operation. (See the first two paragraphs of the section "Progress in the Physics of Particles and Fields, 1948-1978.") Surely, the theory deserves a better foundation, which probably would still yield the same operational results. The search has been on since 1948, and more intensely since the mid-1950s, but the end is still not in sight.

Another such gap exists between gravitation and the other interactions (or forces) in nature: electromagnetism, the Fermi (short-range) weak interactions (such as radioactive beta decay) and the strong interactions (also short ranged, responsible for the glueing together of nucleons in the nucleus). This is related to Einstein's last and unfinished program. We have recently witnessed (1960-73) what may possibly be the ad hoc operational merger of weak and electromagnetic interactions into a coherent "electroweak" interaction. (See subsection "The Weak Interactions" in the section "Progress in the Physics of Particles and Fields, 1948-1978, for an overview, and Part V for a full discussion of the present issues.) The evidence at present is highly favorable, but the crucial test will be available around 1982. Again this is a very ad hoc wedding, with none of the aesthetic impact of Einstein's work. (For one recent speculative attempt to supply an aesthetic foundation, see Y. Ne'eman, *Physics Letters* **81B**, 190-94, 1979.) This success has encouraged many of us to renew the search for the other missing connections between interactions. (See Part VI, "Grand Unifications.") In particular, many of us are studying a possibly new link between gravity and the other forces, which may at the same time resolve the incompatibility of general relativity and quantum mechanics. This approach, called supergravity, has intrinsic beauty, but it has not yet led to the correct operational wedding, so to speak; the search goes on. (See D. Z. Freedman's contribution in Part III, and contributions by Y. Ne'eman and P. van Nieuwenhuizen in the Princeton volume.)

INVARIANCE PRINCIPLES AND GEOMETRIZATION

One tool that was used by Einstein in both special and general relativity is the application of invariance principles. E. Whittaker has named them "postulates of impotence," though they are highly potent in their results. "It is impossible to pick a preferred frame of reference in space-time" is the wording of the Einstein invariance postulates, first for "inertial frames" only (special relativity), then "noninertial" (accelerated) frames for general relativity. The first resulted in kinematical restrictions that are expressed by ten conservation laws (energy-momentum and generalized angular momentum). Indeed, a theorem proved by E. Noether, a gifted student of Hilbert and later his colleague (though the Göttingen faculty refused to grant her a professorship, finally using the argument that there was only a men's room in the senate building), states that for every invariance (or symmetry) there will be a conservation law and vice versa.

The negative content of the postulate of impotence is in fact the bold step of generalization taken by the new theory. To achieve coherence one has to look for a more general foundation, that is, for simplicity. Simplicity is then achieved by this indistinguishability. One frame of reference is as good as another: here is the simplicity. However, for the laws of nature to obey such sweeping generalizations, tremendous sophistication may be needed in the dynamics. In general relativity — and in all the more recent "gauge theories" inspired by it — this is indeed the situation. These are stronger, local invariance principles — local in the sense that we may even change our preferred frames of reference at different points in space-time. (See C. N. Yang's contribution in Part I.) Any such local invariance postulate then requires the existence at every point of a new "field" that compensates for and cancels completely the changes we are producing in arbitrarily applying our freedom of choice at each point. Such is the role of the gravitational field, and a similar part is played by the electromagnetic field — or perhaps the electroweak, combining electromagnetism and weak interactions. We experience these forces directly, but the justification for their presence, and the understanding of their full action, required this identification of their role as guardians and preservers of simplicity through a Leibnitzian relativity (or indistinguishability) applied locally in space-time.

The other key feature introduced by Einstein was geometrization. In his theory of gravitation, all of physics could be represented as motions in curved space-time. The curvature is another embodiment of the same stresses and compensating forces mentioned in the previous paragraph. Indeed, we have recently learned that such a picture is in the background of all our new gauge theories. Einstein had set up this ideal of a geometrized theory, and for some sixty years it looked as if only gravitation would be affected — with all other forces moving in different directions. Suddenly, the light on the scene has changed. It seems that we are all heading for geometry. (See C. N. Yang's contribution in Part I, and chapters by S. S. Chern and by T. Regge in the Princeton volume.)

ABSTRACTION

A physicist's mind undergoes a constant struggle between the wish to remain close to the perceptions and observations, and the need to look for abstract formulation, required by generalization. If one had to provide one theory for each observation, one could stay at that level. For a theory to encompass the largest possible set of observations, it has to abandon the actual observables and move to a deeper and more abstract level. Aristotle was trying to avoid abstraction in his physics and studied motions that were all taken from daily experience and thereby entailed forces of friction. In such a milieu, force is proportional to the "final" velocity, as was calculated in the first half of the last century by G. G. Stokes. Aristotle thus postulated such a law — which was superficially simple but described a nonsimple situation. This law of force-velocity proportionality could not be generalized, and obstructed progress.

Galileo and Newton imagined a vacuum and stated their laws in this "unnatural" background. A simple formalism fitted this simple picture — and Newton's abstraction won. "Force is proportional to acceleration" is a generalizable statement and does not involve one particular force (friction in Aristotle's case). Einstein followed the same path and started from a highly abstract picture of a generalized manifold. His theory did achieve the aims he had set, including Mach's principle, but its abstract structure made it capable of going even beyond that image. Einstein's theory handles situations (such as Godel's "universe," a universe in which time is cyclic) that go beyond Einstein's intuitive goals (including non-Machian universes). Indeed, Einstein himself started this testing of his theory as a theory of the universe (cosmology), beyond its role as a theory of gravitation. (See J. D. Bekenstein's contribution in Part II and chapters by J. A. Wheeler and by S. W. Hawking in the Princeton volume.)

In this abstractive jump, the creative scientist sometimes has to free himself — and us — from some preconceived notion, whether based on sense perceptions or on conditioning and prejudice. It has happened many times — for example, accepting that the world is round (an ancient Greek discovery, forgotten in medieval times until it was reestablished in the Age of Discovery). Copernicus and Galileo had to struggle hard to replace the earth by the sun at the center of the system of planets. Darwin had to fight prejudice when he announced the common ancestry of men and apes. Einstein's original shock action was his dethronement of Chronos, immutable time. To this very day you will find some partisans of "commonsense" time who still have not absorbed the message after a huge amount of direct observational evidence. Remember, indeed, that although the good theory requires abstraction, it also has to be experimentally testable, "falsifiable" in Popper's words. A theory that can never be tested is devoid of any scientific value. Einstein's gravitational theory, general relativity, was accepted only after it had passed several such tests: the deflection of light when passing near the sun, the advance of the perihelion of Mercury, and so on. (See chapters by I. I. Shapiro and by D. T. Wilkinson in the Princeton volume.)

Does the scientific abstraction have anything to do with the artistic one I mentioned earlier? It may require similar imaginative capabilities, although its motivation and effects are very different. Somehow, the one resonates in the subconscious whereas the other is tested by precise computation. However, that aesthetic quality of the great theories affects us in the subconscious, too.

INVOLVEMENT

Traditionally the scientist resides in his ivory tower. Undoubtedly a certain disconnection is essential at the conceptual stage, and the mathematician, the theoretical physicist, and perhaps many others could not function if they were to take part in the daily commotion. But one can partake from time to time of the action and worries of one's surroundings. In many cases, the process is irreversible. However, both the decision to leave the ivory tower and that of returning to it are the individual scientist's personal problem.

With Einstein, the sense of justice and worry about his fellow men were strong motivations that led him out of his personal tower after 1914. The outbreak of the First World War caused him to lend his name and energies to pacifism. He could see no sense in that war, which was fought mostly for "honor," or because of the assassination of an archduke. From Berlin itself, he joined George Nicolai in a manifesto to the "Europeans," transcending the various nationalisms.

Around 1920 he involved himself in Zionism. I shall discuss this aspect in more detail, having been personally interested in it for obvious reasons. To come back to pacifism, the outbreak of the Second World War brought a different response. Nazi Germany was our nemesis; it was extending its domination over the whole of Europe and bringing in its wake prejudice and persecution, mostly racial. It was every free man's duty to help stop Nazism and (we hoped) break its power. At the suggestion of Szilard and Wigner he sent his famous letter to Roosevelt, which influenced the decision to start the American nuclear project. There is no doubt that he was right in fearing for the fate of the world in case the Nazis should succeed with their own atomic energy program. Hitler had the right intuition about the development of weapons, to the point where one new weapon could win the war. (I shall always remember the speeches in which he mentioned Germany's "secret weapon.") It turned out that he erred in the actual choice, putting all his weight behind rockets instead of nuclear weapons.

After the fall of Nazi Germany, Einstein's sense of human responsibility made him an initiator of nuclear disarmament, an aim that led him to the Russell–Einstein manifesto.[8] It is invigorating to see how he could be moral and at the same time pragmatic. He was not a starry eyed intellectual dealing with an unfamiliar world. He understood the realities and reassessed the situation at each stage, rather than taking one dogmatic stand throughout.

Einstein and the Political Life

He could be courageous and daring in his decisions, as exemplified by his stand against Joseph McCarthy. There was a time when Einstein's was one of the relatively few dissenting voices to be heard in the United States.

We now come to Zionism.[9] Einstein espoused this cause between 1914 and 1919, coming to the conclusion that the creation of the Jewish State in Palestine was indeed the only way of saving his fellow Jews. To be sure, the League of Nations mandate to Great Britain, with the explicit aim of making Palestine the Jewish "national home," made him very happy.[10] "What pleases me the most is the realizing of the Jewish State in Palestine," he wrote to Ehrenfest. His first visit to America, in 1921, was as a Zionist emissary, together with Chaim Weizmann, the future first president of Israel. Einstein's role was to collect the people and the means for the creation of the Hebrew University in Jerusalem. He was later the first chairman of its board of trustees.

After 1921 Einstein took an active part in the general Zionist effort by writing a large number of articles and addresses. Beyond physical salvation for the persecuted, which finally came, but too late for six million Jews of Central and Eastern Europe, he was considering the effect on the Jew of the Diaspora: "I am a national Jew in the sense that I demand the preservation of the Jewish nationality as of every other. I look upon Jewish nationality as a fact. . . . I regard the growth of Jewish self-assertion as being in the interests of non-Jews as well as Jews. That was the main motive in my joining the Zionist movement," he said in a speech made in the United Kingdom in 1921 upon his return from the American trip. Again, in the same speech: "I have always been annoyed by the undignified assimilationist cravings and strivings which I have observed in so many of my friends."

In 1923, Einstein visited Palestine. He kept a diary that reflects his emotions. In Jerusalem he gave a lecture and laid the cornerstone of the future university. He was asked to stay and settle in

Jerusalem and noted in his diary "the heart says yes, but the mind says no." Retrospectively, I am not sure he made the right decision at that point.[11] True, Palestine was still an intellectual desert as far as the sciences were concerned, but his own immigration — perhaps by the 1930s — might have accelerated that development and still enabled him to continue his work.

The Temptations of Leadership

In Tel Aviv he was taken by the local enthusiasm and joined a group of young German Zionists who were laying bricks in an apartment building. He enjoyed this experience of creative manual labor, though he later often wondered whether the Jewish return to the land would not inhibit the intellectual capacities developed in the Diaspora.

During the 1930s and 1940s he was active in fund raising but was also interested in the political developments. In 1932 he said in a speech in Los Angeles that "the Zionist goal gives us an actual opportunity to put into practice, through a viable solution of the Jewish–Arab problem, those principles of tolerance and justice that we owe primarily to our prophets." Often he pressed for compromise — but there was no response from the other side, which considered Zionism as an intrusion. In the onslaughts of 1936–39 and in the six-nation Arab invasion of 1948, he stood for firmness and the absolute necessity of defeating aggression.

In 1952, upon Weizmann's death, it was only natural that Einstein should be asked to succeed him as president of Israel. He had indeed become a symbol to all Jews and Israelis, a unifying personality and a central link. He declined the offer, fearing overinvolvement, and wrote: "I am the more distressed over these circumstances because my relationship to the Jewish people has become my strongest human bond, ever since I became aware of our precarious situation among the nations of the world."[12]

Progress in the physics of particles and fields (1948–1978)*

Quantum electrodynamics (QED, 1948) is the first successful relativistic quantum field theory (RQFT), following the classical field theories of gravitation (Newton's and Einstein's) and Maxwell's electrodynamics. It provides the means of calculating all electromagnetic transition amplitudes in terms of a perturbative expansion in the electric coupling between current and potential, that is, the electron charge e or better the rationalized fine-structure constant $\alpha = e^2/4\pi \hbar c$, whose value in "natural" units ($c = \hbar = 1$) is $\alpha = 1/137$. This small value of α is thus essential to the effectiveness of the method.

Construction of QED had taken twenty years, spent in the search for a way of avoiding "ultraviolet" divergences (infinities arising in integrations over the larger values of the momenta). These were finally segregated and removed by the method of renormalization, yielding finite answers whose fit with experiment is of the highest precision encountered in physics. QED also embodies a geometrical invariance idea due to H. Weyl, namely $U(1)$ gauge invariance: the invariance of the theory under locally (space-time) dependent transformations of the quantum phase of the electron wavefunction, somewhat similar to the invariance of general relativity both under locally dependent general coordinate transformations and under the even more gaugelike locally dependent Lorentz transformations of a local "tangent" frame ("tetrads") in curved space-time.

*Republished by permission, from Physics Bulletin, London **29**, pp. 422-24 (1978).

Weyl gauge invariance[13] was essential to renormalizability, as a constraint relating the iterative self-screening of the charge to the self-dressing (for example, by self-repulsion) of the electron wavefunction (the Ward–Takahashi identity). Weyl's gauge was later generalized[14] to a non-Abelian group such as the $SU(2)$ of isospin (relating the proton p^+ and neutron n^0 wavefunctions as one formal spinor), leading to an interaction mediated by a multiplet of vector potentials A_μ^i ($\mu = 0 \ldots 3$ is the space-time index, i the internal symmetry, e.g. $i = 1 \ldots 3$ for $SU(2)$ and $i = 1 \ldots 8$ for $SU(3)$) or "1^-" mesons, where J^p stands for spin J and parity p. In such gauges, the mediators themselves carry some of the non-Abelian charge (whereas the A_μ of QED is uncharged) and thus generate an additional self-interaction term.

After the success of 1948, there remained at least three interactions to be taken care of by appropriate theories, namely quantum gravitation and the two short-range (nonclassical) nuclear forces: the strong interactions (the nuclear "glue," including the exchange of π and K mesons) and the weak (Fermi) interactions (neutron β-decay, muon decays, and so on). Within a few years, however, the task appeared hopeless. Strong interactions, if mediated by π–K mesons and the like (0^- mesons) would still be renormalizable, but with α replaced by $\alpha_S = g^2/4\pi\hbar c = 14$, thus effectively doing away with the idea of an expansion. This led to approaches based on analytical continuation in energy, momentum transfer, angular momentum, and so on. "S-matrix theory" or dispersion relations "replaced" RQFT throughout the 1950s and 1960s, yet providing at best a parametrization of the domain, and as a result, an aid to classification ("Regge trajectories") and indications of a "string" structure, but no true dynamical theory.

The Weak Interaction

For the weak interaction (see Part V), the Fermi coupling is $G = 10^{-5}/m_N^2$, m_N being the nucleon mass. This is a "small" number but with dimensions of an inverse mass squared. As a result, the theory itself was nonrenormalizable. (The same is true of gravitation in the presence of matter.) It was, however, suspected from the very beginning that the (mass)$^{-2}$ in G might just be the mass of a pair of $W_\mu^\pm$ charged 1^- bosons mediating that interaction. In some processes such as beta decay $n^0 \to p^+ + e^- + \bar{\nu}_e$, the intermediate stage $n^0 \to p^+ + W_\mu^-$, $W_\mu^- \to e^- + \bar{\nu}_e$ involves an off-mass shell "virtual" W^-, just as the Coulomb force between p^+ and e^- involves a virtual photon exchange, $p^+ \to p^+ + A_\mu$, $A_\mu + e^- \to e^-$; note that in the even more analogous conjugate process of a proton p^+ emitting a photon A_μ which makes a pair $A_\mu \to e^+ + e^-$, the photon goes off mass shell in the opposite direction! If indeed one should regard $G = \alpha_W(1/M_W^2)$, α_W might even be the same as the electromagnetic α, which would imply $M_W^2 \sim 10^3 \, m_N^2$. With various factors that I have disregarded, $M_{W^\pm} \sim 50$ GeV/c^2.

In the late 1950s, R. Marshak and E. C. G. Sudarshan and R. Feynman and M. Gell-Mann[15] noted the vector nature of the weak current (though left-handed, or parity-breaking), implying mediation by a vector potential. S. L. Glashow in 1961 showed that an $SU(2)_{\text{left}} \times U(1)$ gauge group would produce just such a combination, provided there existed an additional neutral weak current resembling the electric current. Electric charge is then a linear combination between the $U(1)$ and a component of the $SU(2)_{\text{left}}$. The stage was thus set for a unification of weak and electromagnetic interactions, except that there existed no credible (renormalizable) theory for such an interaction involving non-Abelian (and massive, or short-ranged) vector potentials. S. Weinberg and A. Salam nevertheless went ahead[16] and in their optimism about RQFT perfected the as yet unutilizable model. They adopted the Glashow group as a gauge group à la Yang–Mills, that is, generating a set of four vector potentials $W_\mu^\pm$, W_μ^0 and A_μ (of QED), and further introduced a "spontaneous" symmetry breakdown (like a ferromagnet setting on some spins orientation) to allow for the appearance of masses in $W^\pm$ and W^0, ($M_{W^0} \sim 70\text{-}90$ GeV/c^2), but not in A_μ. That kind of symmetry breakdown[17] had originally been studied by P. Higgs, T. W. Kibble, R. Brout, F. Englert, G. S. Guralnik and C. R. Hagen.

The existence of a W^0 and of the corresponding neutral weak current had first appeared to contradict the observed absence of neutral decays such as $K^+ \rightarrow \pi^+ + e^+ + e^-$. These decays however involve a change of strangeness ($K \rightarrow \pi$), and in 1970, S. L. Glashow, J. Iliopoulos, and L. Maiani ("GIM") showed[18] that the existence of a fourth "charmed" quark c (see the following paragraphs) would result in just such a cancellation of neutral strangeness-changing currents. By 1970, the Schwinger-Glashow guess in its final Salam-Weinberg form was ready for both a theoretical check (renormalizability) and experimental ones (the observation of a strangeness-preserving weak neutral current with precisely predictable transitions, and the further direct observation of the $W^\pm$ and W^0 at their predicted masses).

The early 1970s indeed provided the fullest support to this hypothesis. On the theoretical side, work on renormalization had been pursued "in the background" throughout the 1960s by R. P. Feynman, B. S. DeWitt, L. Faddeev and V. N. Popov, E. S. Fradkin, and I. V. Tyutin and M. Veltman.[19] These workers succeeded in paving the way for the proof of renormalizability of a Yang-Mills-type gauge theory. The remarkable accomplishment was completed by G. 't Hooft (1971), who also proved that renormalizability even survives spontaneous symmetry breakdown;[20] that is, it holds for massive W^+ and W^0 in the $SU(2)_{\text{left}} \times U(1)$ model.

On the experimental side, the discovery of neutral currents by F. J. Hasert and his coworkers[21] came in the summer of 1973. At CERN, $\nu_\mu + e^- \rightarrow \nu_\mu + e^-$, ν_μ + nucleon $\rightarrow \nu_\mu$ + hadrons and $\bar{\nu}_\mu$ + nucleon $\rightarrow \bar{\nu}_\mu$ + hadrons were observed, soon to be confirmed by experiments at the Fermi National Accelerator Laboratory in the United States. The effect has since been seen and measured in a variety of cases, with all but one fitting the Salam-Weinberg model's precise predictions. The only result that does not fit involves indirect conclusions from an atomic physics (Bi) experiment, and errors may still exist in the interpretation. (The recent SLAC e-d polarization experiment appears to have settled this issue in favor of the Weinberg-Salam model.) As to the direct observation of the W mesons, the appropriate accelerators will probably be available in the 1980s. However, it would be fair to say that the unified theory of weak and electromagnetic interactions already appears to have provided yet another jewel for the crown of physics, the field theories. It does as usual leave a new problem in its wake, namely that of the inner structure of the spontaneous breakdown mechanism. Does it represent yet another field and particle, and if so does it fit into some peculiar order?

Strong Interactions: ψ and Charm, QCD?

I now return to the strong interactions (see Part IV). The key development had been provided in the early 1960s by the discovery of $SU(3)$ symmetry (Y. Ne'eman and M. Gell-Mann) involving a classification of the hadron spectrum of states and the behavior of the symmetry-breaking term.[22] The nucleon appeared in an octet, not in the simplest defining representation. It led us (H. Goldberg and Y. Ne'eman) to postulate the existence of a fundamental field with atomic mass number $A = 1/3$, so that a nucleon would be made of three such elements.[23] This was soon perfected in the quark model with quarks u,d,s and $p^+ = uud$, $n^0 = udd$, and so on.

The model was highly successful in describing all hadron interactions (strong-interaction branching ratios, mass formulas, particle electric and magnetic moments, weak amplitudes, and so forth) with the quarks appearing unaffected by their supposed strong binding. Note, however, that to preserve the spin-statistics correlation, yet another quantum number had to be assumed. For example, the theoretical result $\mu_p/\mu_n = -3/2$ (where μ_N is the nucleon magnetic moment), which fits rather nicely the measured -1.46 (for a symmetry with an explicit breaking term of at least 10 percent), involves spin and $SU(3)$, with the wavefunctions appearing as symmetric in both spin and $SU(3)$ as well as in space. For the nucleon to be a fermion, there thus has to exist a further property in which the wavefunction is antisymmetric. O. W. Greenberg in 1964 suggested parastatistics, and M. Han and Y. Nambu in 1965 yet another $SU(3)$ (now known as "color" $SU(3)$), which was soon shown to be effectively equivalent to parastatistics.[24] Total antisymmetry in that $SU(3)$ explains saturation at three quarks. Color has since been checked out in various other experiments, such as the colliding beams $e^+ + e^- \rightarrow$ hadrons$/(\mu^+ + \mu^-) = R$ where the cross-sections ratio $R = 2/3$ if there be no color, and $R = 2$ color (for example, three types of 3 quarks). Experimentally, $R \sim 2$ up to a center of mass energy of 3 GeV.

In 1968-9, experiments at the Stanford Linear Accelerator Center with deep-inelastic scattering of e^- on nucleons showed the nucleon to be made of two ingredients: an electrically neutral "glue" (50 percent) and three pointlike objects that most probably carry exactly the quark quantum numbers. (These experiments have been compared with Rutherford's probing of the atom.) The pointlike structure produces a "scaling" effect in which energy-independent structure functions are observed.

It was pointed out by H. O. Politzer and by D. Gross and F. Wilczek in 1973 that such a result should be expected as a logarithmic approximation from a Yang–Mills gauge theory of the strong interactions,[25] and such theories have since been suggested ("quantum chromodynamics" or QCD), involving $SU(3)_{color}$ as the gauge group (with eight 1^- gluons, massless in one version), but the implications with respect to the possible "confinement" of quarks are still not fully understood. Note that there is as yet no corroborated evidence for the direct observation of a free quark (the SLAC experiments can count as direct observation of the quarks inside the nucleon itself). The main evidence for a color $SU(3)$ "QCD" gauge theory is its prediction of a sharp weakening of the effective coupling (color charge) at short distances, that is, quasi-free floating quarks in the center of a hadron. This "asymptotic freedom" would explain the successes of the "quasi-free" behavior of the quarks.

Prizewinning Experiments

At the London conference of 1974 everything seemed to be going well for the quark model, except that beyond 3 GeV the value of R suddenly increased to 4 or 5. The mystery was resolved when in the autumn of 1974 the $\psi(3.2$ GeV$)$ meson ($J^p = 1^-$), and its system of excitations (at 3.7, 4.1 GeV, and so on) were observed in two independent experiments: $e^+ + e^- \rightarrow$ virtual photon $\rightarrow \psi$ (B. Richter and colleagues at SPEAR) and $p^+ + Be \rightarrow$ nucleus $+ \psi$, $\psi \rightarrow$ virtual photon $\rightarrow e^+ + e^-$ (S. C. C. Ting and colleagues at Brookhaven Laboratory).[26] The ψ width is 67 ± 12 KeV, corresponding to a long lifetime ($\sim 10^{-20}$ sec) for a hadron (which it was proved to be). This is similar to the inhibited decay of another 1^- meson, the $\phi(1.02$ GeV$)$: $(\phi \rightarrow 3\pi)/(\phi \rightarrow K + \bar{K}) = 16$ percent. The actual inhibiting factor is of the order of 50, as the 3π decay mode is energetically favored. In the quark picture, the ϕ is made of the third (strange) quark and its antiquark $\phi = s\bar{s}$. Making $K + \bar{K}$ involves the creation of an additional $u\bar{u}$ or $d\bar{d}$, as $K^+ = u\bar{s}$, $\bar{K}^- = \bar{u}s$, $K^0 = d\bar{s}$, $\bar{K}^0 = \bar{d}s$. On the

Introduction

other hand, $\pi^+ = u\bar{d}$, $\pi^- = \bar{u}d$, $\pi^0 = (\bar{u}u - \bar{d}d)/2^{1/2}$ so that $\phi \to 3\pi$ requires annihilating the $s\bar{s}$ of the ϕ itself and creating the entire new hadron (or three pairs of quarks). In QCD, this implies a three color-gluon intermediate state, which is suppressed because of asymptotic freedom. If the $\psi = c\bar{c}$, c being a new type of quark (with a new quantum number, "charm"), the same reasoning would explain the slow decays.

Its excitation system (eight states observed to date) fits such a "charmonium" picture rather well, with the narrow states stopping at 3.7 GeV. This would indicate that some states containing $c\bar{c}$ can be produced above 3.7 GeV, thus removing the inhibiting factor. Indeed, in the fall of 1976, G. Goldhaber and coworkers discovered the $D^+ = c\bar{d}$ and $D^0 = c\bar{u}$ at 1.86 GeV, in complete agreement with the foregoing deduction.[27] The discovery of these charmed states (since D^0 has $C = +1$, for instance) also completed the experimental proof of the existence of charm as such, and of the original GIM deduction from the absence of strangeness-changing neutral currents. Figure I.1 shows a six-prong (overdetermined) decay of ψ'' (4.1 GeV) into charmed particles:

$$e^+ + e^- \to \psi'' \to D^0 + \bar{D}^0$$

$$D^0 \to K^-{}_1 + \pi^+{}_1 \text{ (involves a ``weak'' transition } c \to s)$$

$$\bar{D}^0 \to K^0{}_{s2} + \pi^+{}_2 + \pi^-{}_2 + \pi^+{}_2 + \pi^-{}_2 \text{ (involves } \bar{c} \to \bar{s})$$

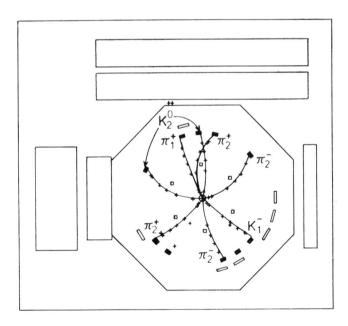

Fig. I.1 A six-prong (overdetermined) decay of ψ'' (4.1 GeV) into charmed particles.

(Stanford Linear Accelerator Center-Lawrence Berkeley Laboratory magnetic detector at SPEAR). The extra trigger and shower counters have been fired by secondary interactions. This is deduced from the time-of-flight information that indicates that they were not fired by prompt particles.

Presumably a fifth type of quark "b" has since made its appearance[28] through the discovery by L. Lederman and colleagues of the Υ(9.4 GeV) 1^- meson, in an experiment similar to the Ting production of ψ. Υ has since been observed at PETRA in an $e^+ + e^-$ analogue of the Richter ψ (from weak-interaction theory we can now predict the existence of a sixth quark "t" somewhere higher up). The appearance of $c\bar{c}$ at 3.1 GeV has also settled the R mystery, since the threshold for that new quark raises the quark model predictions to 10/3, and that of the "b" to 11/3. Moreover, Perl and coworkers[29] found in 1975 a new lepton, the τ, at 1.9 GeV so that the sequence of leptons is now (e^-, ν_e), (μ^-, ν_μ), $(\tau, \nu_\tau?)$ and their antiparticles. The τ raises R by another unit.

The last decade has thus witnessed highly significant experiments: neutral currents, the ψ and Υ, charm, a new lepton. Theory now disposes of one more RQFT as a tool: that of the Yang-Mills gauge, with or without spontaneous symmetry breakdown. Most probably we have the foundations of a unified theory of weak and electromagnetic interactions, and perhaps that part of the strong interaction that is responsible for interquark forces, with ordinary meson exchange appearing as a van der Waals-like higher-order effect.

Acknowledgments

I would like to thank J. Richardson, editor of the UNESCO journal *Impact of Science on Society,* and the journal's editorial board, for permission to reproduce "Coherence, abstraction and personal involvement: Albert Einstein, physicist and humanist" (Vol. 29, No. 1, pp. 17-25, 1979).

I would also like to thank Malcolm Clarke and the editors of *The Physics Bulletin,* Journal of the Institute of Physics (UK), for permission by the Institute of Physics to reproduce "Progress in the physics of particles and fields" (Vol. 29, pp. 422-24, 1978).

NOTES

1. A. Einstein, *Out of My Later Years* (Philosophical Library, New York, 1950), p. 31.
2. A. Einstein, The World as I See It (Philosophical Library, New York, 1949), p. 90.
3. A. Einstein, Letter to "Falastin," 16 January 1929.
4. Excerpted from R. W. Clark's *Einstein: The Life and Times* (T. Y. Crowell, New York, 1971), p. 95.
5. Y. Ne'eman, "Coherence, abstraction and personal involvement: Albert Einstein physicist and humanist," Impact of Science on Society **29**, 17-25 (1979).
6. Y. Ne'eman, "Progress in the physics of particles and fields, 1948-1978," Phys. Bull. (London) **29**, 422-24 (1978).
7. Let us note however that both Poincaré and Lorentz (and Michelson!), though they confirmed the mathematical content of special relativity, stubbornly rejected to their last days Einstein's conceptual breakthrough, especially in the matter of time and the breakdown of simultaneity.
8. See Impact of Science on Society **26**: Science and War, 15 (1976).
9. The quotations from Einstein's speeches and diary are taken from B. Hoffmann's paper in *General Relativity and Gravitation* (GR7) edited by G. Shaviv and J. Rosen (John Wiley, New York, 1975), pp. 233-42.

10. The Balfour Declaration (1917) pledged support to the Jewish people in their efforts to establish a "national home" in Palestine, provided that the "civil and religious rights" of non-Jewish communities there would be respected. See also *The Universal Jewish Encyclopaedia* (Ktav, New York, 1948), vol. 8, pp. 376–78.

11. See Einstein's own remark in that context, in N. Rosen's reminiscences at the Jerusalem symposium in the nonphysics volume.

12. To delve more deeply, see: R. Clark, *Einstein, the Life and Times* (World Publishing, New York, 1971); C. Lanczos, *The Einstein Decade (1905-1915)* (Academic Press, New York, 1973); R. Maheu, et al., *Science et Synthese* (Gallimard, Paris, 1967).

13. H. Weyl, Sitzber. Preuss. Akad. Wiss. 1918, p. 465; H. Weyl, Math. Z. **2**, 384 (1918); H. Weyl, Z. f. Physik, **56**, 330 (1929).

14. C. N. Yang and R. L. Mills, Phys. Rev. **95**, 631, and **96**, 191 (1954); T. Shaw, thesis, Cambridge University.

15. E. C. G. Sudarshan and R. E. Marshak, Proceedings of the Conference Padua-Venice Conference on Mesons (1957); R. P. Feynman and M. Gell-Mann, Phys. Rev. **109**, 193 (1958).

16. S. Weinberg, Phys. Rev. Lett. **19**, 1264 (1967); A. Salam, in *Elementary Particle Theory,* edited by N. Svartholm (Almquist and Wiksells, Stockholm, 1968), p. 367.

17. P. Higgs, Phys. Lett. **12**, 132 (1964); G. Guralnik, C. R. Hagen, and T. W. Kibble, Phys. Rev. Lett. **13**, 585 (1966); F. Englert and R. Brout, Phys. Rev. Lett. **13**, 2386 (1964).

18. S. L. Glashow, J. Iliopoulos, and L. Maiani, Phys. Rev. **D2**, 1285 (1970).

19. R. P. Feyman, Acta Phys. Polon. **26**, 697 (1963); B. S. DeWitt, *Dynamical Theory of Groups and Fields* (Gordon and Breach, New York, 1965).

20. G. 't Hooft, Nucl. Phys. **B33**, 173 (1971), and **B35**, 167 (1971).

21. F. J. Hasert et al., Nucl. Phys. **B73**, 1 (1974); B. Aubert et al., Phys. Rev. Lett. **32**, 1954 (1974); A. Benvenuti et al., Phys. Rev. Lett. **32**, 800 (1974).

22. Y. Ne'eman, Nucl. Phys. **26**, 222 (1961); M. Gell-Mann, Caltech report CTSL-20 (1961).

23. H. Goldberg and Y. Ne'eman, Nuovo Cim. **27**, 1 (1963), and AEC report 1A-725 (1962); M. Gell-Mann, Phys. Lett. **8**, 14 (1964); G. Zweig, CERN report TH 401, 412 (1964).

24. O. W. Greenberg, Phys. Rev. Lett. **13**, 598 (1964); M. Han and Y. Nambu, Phys. Rev. **139**, 1006 (1965).

25. H. D. Politzer, Phys. Rev. Lett. **30**, 1346 (1973); D. Gross and F. Wilczek, Phys. Rev. Lett. **30**, 1343 (1973).

26. J. J. Aubert et al., Phys. Rev. Lett. **33**, 1404 (1974); J. E. Augustin et al., Phys. Rev. Lett. **33**, 1406 (1974).

27. G. Goldhaber et al., Phys. Rev. Lett. **37**, 255 (1976); E. H. S. Burhop et al., Phys. Lett. **65B**, 299 (1976).

28. L. M. Lederman, *Proceedings of 1978 (Tokyo) International Conference on High Energy Physics,* edited by S. Homma, (Physical Society of Japan, 1979), pp. 706-21.

29. M. Perl et al., Phys. Rev. Lett. **35**, 1489 (1976); M. Perl et al., Phys. Lett. **63B**, 766 (1976).

To Fulfill a Vision

Jerusalem Einstein Centennial Symposium on
Gauge Theories and Unification of Physical Forces

PART I: UNIFICATION: AIMS AND PRINCIPLES

This part comprises two chapters touching upon the general program of unification: C. N. Yang's, stressing the role of geometry, and P. G. Bergmann's, sketching Einstein's aims and attempts beyond general relativity. The third chapter, by Gürsey, develops the geometrical theme and throws some further light on Einstein's influence in mathematics.

The year 1979 was not only the centennial of Einstein's birth, it was also that of Maxwell's death, at the relatively young age of forty-eight. Yang reminds us that it was in Maxwell's mathematical rendering of Faraday's phenomenology — with the concomitant discovery of a missing term, the displacement current — that special relativity imposed itself in physics. Indeed, Maxwell's equations display Poincaré (or Lorentz) invariance, rather than Galilean. Einstein's discovery of special relativity consisted in identifying that symmetry in the equations, and realizing that it represented a fundamental feature of space-time.

Yang's message in the matter of phase (gauge) fields is highly relevant. It has turned out that gauge theories are equivalent to geometric manifolds of the fiber bundle type. In particular, the gauge field itself (with no matter fields as sources) is represented in a principal bundle. This should not be treated as a sidelight. The search for solutions (monopoles, instantons, merons, and so forth) has shown that it is indeed by assuming that we are in the geometry of a principal bundle that we get these results. What Yang is stating is that just as it was important for the development of special and general relativity to assume that space-time *is* indeed Minkowskian or pseudo-Riemannian, rather than that it manages artificially to appear "as if" it were non-Euclidean, so is it also important now for gauge theories to view space-time as a section in a principal bundle. This is geometrodynamics of the purest type!

Yang's lesson, aside from the aforementioned treatment of solutions, has recently been vindicated by the geometric identification of the ghost fields, necessary for the renormalization of the Yang–Mills interaction,[1] and of Goldstone–Nambu fields[2] required by spontaneous symmetry breakdown.

Bergmann's chapter explains Einstein's motivation, which included a hope for a deterministic theory that would replace quantum mechanics, aside from the attempt to explain all interactions and the particle spectrum. The main lines used by Einstein and by his successors among workers in general relativity are traced to the natural evolution of the theory, beyond the elucidation of the

---

Yuval Ne'eman (ed.), To Fulfill A Vision: Jerusalem Einstein Centennial Symposium on Gauge Theories and Unification of Physical Forces
ISBN 0-201-05289-X

Copyright © 1981 by Addison-Wesley Publishing Company, Inc., Advanced Book Program.
All rights reserved. No part of this publication may be reproduced, stored in a retrieval system, or transmitted, in any form or by any means, electronic, mechanical photocopying, recording, or otherwise, without the prior permission of the publisher.

equations of motion by Einstein, Infeld, and Hoffmann in 1938. Notice how difficult the actual calculations are: it took forty-five years to go from the Schwarzschild solution (static sphere) to the Kerr solution (rotating sphere). Bergmann's last comments introduce supergravity (see D. Z. Freedman's chapter in Part III), viewed as a unified field theory. His analysis of the role of complexification and hypercomplexification also relates to Gürsey's paper.

Gürsey's text brings out another interesting insight about the interlocking roles of mathematics and physics. Not only do we exploit Einstein's physics influence in our present interest in gauge theories, born out of Weyl's attempt at generalizing general relativity to include electromagnetism: the mathematical treatment of the solutions exploits Einstein's influence in mathematics.

Our requirements for a unified field theory, stated in the light of present-day physics, would consist of the following:

1. Relating all couplings (strong, weak-electromagnetic, CP violation, gravity) through some symmetry structure, exact under special conditions.

2. Explaining their differences in terms of some symmetry breakdown mechanism. This entails introducing some mass-scales (such as the mass of the weak intermediate boson, which explains the departure of G_w from α). In particular, the Planck mass (10^{19} GeV) comes in when comparing gravity with the strong interactions.

3. Explaining the double group structure, that is, the departure of the mass spectrum symmetries (flavor, such as $SU(3)$ and so on) from the (weak-electromagnetic) asthenodynamic gauge group. This is represented by a set of generalized Cabibbo angles (including perhaps CP violation).

4. Constraining (group theoretically) the symmetry breakdown mechanisms (spontaneous?) in (2) and (3). Most present attempts are very lax on that point, and introduce arbitrary numbers of Higgs fields.

5. Explaining the spin parity structure: why $J = 2^+$ for gravity, $J = 1^-$ for QCD, $J = 1$ with QAD, $J = 0$ for spontaneous breakdown fields and so forth, and removing any difficulties relating to those spins.

In discussing actual theories, we should refer to this list of requirements.

NOTES

1. J. Thierry-Mieg, to be published in J. Math. Phys.; J. Thierry-Mieg, to be published in Nuovo Cimento A; J. Thierry-Mieg, Doctorat d'Etat, thesis; Y. Ne'eman, *Proceedings of the XIX International Conference on High Energy Physics* (Tokyo 1978), edited by S. Homma et al. (Physical Society of Japan, 1979), 552. Y. Ne'eman and J. Thierry-Mieg, Ann. Phys. (N. Y.) **123**, 247 (1979).

2. Y. Ne'eman, Phys. Letters **81B**, 190 (1979); Y. Ne'eman and J. Thierry-Mieg, *Proceedings of the Eighth International Colloquium on Group Theoretical Methods in Physics* (Kiryat Anavim, 1979), edited by L. Horwitz (Ann. Isr. Phys. Soc.) 1980.

1. GEOMETRY AND PHYSICS
Chen Ning Yang

It is an honor for me to contribute to this volume celebrating the centennial of the birth of the greatest physicist of the twentieth century. I consider it especially meaningful that the centennial celebration took place at the Israeli Academy of Sciences, for no declaration nor historical analysis could better underline the important contributions of the Jewish people to the common heritage of mankind.

My subject is geometry and physics. Geometry is a science that originated from the quantitative study of forms and sizes. Mankind has been fascinated with geometry in almost all cultures. This perhaps is a basic reason why geometrical constructions have assumed important roles in art. That geometrical constructs also play important roles in the laws of physics is a well-known fact, although the precise reason for this is not really understood. I should like to delineate some of the intricate roles geometry is playing in contemporary physics.

In his autobiographical notes of 1946 Einstein said: "The most fascinating subject at the time that I was a student was Maxwell's theory. What made this theory appear revolutionary was the transition from forces at a distance to fields as fundamental variables."[1] Maxwell's equations, in today's notations, are as follows:

$$\nabla \cdot \mathbf{E} = 4\pi\rho \qquad \text{Coulomb's law}$$

$$\nabla \cdot \mathbf{H} = 0 \qquad \text{Gauss's law}$$

$$\nabla \times \mathbf{H} = 4\pi\mathbf{j} + \dot{\mathbf{E}} \qquad \text{Ampere's law}$$

$$\nabla \times \mathbf{E} = -\dot{\mathbf{H}} \qquad \text{Faraday's law}$$

The empirical bases of these laws were derived from fundamental experiments concerning electricity and magnetism carried out by physicists over approximately 50 years starting from the second half of the eighteenth century. Familiarity with these laws allowed Faraday (1791-1867) to create the geometrical concept of lines of forces (Fig. 1.1) and the concept of the field. Faraday was an experimental physicist with profound instincts, but he was not very learned in mathematics

---

Yuval Ne'eman (ed.), To Fulfill A Vision: Jerusalem Einstein Centennial Symposium on Gauge Theories and Unification of Physical Forces
ISBN 0-201-05289-X

Copyright © 1981 by Addison-Wesley Publishing Company, Inc., Advanced Book Program.
All rights reserved. No part of this publication may be reproduced, stored in a retrieval system, or transmitted, in any form or by any means, electronic, mechanical photocopying, recording, or otherwise, without the prior permission of the publisher.

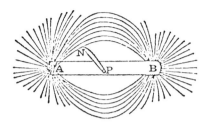

Fig. 1.1 Illustration of lines of force by Faraday in the Philosophical Transactions (1832).

and did not put down in mathematical language his concepts about electricity and magnetism. About him Helmholtz has written: "It is in the highest degree astonishing to see what a large number of general theorems, the methodical deduction of which requires the highest powers of mathematical analysis, he found by a kind of intuition, with the security of instinct without the help of a single mathematical formula."[2]

Although Faraday invented the field concept, the honor of inventing the term "field theory" belonged to Maxwell (1831–79). In 1865 Maxwell published a paper that is deservedly considered the greatest paper in physics in the last century. The title was "A Dynamical Theory of the Electromagnetic Field." Two sentences of it read as follows: "The theory I propose may therefore be called a theory of the Electromagnetic Field, because it has to do with the space in the neighborhood of the electric or magnetic bodies and it may be called a dynamical theory, because it assumes that in that space there is matter in motion by which the observed electromagnetic phenomena are produced." . . . "The Electromagnetic Field is that part of space which contains and surrounds bodies in electric or magnetic conditions." By its explicit statement, Maxwell's program was to write in mathematical formulas what Faraday had already conceived as physical ideas. In the process of working on his paper, once he materialized the empirical laws into equation form, he found that there were inconsistencies that could only be removed by adding the "displacement current." This development was of great importance and illustrates how instinct is oftentimes not quite enough. A detailed mathematical formula is decisive because with it one is able to manipulate, using well-developed formal tools of mathematics.

After he added the displacement current, Maxwell went on to demonstrate that his equations allowed for wave solutions. He deduced the velocity of the waves and compared it with the observed velocity of light, coming to the dramatic conclusion that light propagation is just electromagnetic propagation.

The excitement that he felt can be sensed from the following quote from a letter that he wrote to Lord Kelvin in 1861: "I made out the equations in the country before I had any suspicion of the nearness between the two values of the velocity of propagation of magnetic effects and that of light, so that I think I had reason to believe that the magnetic and luminiferous media are identical."

Maxwell had informed Faraday, who was forty years his senior, of his efforts to express Faraday's physical ideas in mathematical form. Faraday admired young Maxwell's efforts but was not entirely at ease: he was a little bit afraid, as a true experimentalist should be, of too much mathematical formalism. He was afraid that it would spoil his physical ideas. His attitude is vividly revealed in the following letter he wrote to Maxwell in 1857: "My dear sir, I received your paper, and thank you very much for it. I do not say I venture to thank you for what you have said about "Lines of Force," because I know you have done it in the interest of philosophical truth; but you must suppose it is work grateful to me, and gives me much encouragement to think on. I was at first almost frightened when I saw such mathematical force made to bear upon the subject, and then wondered to see that the subject stood it so well." Many experimentalists today have the same hesitations about accepting the very sophisticated mathematics that is becoming increasingly prevalent in the language of the theorists.

Maxwell talked about the luminiferous medium — that is, the ether. That he did so was quite natural since mechanical models were considered essential in Maxwell's time and a mechanical model must be constructed, of course, in a medium. "It seems to me that the test of 'Do we or do we not understand a particular point in physics?' is 'Can we make a mechanical model of it?'" wrote a contemporary of Maxwell's, Lord Kelvin.[3] Maxwell, following the beliefs of the time, made very elaborate mechanical models in order to explain the equations he had already obtained. His attitude about these models was ambivalent. There were places where he said explicitly that these models are unessential and merely pedagogical, yet there were other places where he said explicitly the medium must be there.

Einstein in his autobiographical notes referred to these matters and asked himself why it was so difficult for physicists to shed the concept of the stationary ether. His conclusion was that, in the nineteenth century, propagation had to be supported by a medium and the "vacuum" was considered just a special case of a "medium."

Maxwell's equations were studied by many physicists. Among them there were Hertz (1857-94), Lorentz (1853-1928), and Poincaré (1854-1912). A particularly important development was the discovery of the Lorentz transformation, a mathematical substitution that leaves Maxwell's equations invariant. But it required the genius of Einstein to tell the physicists the precise meaning of that transformation.

Brilliant as Poincaré was, learned as Lorentz was, neither dared to take the decisive and revolutionary step to reexamine our concept of simultaneity, a concept perhaps not just learned from our parents, but inherited in our genes through a million years of evolution.

To summarize:

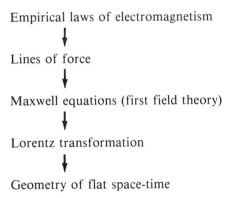

Thus, in the historical developments outlined the empirical laws of electromagnetism led to the concept of lines of force of Faraday, which, once put into mathematical form, gave the Maxwell equations. The latter gave birth to field theory, which is still the central theme of particle physics today. Maxwell's equations led to the concept of Lorentz transformations, which, through the work of Einstein, revealed the geometry of flat space-time.

A particularly important conclusion to draw, which was emphatically drawn by Einstein, was that symmetry plays a very important role. Before 1905 equations were derived from experience, and symmetries from the equations. Then — Einstein said — Minkowski made the important contribution of turning things around: first you declare the symmetry, and second you look for those equations that are consistent with it.

This idea worked deeply in Einstein's mind and from 1908 on he wanted to exploit it by enlarging the symmetry itself to a larger one. He wanted general coordinate symmetry, and that was one of the motivating forces that led to the general theory of relativity. The other motivating force was the idea of the equivalence principle. Putting these two together and struggling with them over seven years, he gave the world finally, in 1915, the geometry of curved space-time and the theory of general relativity.

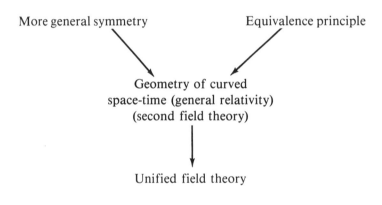

Fig. 1.2

General relativity is a field theory. Einstein was profoundly impressed with this singular creation of his and emphasized that one should build on it and embrace more of the world of physics, as shown in the accompanying diagram (Fig. 1.2). At that time the only other well-known interaction besides gravitation was the electromagnetic interaction, and therefore he wanted a unified field theory of these two.

On this project he labored most of the rest of his life. In the last editions of *The Meaning of Relativity* he added an appendix in which he proposed a unified theory with a nonsymmetrical $g_{\mu\nu}$ tensor. The antisymmetrical part was to be identified with the electromagnetic field tensor $F_{\mu\nu}$. This effort was not particularly successful, and there has been for some time among some people the impression that the idea of unification was a kind of obsession affecting Einstein in his old age. Yes, it was an obsession, but an obsession with an *insight* of what the fundamental structure of theoretical physics should be. And I would add that that insight is very much the theme of the physics of today.

This emphasis on unification produced something at once. It led many distinguished mathematicians, including Levi-Civita, Cartan, and Weyl, to look more deeply into possible additions to the mathematical structure of space-time.

Weyl made an effort of incorporating electromagnetism into gravitation. His idea led to what is called "gauge theory." This development dates back to 1918 and 1919. Since the proper treatment of coordinate invariance had produced gravity theory, Weyl thought that a new geometrical invariance could be tied to electromagnetism. His proposal was scale invariance: If x^μ and $(x^\mu + dx^\mu)$ are two space-time points in the neighborhood of each other and f describes a field such that it is f at x^μ and

$$f + \frac{\partial f}{\partial x^\mu} dx^\mu$$

at $(x^\mu + dx^\mu)$, Weyl considered the space-time-dependent rescaling of f by a scale factor that is given in the third row of the following diagram:

Coordinate	x	$x^\mu + dx^\mu$
f	f	$f + \frac{\partial f}{\partial x^\mu} dx^\mu$
Scale	1	$1 + S_\mu dx^\mu$
Scaled f	f	$f + (\partial_\mu + S_\mu)f dx^\mu$.

(1.1)

Notice particularly the scale factor, $1 + S_\mu dx^\mu$.

Now Weyl observed two things. First, S_μ has the same number of components as the electromagnetic potential A_μ. Second, by further developments, he proved that by requiring the theory to be invariant under the scale change, only the curl of S_μ occurred and not S_μ itself. That is a feature of A_μ also. So he identified S_μ with A_μ.

This idea did not work, however. It was discussed by several people, including Einstein, who demonstrated that Weyl's theory cannot possibly describe electromagnetism, and Weyl gave the idea up.

Then 1925 came and quantum mechanics was invented, completely independently of this development.

We all know that in classical mechanics it is not the particle momentum p that occurs, but, in presence of electromagnetism, it is always the combination

$$\pi_\mu = p_\mu - \frac{e}{c} A_\mu .$$

In quantum mechanics this is to be replaced by

$$-i\hbar(\partial_\mu - \frac{ie}{\hbar c} A_\mu) .$$

(1.2)

This was pointed out by Fock in 1927.[4] Immediately afterward London[5] compared expression (1.2) with the increment operator $(\partial_\mu + S_\mu)$ in (1.1) and concluded that S_μ is to be identified not with A_μ but with $-ieA_\mu/\hbar c$. So the only important new point is just the insertion of an imaginary unit i. This has the far-reaching consequence that the scale factor in (1.1) becomes:

$$1 - \frac{ie}{\hbar c} A_\mu dx^\mu \cong \exp\left(-\frac{ie}{\hbar c} A_\mu dx^\mu\right), \tag{1.3}$$

which is a *phase* change, not a *scale* change. Therefore, local *phase invariance* is the correct quantum mechanical characterization of electromagnetism.

Weyl himself had called his idea *Mass-stab Invarianz* at first but later changed the name to *Eich-Invarianz*.[6] In the early 1920s it was translated into English as gauge invariance.

If we were to rename it today, it is obvious that we should call it *phase invariance,* and gauge fields should be called *phase fields.*

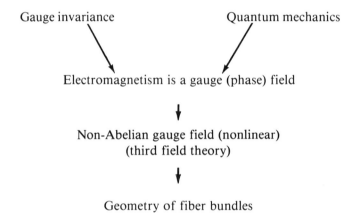

Fig. 1.3

Once one has understood that gauge invariance is phase invariance, one finds that the key idea is a *nonintegrable phase factor*. The substitution for the simple phase of complex numbers with a more complicated phase, namely an element of a Lie group, leads one to non-Abelian gauge theories.[7] (See Fig. 1.3.)

However, historically, the concept of non-Abelian gauge fields did not originate this way.[8] It instead sprang from two roots.

One root was the concept of local symmetry, which goes very simply this way: Protons and neutrons were known to be very similar; they form an isotopic spin doublet. They are not quite the same because one is charged and the other is not, and because there is a slight mass difference. Yet everybody believed that, if one could switch off the electromagnetic interaction, they would become truly identical. So let us go into a world where that is the case. We could call one of the nucleons proton and the other neutron, but which is which is a matter of convention. In fact, we can take any linear combination of the two states and adopt the convention of calling it the proton, and call the state orthogonal to it the neutron. A question arises whether this choice of convention in this room is binding for a person making experiments in the next room — that is, whether each observer in a different space-time point can choose the convention independently.

If you believe the convention should be independently chosen, you are obliged to construct a theory that allows for a point-to-point convention-freedom. You then naturally arrive at an SU_2 non-Abelian gauge theory.

The second motivation for non-Abelian gauge theories is the following: We all know that in the 1950s physicists had been discovering all sorts of mesons with different masses, spins, parities, and isotopic spins and were busily trying to write down interactions for them. Given only the requirement of consistency with Lorentz invariance and global internal invariance (isotopic spin), there were large numbers of allowed interactions and there was *no general principle* to choose among them. On the other hand, the electromagnetic interaction of any new particle was always prescribed in a unique way through the $p - (e/c)A$ recipe. That is, it was realized that the gauge principle allows you to write down interactions in a unique way once you have a conserved quantity (in the electromagnetic case, the charge). Since there were other conserved quantities besides the electric charge — for example, isotopic spin — the question was: "Can we duplicate exactly the same process for isotopic spin?" Trying to do that, one arrived at general non-Abelian fields.

It was immediately obvious that a non-Abelian gauge field, in contrast to the Maxwell field, has the very desirable property of being *nonlinear*. It therefore can generate itself without external sources. That we need field theories that are nonlinear was repeatedly emphasized by Einstein.[9]

Another attractive feature of non-Abelian gauge theories is related directly to the second motivation just discussed. It allows one to write down interactions with *one* coupling constant, and that is the central theme today of efforts to unify all strong, electromagnetic, and weak interactions.[10]

We are today very far from that unification, but experiments of the last six years have shown amazing agreement with a specific model of unification of weak and electromagnetic interactions: the Weinberg–Salam model.[11] This model has two conceptual ingredients: one is the non-Abelian gauge theory; the other is broken symmetry.

The concept of broken symmetry is very handy.[12] One has to cope with the fact nature is not completely symmetrical; masses and coupling constants are not all the same. The broken symmetry mechanism allows one to have a theory where the interaction Lagrangian possesses the full symmetry and yet it allows for experimental manifestations that are not fully symmetrical. An example of broken symmetry is found in the crystal. The Hamiltonian of a crystal has full rotational symmetry. That symmetry is manifest in the liquid phase, but at low temperatures the system becomes a solid that physically is not a rotationally symmetrical system. In other words, a symmetrical system of equations may, through the symmetry-breaking mechanism, give rise to unsymmetrical experimental manifestations.

Combining the idea of gauge fields and symmetry breaking, Weinberg in 1967 and Salam in 1968 constructed a model that unified weak and electromagnetic interactions.[13] It is remarkable that it turned out to agree with experiments very well, although from the theoretical viewpoint it is one of many possible models.

I want to emphasize that the lack of full symmetry in a broken symmetry model is a low-temperature effect, as in statistical mechanics. In this view, doing usual *high-energy* experiments we are always probing broken symmetry gauge theories at *zero temperature*. No matter how high the energy is, the initial states are always *prepared* at zero temperature. Therefore, the full symmetry is never realized at any energy. However, if we could reach a sufficiently *high temperature* regime, we would see the full symmetry restored. This is satisfying because it means that the full symmetry can in principle be directly tested experimentally.[14]

There have also been efforts to develop gauge theories of strong interactions, about which there are many discussions later in this volume.

In the diagram about the historical origin of gauge fields presented earlier (Fig. 1.3) I illustrated the development of concepts that physicists found necessary for the description of experimental laws. The concepts were motivated by knowledge of physical phenomena. It was therefore very surprising[15] that the concept of gauge field proved to be identical to a *geometrical* concept called fiber bundles that was developed by mathematicians entirely independent of any relation it might bear to physical reality.[16]

Not only is it true that gauge field is a geometrical concept, but it turns out that *topological* complexity is important for gauge fields. Appreciation of this point came with Dirac's magnetic monopoles, 't Hooft's monopoles, and the pseudoparticle solution of Belavin et al.[17] In fact, it is through the topological complexities, for example, necessitated by the Dirac monopole, that it became absolutely convincing for me that gauge fields, such as Maxwell fields, are not just expressible in the geometrical language of fiber bundles, but *must* be so expressed to bring out their full meaning.[18]

The development of fundamental concepts in physics in this century has turned out to be framed in deep concepts of mathematics:

Special relativity	Four-dimensional space-time
General relativity	Riemannian geometry
Quantum mechanics	Hilbert space
Gauge theories	Fiber bundles (with topological complexities)

This list does not include uses of mathematics as a tool to solve mathematical problems in physics. It would be wrong, however, to think that the disciplines of mathematics and physics overlap very much; they do not (see Fig. 1.4). And they have their separate aims and tastes. They have distinctly

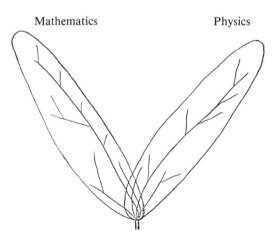

Fig. 1.4 Overlap of mathematics and physics.

different value judgments, and they have different traditions. At the fundamental conceptual level they amazingly share some concepts, but even there, the life force of each discipline runs along its own veins.

NOTES

1. A. Einstein, "Autobiographical Notes," in Albert Einstein, *Philosopher-Scientist,* edited by P. A. Schilpp (Library of Living Philosophers, Evanston, Ill., 1949), pp. 1-95.
2. Helmholtz lecture (1881) quoted in A. Koestler, *The Act of Creation* (Dell, New York, 1964).
3. Quoted in S. P. Thompson, *The Life of Lord Kelvin* (Macmillan, London, 1910), p. 8301.
4. V. Fock, Z. f. Phys. **39**, 226 (1927).
5. F. London, Z. f. Phys. **42**, 375 (1927).
6. For this history see C. N. Yang, Ann. New York Acad. Sci. **294**, 86 (1977).
7. See ibid.
8. See ibid.
9. Einstein, op. cit. in n. 1, p. 89.
10. See A. Salam in *Proceedings of the 19th International Conference on High Energy Physics, Tokyo 1978* (Physical Society of Japan, 1979). There is much more about this topic later in the present volume.
11. S. Weinberg, Phys. Rev. Lett. **19**, 1264 (1967); A. Salam, in *Elementary Particle Physics,* edited by N. Svartholm (Almquist and Wiksells, Stockholm, 1968).
12. Many people contributed to bringing broken symmetry into particle physics. Among them were Nambu, Goldstone, Anderson, Higgs, Salam, and Weinberg. See the review article by S. Weinberg, Rev. Mod. Phys. **46,** 255 (1974).
13. Weinberg, op. cit. in n. 11; Salam, ibid.
14. I was not convinced of the relevance of the idea of broken symmetry in elementary particle physics in the 1960s. I argued that the rotational symmetry of the Hamiltonian of a crystal can be tested experimentally by considering a case of only a few atoms, for which rotational symmetry is precise. However, with elementary particle physics that is not possible; the symmetry cannot be precisely tested, I argued at that time. [See discussions between Nambu and me after R. E. Marshak's talk in *Proceedings of the International Conference on Elementary Particles, in Commemoration of the 30th Anniversary of Meson Theory,* edited by Y. Tanikawa (Yukawa Hall, Kyoto University, 1966).]
15. See Yang, op. cit. in n. 7.
16. The mathematical concept of fiber bundles was first discussed by H. Hotelling in 1925, H. Seifert in 1932, and H. Whitney and E. Stiefel in 1935. It assumed great importance starting with the mathematical developments of the 1940s. See A. Weil in *Shiing-shen Chern, Selected Papers* (Springer-Verlag, New York, 1978).
17. P. A. M. Dirac, Proc. Roy. Soc. **A133,** 60 (1931); G. 't Hooft, Nuclear Phys. **B79,** 276 (1974); A. Belavin et al., Phys. Lett. **59B,** 85 (1975).
18. T. T. Wu and C. N. Yang, Nuclear Phys. **B107,** 365 (1976).

2. THE QUEST FOR UNITY: GENERAL RELATIVITY AND UNITARY FIELD THEORIES

Peter G. Bergmann

A centenary jubilee is a good time to evaluate an outstanding individual's intellectual contributions. Enough time has elapsed to free us of the fashions of the moment, but that time is sufficiently short so that some of us who have come under that person's influence are still alive. The memory of Albert Einstein, who had earned the greatest fame of all physicists in his own lifetime, has come to serve as a focus for a vast range of endeavors, both humane and scientific, on the occasion of the one-hundredth anniversary of his birth. Having had the privilege of being associated with Einstein in my youth, I am very happy to have this occasion to express my deep gratitude for the instruction and stimulation that I received from him. Albert Einstein's memory will last as long as there are human beings who strive for a more perfect society, and for a deeper comprehension of the physical universe.

Einstein's contributions to physics are many. I shall address myself to but one: his quest for unity in science, which found its expression in his formulation of the theory of relativity and in his search for a unitary field theory that would lead beyond it.

Let me begin with the special theory of relativity. At the turn of the century there was among the many puzzles confronting physicists one that touched the very foundations of all natural science; it concerned the nature of space and time. Most of physics was then dominated by mechanics, which dealt with the interaction of physical bodies with each other. The crowning achievement of mechanics had been the complete and quantitative explanation of the workings of the solar system, so that astronomers were able to predict such events as eclipses decades and centuries away, with great accuracy and complete reliability. The laws of mechanics, formulated by Isaac Newton, concerned the *accelerations* of the interacting bodies, determined by the forces of interaction, which in turn depended only on the (instantaneous) configuration. If Newton's laws were valid, then it followed that in our universe there is no possibility of identifying a state of rest or, for that matter, of absolute motion. As far as absolute properties of space and time were concerned, the laws of mechanics called for a set of states of nonrotational uniform rectilinear motion, all of equal stature, which are usually referred to as inertial frames of reference.

The then new physics of the electromagnetic field, brilliantly formulated by Faraday, Maxwell, and Lorentz, differed from the laws of mechanics in that they introduced the notion of the pervasive field, which was to fill the space between the particles. The laws of the field, however, involved a *velocity,* the speed with which any electromagnetic disturbance would spread in empty space; today we call this the speed of light. The electromagnetic laws would seem to single out one state, the state of absolute rest, that state in which in the absence of matter the speed of propagation of electromagnetic waves is isotropic. As everybody knows, the search for that state, or frame, of absolute rest was unsuccessful: it appeared that the electromagnetic field is totally insensitive to the absolute motion of the earth through space. This experimental fact, confirmed in the meantime in all manner of ways, apparently presented an internal inconsistency, unless of course you assumed that the earth represented the state of absolute rest. And that would have been a regression to Ptolemaic ideas, unacceptable to nineteenth-century scientists.

In this situation increasingly tortured proposals were considered by the outstanding theorists of that time. Einstein's contribution was revolutionary because it was formally simple, and deep at the conceptual level. Aside from permitting observers in different states of motion to have different scales of distance and of time, Einstein demonstrated that the simultaneity of distant events would be observer- or frame-dependent if one accepted the proposition that the speed of light cannot be exceeded by any signaling device. By an intricate argument, into which I shall not enter here, but which is intricate not because of abstruse mathematics but because of a very delicate analysis of experimental procedures, he showed that once the notion of absolute time marks is dropped, two moving observers each can perceive the other's clock to be slow, and each can perceive the other's yardsticks to be contracted. The paradox was resolved by a profound modification of classical space and time concepts.

A very few years later Minkowski discovered the natural mathematical formulation of Einstein's new physics: the four-dimensional space-time model. The relation between space and time measurements of two observers moving differently was analogous to a rotation in four dimensions, except for a few signs that differed from an ordinary rotation.

Through his revision of the space-time concepts Einstein had succeeded in removing from physics the apparent contradiction between the (classical) principle of relativity of mechanics and the laws of electrodynamics. To this extent unity was restored, but a new contradiction was created. Newtonian mechanics involved at its foundations the notion of absolute simultaneity; the forces between distant bodies, for instance between the sun and the earth, depended on their instantaneous distance from each other, which in relativity would differ for different observers. If the new theory of space and time was to prevail, mechanics needed to be modified.

Relativistic mechanics was designed to bridge the gap partially, by making the mass velocity-dependent — hence the proportionality between mass and energy — and by modifying the force. These changes had, however, no effect on the dependence of the action at finite distances on absolute simultaneity. This could be accomplished only by replacing the Newtonian action by the intermediary of fields. Thus the need arose for a relativistic gravitational field.

It is possible to introduce relativistic field equations for a gravitational field with relatively little effort. Einstein was, however, troubled by two considerations. One was that there were more ways than one of doing so, and there seemed to be very little grounds for choosing one over the others. The other was a peculiar property of gravitation, the universality of gravitational acceleration. In a gravitational field all bodies undergo the same acceleration, on the surface of the earth for instance, 9.8 meters per second per second.

In an electric field the force acting on a body depends on its electric charge, and the acceleration on the ratio of its charge to its mass, e/m. No analogous parameter enters into the expression for acceleration caused by a gravitational field. This fact had already been ascertained by Galileo, and certainly had been recognized by Newton; but it remained a curiosity. It was Einstein who recognized the implications. If gravitational acceleration is the same for all bodies, then it vanishes, locally, for an observer who himself undergoes the same acceleration. One is led naturally to the notion of a free-falling frame of reference, rather than the inertial frame of reference. The difference between the two concepts is this: Whereas an inertial frame of reference presumably extends over the whole universe, a free-falling frame is defined only locally, in a sufficiently small region. An astronaut, or cosmonaut, will perceive no gravitational field in his free-falling vehicle, but distant objects appear to be accelerated relative to himself. Thus the local uniformity of gravitational acceleration precludes the determination of inertial frames of reference by local means, replacing these frames by constructs that cannot be extended globally.

This line of reasoning leads to the general theory of relativity, Einstein's theory of the gravitational field. When the new theory was completed, some sixty years ago, it replaced the space-time of the special theory of relativity by a yet more general geometric concept, that of a Riemannian space-time, which locally has properties resembling those of the special theory but on a larger scale is much more involved, being a curved manifold.

I do not wish to give you the impression that the progress from the special to the general theory was straightforward, or logically inescapable. Far from it. If inertial frames cannot be determined by local observations, it might be possible to preserve the concept by relying on observations of distant objects. This is in fact what astronomers do. But reliance *in principle* on distant objects runs counter to the spirit of a field theory, which relies on physical interaction of fields in the neighborhood of the particle, not at a distance, thus circumventing the embarrassment of instantaneous action at a distance. From a logical point of view, the progress toward general relativity depended on a number of choices to be made; its eventual adoption, first by Einstein himself and later by the community of physicists, depended on the esthetic appeal of the finished theory, and on its confirmation by experiment and observation.

As for experimental confirmations, the universality of gravitational acceleration has been confirmed to an accuracy beyond 10^{-11}. As for relativistic effects — that is, gravitational effects that deviate from the predictions of classical mechanics and of the special theory of relativity — all quantitative observations that can be performed with today's technology have confirmed Einstein's theory well within the bounds of instrumental error, including such cases where competing modern theories predicted different results. This is an ongoing enterprise.

The issue of esthetic appeal of general relativity is closer to the principal theme of this essay. Once more general relativity had restored a measure of unity to physics by modifying our ideas of space and time, which lie at the foundations of any dynamical conceptual construction. The new framework accommodated gravitation. Its essence was to be sought not in the properties of the single local free-falling frame of reference but in its relation to free-falling frames in adjacent regions. These relations were subject to field laws that were chosen according to principles of formal simplicity and the requirement that for weak fields the classical results should agree in lowest approximation with those of the new theory.

One major conceptual difficulty was removed from the new theory some twenty years after its inception: the interaction of the local field with a particle. Every mass serves as a source of the gravitational field, just as each charge is a source of the electromagnetic field. At the site of a particle the field becomes very large. If the particle is conceived of as a mass point, the field becomes infinite. But the force that affects the kinematic behavior of the particle is determined by the surrounding field. What if that field is finite? The first response, historically, was a holding operation. If the particle itself was small, if its mass was slight, then one could imagine the field as it would be if the particle under consideration did not exist. Einstein then postulated that such a small particle would travel on a so-called geodesic, a curve in space-time that corresponds to unaccelerated motion in special relativity, or in the local free-falling frame. This assumption was in fact the point of departure for the geometric interpretation of the fact of uniform gravitational acceleration.

But what if the particle was not so small? How would one deal with the problem of a double star, for instance, in which the field caused by the "other" star could not reasonably be assumed to be larger than the field caused by the star under scrutiny? Eventually Einstein, Infeld, and Hoffmann developed an approach that was applicable in such cases. They found that the field laws outside a particle could not be satisfied unless the particle itself behaved "properly." Viewed from that angle, the behavior of sources of the gravitational field was determined by the field laws themselves. This was a property not shared by other field laws, and certainly not by those of electrodynamics. Thus general relativity turned out, after all, to be conceptually more nearly of one piece than any physical theory then known.

With the laws of motion of particles having been absorbed into the logical structure of the field laws, mechanics, once the dominant structure of theoretical physics, was all but eliminated from it. The quest for unity had apparently reached its objective. But there were several hairs in that ointment.

Atomic physics, we know, is not governed by classical laws but obeys quantum rules. General relativity, however, is nonquantum. It satisfies essentially strictly deterministic laws, whereas quantum laws are essentially statistical. Einstein could never bring himself to accept statistics as the definitive form of the laws of nature, even though as a young person he had made major contributions to quantum theory and to statistical mechanics. He always considered statistical approaches preliminary to a better understanding, which would be strictly causal.

The second drawback of general relativity was that it treated particles as singularities of the field, infinities, and failed to explain their structural properties, such as the masses, charges, and other characteristics. Finally, as nature is not purely gravitational but allows for other forces as well, the gravitational and the nongravitational fields appear to be essentially different. From the point of view of general relativity, gravitation is needed in order to give space and time their geometric structure; all the other forces are gratuitous.

Unitary field theory was intended to remedy all these blemishes.

From the early twenties to the end of his life Einstein developed ever new approaches to unitary field theory. At the time of his death, he was working, together with Bruria Kaufman, on the so-called asymmetric theory.

Riemannian geometry in four dimensions is a well-defined and fairly rigid structure, which admits very little variation in the proposed dynamical laws. Somehow this mold must be broken if more physical fields than gravitation are to be accommodated within the geometric framework. Before I discuss a few of these attempts, permit me to address myself to a preliminary question: What is geometry?

I suspect that there is no answer to this question that will satisfy everybody. Basically, geometry might be considered any kind of mathematical structure that begins with the construction of a set of points that satisfies the minimal properties of continuity that justify one in speaking of a space. A space may, but need not, involve such concepts as volume and distance; it may, but need not, involve the existence of vector fields and the possibility of defining when two vectors at distinct locations are to be considered parallel to each other.

These are but examples of properties that geometric spaces might possess. Many more have been investigated; in fact many more have been used by physicists in their pursuits and endeavors to understand nature. Depending on the properties ascribed to a new model for space-time, its structures might lend themselves to interpretations that are reminiscent of fields known to physicists. How does such a "geometrization" contribute to unification? Einstein has stated repeatedly that he did not consider geometrization of physics a foremost, or even a meaningful objective, and I believe that his comments remain valid today. What really counts is not a geometric formulation, or picturization, but a real fusing of the mathematical structures intended to represent physical fields.

How can we visualize such a fusing? One possibility, suggested by the history of relativity itself, is that the decomposition of fields into gravitational, electromagnetic, "strong," and "weak" nuclear forces might depend on the frame used for their description, that, for instance, a field that appears in one frame to be purely gravitational is mixed gravitational and electromagnetic in another frame. This is possible if the variety of equivalent frames, or modes of description, is sufficiently large.

There are other possibilities. Some fields might require additional fields complementing them before any meaningful differential operations can be defined. This situation obtains, for instance, in Weyl's geometry, on which I shall comment.

To formulate, and to survey, such possibilities a geometric formulation often is a real help. Essentially, mathematicians and physicists too proceed intuitively when they endeavor to create new concepts and relations. Geometry often helps to "think in images." Thus, geometry may serve as a heuristic device. That might not exhaust its role, but it is a major part of it.

I cannot give you a complete listing of all attempts, by Einstein and by many others, to create generalizations of the four-dimensional Riemannian model of space-time. Though I have worked on unitary field theories myself, I cannot claim any comprehensive knowledge. One whole class of attempts may be characterized as maintaining the four-dimensionality of space-time but modifying or enriching the Riemannian structure. In this class belongs, for instance, Weyl's geometry. Weyl weakened Riemann's idea of an invariant distance at the infinitesimal level; he replaced it by the notion of relative distance. Only the ratio of two distances would have any invariant (frame-independent) meaning. With this weakening of the metric concept, one cannot form differential structures without introducing a pseudo-vector field that looks like the potentials of the electromagnetic field.

Another enrichment, suggested originally by Cartan, generalizes the notion of parallel transport of vectors. In Riemannian geometry if you introduce a free-falling frame of reference, then a vector is parallel to a vector if, in that frame, the components are the same. In Cartan geometry, they may be rotated.

Finally, in Einstein's asymmetric theory the dot product of two vectors (at the same point) is not symmetric in the two vectors, $\mathbf{A} \cdot \mathbf{B} \neq \mathbf{B} \cdot \mathbf{A}$. In all three of these examples the minimal geometric structures are richer than in Riemannian geometry, so they are capable of accommodating a greater variety of physical fields.

In Weyl's geometry the gravitational and the electromagnetic structures are distinct in that they are not being converted into each other under changes of frame, but they are both required to produce a harmonic whole. In Einstein's asymmetric theory there exists one type of change of frame that mixes the Riemannian with the other parts of the geometry; in the Cartan geometry I do not see that kind of fusing, but from a somewhat different point of view Cartan's geometry also hangs all together.

How can any enlargement of the geometry lead to an understanding of the properties of particles? That is a very difficult question to answer. The occurrence of singularities in a field theory represents a sort of breakdown of that theory: The field equations admit of solutions that go out of control of those equations and ruin the causal character of the field laws. And there seems little doubt that general relativity as we know it today leads to singularities under a variety of circumstances. There are no solutions that might be interpreted as particles that are everywhere finite. Once you are dealing with different field equations, you can hope that such solutions might exist. I might add that the theorems concerning the unavoidability of singularities in the standard theory were all discovered long after Einstein's death, mostly by R. Penrose and by S. Hawking. I have not seen their methods of proof extended to any of the unitary theories, but this might well be possible.

If nonsingular solutions should exist, then one might investigate whether these can in some way be related to the properties of particles that occur in nature. There is a way to relate the ratio of charge to mass of an elementary particle to a pure number, of the order of 10^{20}, depending on the kind of particle. A theory of elementary particles should yield at the very least numbers like this one.

Einstein hoped to obtain quantum rules in a similar fashion. If particles interact with each other, it is not likely that singularities can be avoided in the course of time unless the initial conditions are just right. I suspect that few practitioners of unitary field theory today would share these hopes; many of them would feel, I believe, that to achieve successes in other respects is worth their efforts, even if quantum theory will continue to flourish in its present form.

There are other kinds of unitary field theories, including some that today claim a great deal of interest. These utilize, in some way or another, an increase in the number of dimensions of space-time. One famous example is Kaluza's proposal. He increased the number of dimensions to five, without changing the Riemannian character of the model. He was thus able to increase the number of components of the metric so as to accommodate the electromagnetic field as well. He set one extra component equal to a constant, because he had no use for it. To account for the observed four-dimensionality of space-time, he assumed that no field depended on the fifth coordinate.

Strangely enough, Kaluza's field, though conceived of as a single structure, the metric, separated quite naturally into the gravitational and the electromagnetic fields, in a manner that did not at all depend on the frame used. To this extent Kaluza's fusion of fields failed. But his idea continued to intrigue others, and several variants were tried in the course of the years. One, by Einstein, V. Bargmann, and myself, replaced Kaluza's assumption of strict independence from the fifth coordinate by a weaker assumption, that the universe is closed in the fifth dimension, that it looks a bit like a tube, and that the dependence on the fifth coordinate, limited as it must be if the circumference of the tube is sufficiently small, has something to do with quantum phenomena. Alas, the idea did not work out.

Another idea, discovered and rediscovered several times over, was not to kill the supernumerary field component but to retain it and to assign to it such tasks as to serve as a cosmological parameter. Brans's and Dicke's so-called tensor-scalar theory is one of these attempts, though I believe that these authors were initially unaware of the preceding history of that idea.

There are other methods for increasing the dimensionality of space-time. One is to permit the coordinates of space-time to assume complex values. Penrose's twistor formalism is a case in point. Complexification is utilized by some authors as a mere technical device for discovering new solutions of Einstein's equations in the real domain; this is a productive approach, but has little to do with unitary field theory. Others, and I believe Penrose is among them, take complex space-time seriously. They hope to break new ground. Formally, a complex number is a pair of real numbers. A complex four-dimensional space or space-time is in that sense equivalent to a real eight-dimensional manifold. But if the pairing into sets of complex coordinates (or dimensions) is taken seriously, then the rules of algebra and of analysis applied to complex numbers are equivalent to the introduction of an additional invariant structure, the so-called complex structure, which must be reproduced under all changes of frame. Thus the structure of a complex space differs significantly from that of a real space having twice as many dimensions. Penrose hopes that by pursuing this line of inquiry he may succeed in understanding elementary particles and perhaps also the quantum character of nature.

If complex numbers are good, hypercomplex numbers may be better. Hypercomplex numbers are one way of looking at algebras that have at least some of the properties of the algebra of ordinary numbers. Whereas complex numbers are equivalent to pairs of real numbers, hypercomplex numbers involve larger multiplets. Their rules of arithmetic cannot be as simple as those involving real and complex numbers. They will involve noncommutative products ($ij \neq ji$). Most systems of hypercomplex numbers also contain null divisors, nonzero elements whose product with some other nonzero number equals zero.

One particular type of hypercomplex algebras is known as Grassmann algebras. The product of any two of the basic elements of a Grassmann algebra is anticommutative, $ij + ji = 0$. Interest in Grassmann algebras and in fields formed with their help originated with mathematicians and with physicists who were impressed with the possibility of using them in elementary particle physics. It has been observed that there are collections of elementary particles that resemble each other even though some members of the set have integral, others half-odd spin. In quantum theory the state vectors (or wave functions) belonging to particles with integral spin are symmetric with respect to the permutation of particles; those belonging to particles with half-odd spin are antisymmetric. One type obeys Einstein–Bose statistics, the other Fermi–Dirac statistics. Some elementary-particle physicists believe that there must be some changes in frame that change one kind of particle into the other field. Formally, such a scheme can be set up, and most conveniently, with the help of Grassmann "numbers." These endeavors go under the name of supersymmetry. If they involve an attempt at unitary field theory, they are called supergravity.

There is some formal resemblance between complex field theories and supergravity. As for motivation, I am impressed with the seriousness of these novel attempts to draw inspiration from elementary-particle physics, an area in which large numbers of people are obtaining new and exciting insights. Supergravity meets one objection that has been raised against the search for unitary field theory: that it has been purely speculative, without nurture from the findings of experimental physics. Supersymmetry and supergravity are speculative, to be sure; but they are influenced by high-energy physics, and that to me is a very attractive feature. There are many unsolved problems in these attempts, too. I certainly do not wish to give you the impression that I am all sold on supergravity. Rather, I should say, this many years after Albert Einstein's death a new generation of unitary-field theorists is taking up the torch, and they are proceeding along novel lines. They have good contact with other frontier areas of physics; one can only wish them well.

In twentieth-century theoretical physics there have emerged a number of major areas, each dominated by a closely reasoned and closely linked set of laws. These areas have emerged in response to the human quest for understanding, for comprehending the individual event as an instance of an overriding general principle. Albert Einstein created one such area, the theory of gravitation, and he did so by deepening our grasp of the nature of space and time, the scaffolding on which all of physical science takes place. He had hoped to expand and to strengthen this scaffolding so as to take in the physics of the atom and of the subatomic world as well, but this attempt did not succeed in his lifetime.

It behooves each of us to proceed in the manner we each judge best, whether or not it resembles closely Einstein's own way. All our endeavors are supported by what he achieved, and our resolve is strengthened as we perceive Einstein's tenacity, but also his creativity and flexibility. To most of us it is given to contribute but one small step or two toward man's understanding of nature. Let us be content with that. The quest for unity will never be sated. Each achievement will reveal new vistas and mysteries to be conquered.

Acknowledgment

Preparation of this paper was partially supported by National Science Foundation Grant #PHY78-06721.

Discussion

Following the Paper by P. G. BERGMANN

Chairman, A. PAIS, Rockefeller U.

Pais: Professor Bergmann, I have a quotation here from a letter that Einstein wrote to Solovine, one of his old friends. In it he says (and that is almost thirty years ago), "The theories which have gradually been connected with what has been observed have led to an unbearable accumulation of independent assumptions." The quest of Einstein for unification has become much more urgent as the complexity of the phenomena make it obvious to the naked eye that this just can't go on. And in the program that Einstein himself set (and in which he failed), he is not alone anymore. Thirty years later we are still not there, although we would like to think we have made some progress. But not much.

Ne'eman (Tel-Aviv U.): As mentioned by Professor Bergmann, supergravity is a step in this search for unification. From the point of view of unification itself, for instance, it has already produced a very elegant way of unifying electromagnetism with gravitation, better than Einstein's results. In addition, it has produced some useful consequences in other directions. Example: removing some paradoxes relating to spin 3/2 fields. However, it occurs to me that it has even provided a contribution to yet another aspect of Einstein's program, as described by Professor Jammer (see history and philosophy volume of Jerusalem Einstein Centennial Proceedings), an aspect that is less popular with us. Professor Jammer said that in his search for a unified field theory, Einstein was also hoping to solve the quantum aspect. He was hoping to find a better formulation of quantum mechanics, as a by-product.

There is no such solution to date, but it is interesting that supergravity does force one quantum aspect into gravity. Supergravity is a gauge theory based on supersymmetry. It can be regarded as extending space-time with a spinor "region." The introduction of such a spinor submanifold is a result of quantum mechanics. Remember that the only reason for the existence of spin is the fact that the S matrix yields probabilities only by squaring amplitudes, and phases do not count. This allows the appearance of the double covering of the Lorentz group. Supergravity could thus not exist if it were not for quantum mechanics!

Bergmann: I want to comment in one sentence: I agree with everything that Yuval said.

Pais: I would like to make one remark on this point. Many of us, myself included, feel that we subscribe to the Einstein program of a unified field theory, but not to his wish to derive quantum mechanics from underlying geometry. This is wrong, in my opinion. That is how it looks to us in 1979. We may of course be ashamed of this statement someday. Professor Bergmann, can you read us the quotation you just found?

Bergmann: This is from the autobiographical notes: "One must however not legitimize the sin indicated above so far as to assume that distances are entities of a special kind, different from other physical quantities ('reducing physics to geometry', etc.). Man darf aber die erwähnte Sünde nicht so weit legitimieren, dass man sich etwa vorstellt, dass Abstände physikalische Wesen besonderer Art seien, wesensverschieden von sonstigen physikalischen Grössen ('Physik auf Geometrie zurückführen,' etc.).''

3. GEOMETRIZATION OF UNIFIED FIELDS

Feza Gürsey

1. Introduction

Einstein's much maligned later work conjures up key words like *fields, geometry,* and *unification,* which today appeal to us more than to a previous generation. The work itself concerning a unified, geometric theory of gravitation and electromagnetism, the "total field," is forgotten or politely ignored. The title of the present chapter is meant to recall these late Einsteinian themes that could also come in different permutations, such as unification of geometric fields or field theory of unified geometries. I have preferred to put the emphasis on the word *geometry*.

My aim here is not to resurrect the specific theories that left Einstein himself unsatisfied, but to highlight some features of these artful constructions that keep turning up and getting themselves woven into the fabric of our modern field theories.

In his endeavor to understand the global aspects of general relativity and his almost quixotic efforts to construct a unified field theory, Einstein showed extraordinary imagination in inventing some remarkable geometries. After quickly reviewing some of these structures I shall attempt to put our current field theories in an Einsteinian perspective. I also hope to sketch some possibilities and discuss some speculations suggested by a neo-Einsteinian approach to high energy physics that could enable us to understand geometrically some selected features of fields, the structure of the space of internal symmetries (generalized charge space), and problems of unification of fields as well as criteria for the uniqueness of the resulting theories. As a specific example, I shall discuss the geometry of finite Hilbert spaces, which come in real, complex, quaternionic, or octonionic versions.

One general remark before embarking on this brief review: A sense of bewilderment can no doubt result from reading various interesting reports on the present state of physics. In their effort to make some sense of the amazing world of particles, physicists have already discerned remarkable regularities in terms of generations of quarks and leptons with interactions dictated by gauge principles. In spite of these partial successes there are seemingly insuperable difficulties in the way of a truly unified field theory of all interactions including gravity. Brave attempts have been made to overcome these obstacles and many, perhaps too many, ideas have been proposed. We are still very much dependent on experiments utilizing new and future machines to sift these ideas or suggest new theoretical directions. As Pauli once said: "Physicists nowadays have too many ideas and too little conviction." Einstein did have conviction, perhaps too much conviction that made him reject quantum mechanics and made him say: "God does not play dice." In the post-Einstein era we live in an age of doubt and skepticism. We also like to gamble. Here, you are watching contemporary physicists monitoring the casino of distant laboratories and throwing dice in a very un-Einsteinian style while anxiously waiting for the roulette of e^+e^- or p^+p^- rings to stop and tell them who has won and who has lost. Toward the end of this chapter I review some of these dice configurations and restrain myself from hazarding guesses about their relative probabilities.

2. The Neo-Einsteinian Doctrine

Before focusing on the impact of Einstein's concept of unification of fields and the geometries he created for that purpose, let me review the general framework under which we try to formulate fundamental modern physical theories. This framework owes much to Einstein. The tenets of what I will call the neo-Einsteinian doctrine are held by the majority of contemporary physicists, who view them as forming a body of working hypotheses and inspirational guidelines rather than articles of faith. The new pragmatism dictates that they should be discarded as soon as they do not prove useful or conflict with experiment. So far they have proved to be remarkably fertile and resilient.

Einstein held that "physical reality" was the space-time continuum endowed with properties that could be described by a local field satisfying causal differential equations that simultaneously characterize the geometry of space-time. Hence, in the same way that mass and energy coincide we have the convergence of the three concepts: space-time, field, and geometry into the unifying concept of physical reality. The neo-Einsteinian holds a modified view. Physical reality is not confined to space-time and the fields that describe it are not classical but quantum fields. Since quantum fields are operators, they can be defined only in a Hilbert space at each point x of space-time or at the point p of momentum space. The state of a physical system located at x or having the momentum p is further labeled by quantum numbers such as spin, charge, color, flavor, particle number, and so on. It follows that now physical reality has turned into a porcupine with the spines being the Hilbert spaces sticking out of each point of its space-time body. This structure is described by local quantized fields that may also be associated with a new geometry (a quantum generalization of fiber bundle geometry) that is much richer than the simple extensions of Riemannian geometry envisaged by Einstein.

Here are some typical Einsteinian concepts and beliefs as modified by the neo-Einsteinian physicsts whose additions occur between square brackets.

1. *Fields (Local Observables).* The fundamental laws of physics are expressible as a [quantum] field theory in the form of [operator valued] field quantities obeying differential equations. Examples: Maxwell field, gravitational field, [supergravity, Weinberg–Salam theory, chromodynamics].

2. *Symmetry Principle.* The field equations must be determined from invariance principles. This is implemented by deriving the field equations from an action principle where the action that is a functional of the fields is invariant under the symmetry group associated with the invariance principle. Examples: invariance principle: general relativity, associated symmetry group:

diffeomorphisms in space-time, action: the Einstein–Hilbert action, field theory: gravity [supergravity with action invariant under local supersymmetry; Yang–Mills theories with action invariant under a given local compact Lie group].

3. *Geometrical Principle.* Field equations must have a geometrical interpretation in space-time [and in a Hilbert space associated with internal symmetry, provided that the base is space-time or superspace]. Examples: Einstein gravity: the source of the metric is the energy momentum tensor, which is proportional to the modified Ricci curvature tensor (the Einstein tensor). [In superspace the supercurvature tensor is the source. In Yang–Mills fields the field quantities describe the curvature of the fiber bundle.]

4. *Wave-Particle Duality.* a. Field theories must incorporate wave-particle duality. This was Einstein's main motivation when he started constructing unified field theories. [The neo-Einsteinian achieves this aim by canonical quantization or Feynman path integral quantization of the fields starting from the Hamiltonian or Lagrangian formulations of the action principle. The method works for linear as well as nonlinear theories.]

According to Einstein there are some unsolved problems in implementing wave-particle duality. His guidelines were the following: 1) Particle solutions occur only in nonlinear theories. 2) Quantization must be of a topological nature in order to preserve invariance under diffeomorphisms. 3) Particle solutions must be nonsingular stable solutions of the nonlinear field theory classified by a discrete quantum number.

It will immediately occur to the modern physicist that Einstein wanted to explain field quanta in terms of soliton solutions of the field equations. It is true that such solutions can be yielded only by nonlinear field equations. They are also classified by a topological quantum number. Since Maxwell's equations are linear, although the fields can be quantized and yield photons by being turned into operators as in quantum electrodynamics, if we restrict physical reality to space-time by banning the Hilbert space, we are left with only the possibility of nonlinearizing the Maxwell field by embedding it in a larger unified field theory of gravitation and electromagnetism and looking for soliton solutions of the unified theory. This was Einstein's heretic choice. However, when we consider Yang–Mills fields in our enlarged geometrical picture, then the vacuum of the theory becomes quantized in the Einsteinian way with the different vacua being classified by a topological number. This would not be possible if Yang–Mills equations were linear. Hence the alternative quantization invented by Einstein does not describe the wave-particle duality but the rich geometrical structure of the "quantized" vacuum in nonlinear field theories. Similar solutions also occur in gravity (gravitational instantons) and are of importance in quantum gravity.

4b. According to Einstein, quantization must always be in accordance with the principle of general relativity. The usual canonical quantization method does not respect this principle. However, if we start from an action principle for Yang–Mills Fields coupled to gravitation the action is invariant under general coordinate transformations. Then transition to quantum field theory by Feynman path integration over all possible field configurations using the classical action does define, at least in principle, a method that is consistent with Einstein's condition. This is the correct field quantization according to the neo-Einsteinian physicist. However, there are great mathematical difficulties associated with this procedure. In practice, field quantization is often done neglecting coupling with gravity, thereby sacrificing Einstein's quantization principle.

5. *Physical Relevance of Global Properties of Solutions of the Fundamental Field Equations.* Einstein applied this principle to the formulation of Mach's principle that led him to introduce the cosmological background. The same principle led him to the importance of regular particle solutions [monopoles and instantons in our modern parlance]. It is also reflected in his concept of topological quantization associated with the classification of solutions according to their global properties. [The neo-Einsteinian applies this principle to the topological study of the solutions of field equations that cannot be described by perturbation theory. In that sense concepts like the physical vacuum and confinement can be regarded as neo-Einsteinian concepts.]

6. *The Unity Principle (Unification and Uniqueness).* The various fundamental interactions, gravitation, electromagnetism [weak and strong interactions], must be unified within a total field theory, and this unified field theory must be uniquely determined by the invariance principles. More specifically we try to unify gravity, based on local $SL(2,C)$, [or supergravity based on local $OSp(4/1)$] with electromagnetism [and weak interactions based on local $SU(2) \times U(1)$ and with chromodynamics based on local $SU(3)$]. [The uniqueness of the theory may be achieved in two ways:

6a. General relativity in the form of Einstein's theory or supergravity may be compatible with special relativistic quantum mechanics only for a very restricted class of field theories. For example, supergravity as extended by a local internal symmetry group can be constructed only when the internal group is $O(n)$ with $n = 1, \ldots, 8$.[1] There is something unique about this $O(8)$ extended supergravity theory. As shown by Cremmer and Julia,[2] the theory also exhibits a global E_7 invariance (in a noncompact form of it) and can be reexpressed also to have local $SU(8)$ invariance. Although the $O(8)$ gauge group admits the chromodynamical group $SU(3)$ and the electromagnetic gauge group $U(1)$ as subgroups, only its extended version $SU(8)$ is large enough to include also the $SU(2) \times U(1)$ Weinberg–Salam generalized version of $U(1)^{el}$. However, in that case the weak gauge bosons $W_\mu^\pm$ and Z_μ^0 occur as bound states of more fundamental scalar and pseudoscalar fields. It is also not known if the phenomenological 6-quark model can be fitted in this extended supergravity. Finally, since this unified theory is not renormalizable in the usual sense, the consistency of the corresponding quantum field theory cannot be ascertained. Although the final theory does not have the beauty and uniqueness of Einstein's gravitation, it is interesting that this approach limits the internal gauge group to $SO(8)$ or $SU(8)$ and the global internal symmetry group to the unique exceptional structure described by the Lie group E_7.

6b. Instead of restricting the unified geometries by imposing consistency between supergravity and the gauge symmetry of the internal space, we might ask if there are internal symmetry spaces that are uniquely distinguished from the others. Such a criterion would select the internal space independently of gravitation and single out certain grand unified theories based on a purely internal gauge group. To follow this line we recall that finite Hilbert spaces are projective spaces since the ket $|\alpha\rangle$ (that can be regarded as the homogeneous coordinates of a point) represents the same physical state as $\lambda|\alpha\rangle$, λ being an arbitrary complex number. Hence the classification of possible Hilbert spaces is equivalent to the classification of projective geometries. Those are special symmetric spaces that have all been classified by E. Cartan.[3] The projective spaces are certain coset spaces of rank 1 obtained from simple Lie groups sliced with respect to some maximal subgroups. They are associated with orthogonal groups, unitary groups, symplectic groups that leave invariant quadratic forms constructed respectively out of real numbers R, complex numbers C, and quaternions H. They are $SO(n + 1)/SO(n)$ for R (spheres), $SU(n + 1)/SU(n) \times U(1)$ for C (complex projective spaces CP_n) and $Sp(n + 1)/Sp(n) \times Sp(1)$ for H (quaternionic projective spaces HP_n). They correspond respectively to real, complex, and quaternionic Hilbert spaces of dimension $n + 1$. In any of these spaces there is no special dimension that is distinguished from the others. Hence, the internal space can have any dimension. In supergravity the restriction comes from embedding the internal space in a superspace and putting the restriction that there be a unique graviton in the particle spectrum. On the other hand, considering the possibility of constructing a projective space with octonions we obtain one more projective space, namely the Moufang plane $F_4/SO(9)$.[4] Because the octonions are nonassociative[5] this projective space cannot be embedded into similar spaces of higher dimensions. This results in the breakdown of the Desargues property[6] shared by the other projective spaces associated with R, C, and H. Rozental, Fueter, and Tits have generalized projective geometries by extending the underlying composition algebras R, C, H, and Ω (octonions) with their products with another set of commuting algebras R', C', H', and Ω'. Then new geometries that include and extend projective spaces are obtained from simple Lie groups that fit in a square (the "magic square"), as coset spaces with respect to their associated subgroups that fit in another magic square of subgroups.[7] (See Tables 3.1 and 3.2.) The dimensions of the corresponding coset spaces are always given by a power of 2. They are represented by points with two inhomogeneous coordinates that belong to the algebra $A' \times A$, where both A and A' range over R, C, H, and Ω. The first line of coset spaces $SU(2)/U(1)$, $SU(3)/SU(2) \times U(1)$, etc., gives the projective planes while the second line $SU(3)/SU(2) \times U(1)$, $E_6/SO(10) \times SO(2)$ yields the complex planes. The latter are more general than Hilbert spaces and they could describe generalized internal spaces. For instance $E_6/SO(10) \times SO(2)$, in which points are labeled by a pair of complex octonions, is known to be a positivity domain, a property possessed by the complex Hilbert spaces but not by the projective Moufang plane $F_4/SO(9)$. These are unique geometries associated with the exceptional groups and their subgroups that can be good candidates for the description of the internal symmetries alone. Another variant of this method leads to the study of exceptional supersymmetry groups $F(4)$ and $G(3)$,[8] which have remarkable coset spaces with respect to their Lie subgroups $SO(7) \times SU(2)$ and $G_2 \times SU(2)$. In the first case the Lie subgroup also admits $SU(3) \times SU(2) \times U(1)$ as subgroups and the coset space is labeled by two spinor representations of $SO(7)$. Note that the automorphism groups G_2 and $SU(2)$ of octonions and quaternions as well as the norm groups $SO(7)$ and $SO(3)$ of purely

Table 3.1
Groups of the Magic Square

	R	C	H	Ω
R'	$SU(2)$	$SU(3)$	$Sp(3)$	F_4
C'	$SU(3)$	$SU(3) \times SU(3)$	$SU(6)$	E_6
H'	$Sp(3)$	$SU(6)$	$SO(12)$	E_7
Ω'	F_4	E_6	E_7	E_8

Table 3.2
Subgroups of the Magic Square

	R	C	H	Ω
R'	U(1)	SU(2) × U(1)	Sp(2) × SP(1)	SO(9)
C'	SU(2) × U(1)	SO(4) × U(1) × U(1)	SU(4) × SU(2) × U(1)	SO(10) × SO(2)
H'	Sp(2) × Sp(1)	SU(4) × SU(2) × U(1)	SO(8) × SO(4)	SO(12) × SO(3)
Ω'	SO(9)	SO(10) × SO(2)	SO(12) × SO(3)	SO(16)

imaginary octonions and quaternions appear in the subgroups of exceptional groups or exceptional supergroups.]

To sum up, if the neo-Einsteinian physicist is motivated by the uniqueness principle in connection with an algebraic structure that encompasses supergravity, he will be led to the internal symmetry groups E_7, $SU(8)$, and $O(8)$. If he looks for remarkable finite geometries of the internal space alone he will consider exceptional structures for Lie groups as they appear in the magic square or a similar extension to supergroups.[9] In either case most of the structures accommodate the phenomenologically successful symmetry group $SU(3) \times SU(2) \times U(1)$ of strong, weak, and electromagnetic interactions.

3. Neo-Einsteinian Geometries

We have seen that the geometrization of physics initiated by Einstein is not restricted to space-time. It is also useful and suggestive in classifying Hilbert spaces. A discussion of the geometry of Yang–Mills fields or the internal space of symmetries depends on many concepts originally introduced by Einstein. Let me give some examples.

1. *Einstein Spaces.* These are spaces for which the Ricci curvature is proportional to the metric tensor. They were introduced by Einstein[10] in 1917 in connection with the cosmological background. He studied the hypersphere $S^3 = SO(4)/SO(3)$ giving constant spatial curvature to a static cosmological background. De Sitter[11] generalized it to the pseudosphere denoting an expanding universe, namely $SO(4,1)/SO(4)$, which has constant four-dimensional curvature. These examples led Cartan to the classification of all Einstein spaces and their descriptions as symmetric spaces associated with cosets G/H.[12] We have seen that projective spaces and their generalization also fall in this category.

2. *Non-Riemannian Affine Geometries.* In these geometries the affine connection $\Gamma_{\mu\nu}^{\lambda}$ is the fundamental object rather than the metric. Einstein gave an example[13] in 1923. They were subsequently generalized and studied by Eisenhart, Eddington, Schrödinger, and many others.[14]

3. *Spaces of Distant Parallelism.* As proposed by Einstein[15] in 1928 they are characterized by a moving vierbein h_μ^a, which is the fundamental quantity. The metric is obtained from it by $g_{\mu\nu}$

$= h_\mu{}^a h_\nu{}^a$. This was further developed by Cartan in his theory of moving frames. Vierbeins are essential in writing the Dirac equation in curved space or in the formulation of supergravity.[16]

4. *Nonsymmetric Generalization of the Real Symmetric Metric: the Einstein–Kähler Geometries.* In 1925 and later in 1945 Einstein[17] put together a symmetrical tensor $g_{\mu\nu}$ with an antisymmetrical one $F_{\mu\nu}$ into either the Hermitian structure

$$K_{\mu\nu} = g_{\mu\nu} + i F_{\mu\nu} = K^*_{\nu\mu} \; ,$$

or the real nonsymmetrical structure

$$S_{\mu\nu} = g_{\mu\nu} + F_{\mu\nu} \; .$$

Einstein was motivated in this synthesis by the unification of gravity and electromagnetism. He therefore assumed that $F_{\mu\nu}$ derived from a potential a_μ and hence, satisfied the Bianchi identities. Now $g_{\mu\nu}$ determines the line element $ds^2 = g_{\mu\nu} dx^\mu dx^\nu$ while $F_{\mu\nu}$ determines the 2-form $F_{\mu\nu} dx^\mu \wedge dx^\nu$, where the wedge generalizes the vectorial product. When $F_{\mu\nu}$ derives from a potential, the 2-form is closed. The association of a line element and a closed 2-form in a Hermitian structure defines a Kähler–Einstein geometry. Kähler[18] in 1933 studied these structures mainly when $g_{\mu\nu}$ and $F_{\mu\nu}$ are functions of a complex variable. The theory was developed by Kodaira, Chern, Lichnerowicz, and other mathematicians[19] who showed the natural connection of these geometries with the projective spaces CP_n. In the last decade the structure was generalized to HP_n spaces by considering $K_{\mu\nu}$ as a quaternionic Hermitian metric instead of a complex metric. Then the imaginary part is a 2-form with three components. Hence it is a Yang–Mills field. It is covariantly closed since it satisfies $\Delta_\mu {}^* F_{\mu\nu} = 0$; that is, its dual has vanishing covariant derivative. The resulting structure is a quaternionic Kähler structure.[20] It leads to the synthesis of gravitation with Yang–Mills fields instead of Maxwell fields.

Two-dimensional σ models and four-dimensional Yang–Mills gauge theories can be interpreted geometrically in terms of a Kählerian metric.[21] Then instantons appear as smooth mappings of $CP_1 \to CP_n$ or $HP_1 \to HP_n$. Hence Einstein's unified field geometry has made an unexpected comeback in connection with the properties of the vacuum in quantum field theories.

The Einstein–Kähler structures arise quite naturally as generalization of Riemannian geometry in which at a given point coordinates ξ^a can be chosen such that the line element

$$ds^2 = g_{\mu\nu} dx^\mu dx^\nu = \eta_{ab} d\xi^a d\xi^b \; ,$$

where η_{ab} is diagonal with 1 or -1 as its elements. In the locally Euclidean case we have $\eta_{ab} = \delta_{ab}$. Then the line element becomes invariant locally under $SO(n)$ where n is the dimension of the Riemannian space.

Now let us generalize the real coordinates x^μ and ξ^a to be complex or quaternionic. Their complex or quaternionic conjugates obtained by reversing the sign of the imaginary units are denoted by $\bar{\xi}^a$. Then the line element becomes

$$ds^2 = dx^\mu K_{\mu\nu} d\bar{x}^\nu = \eta_{ab} d\xi^a d\bar{\xi}^b \; ,$$

and this is invariant under unitary transformations of ξ^a in the complex case[22] and the symplectic transformations of ξ^a in the quaternionic case. Thus, locally, instead of $SO(n)$ invariance in the tangent space we have invariance under $SU(n)$ or $Sp(n)$. These are Hermitian complex and quaternionic Riemannian spaces. The real part of $K_{\mu\nu}$ is the Riemannian metric and its imaginary (in the complex case) or vectorial part (in the quaternionic case) is the 2-form associated with a Maxwell or a Yang–Mills fields in that space. The extension to octonions is not possible for arbitrary dimensions. Hence, using Einstein's idea we obtain a magic square of possible extensions of Riemannian geometries such that the Tits–Freudenthal geometries hold locally around a point of a manifold. The corresponding curvature tensors are real, complex, quaternionic, or octonionic. In the complex case the transformations of the complex coordinates can be restricted to holomorphic transformations. Then inhomogeneous terms in the transformation laws of many indexed symbols can vanish due to Cauchy–Riemann relations.[23] Thus a restriction to holomorphic transformations in Kähler–Einstein spaces will yield new tensor quantities whose equivalents do not exist in Riemannian spaces. The resulting geometry is then much richer and possesses more structure than the Riemannian geometries. These are the extended geometries that arise naturally for the unification of fields including those associated with internal symmetries. Viewed from this standpoint, Einstein's various proposals towards a unified field theory are of great importance for having created the mathematical arsenal presently needed for an attack on the problem of unification of the forces of nature.

4. Description of Unified Interactions by Some Exceptional Structures

In this section I shall give some examples to illustrate how certain exceptional geometries and their associated field theories can be used to describe in a unified way strong, electromagnetic, and weak interactions. As we have seen, the exceptional structures arise from the application of Einstein's uniqueness criterion to internal space alone. Geometrically, they are connected with nonembeddable structures that can exist only in some definite dimensions, giving a natural reason for the existence of a finite number of degrees of freedom in the generalized charge space of internal symmetries. These degrees of freedom are then naturally associated with the number of leptons, quarks, and basic gauge particles. If this number is not extendable, as is the case for exceptional structures, we have a mathematical argument for the finiteness of the number of fundamental particles. Algebraically, exceptional structures exist in octonionic spaces and it is the nonassociativity of the octonion algebra that limits the number of degrees of freedom. Now, the group G_2 is the automorphism group of octonions and $SU(3)$ is a maximal subgroup of G_2. If the octonion is decomposed into a complex number whose imaginary unit is one of the seven imaginary units of octonions and if one defines a vector product algebra (called a Malcev algebra) for the six additional imaginary units, then $SU(3)$ is the automorphism group of this Malcev algebra.[24] Thus, exceptional structures that give a finite number for color and flavor (the degrees of freedom of basic fermions) also exhibit an intrinsic $SU(3)$ invariance that we can identify with the color group.[25] The sector connected with the complex numbers embedded in octonions is invariant under $SU(3)$. Such basic fermions can be identified with leptons. The coefficients of the remaining six imaginary units decompose into (3) and ($\bar{3}$) representations of $SU(3)$. They can be identified respectively with quarks and antiquarks. To summarize this discussion, Einstein's uniqueness principle, when applied to the Hilbert space of internal degrees of freedom, yields a finite number of leptons and colored quarks.

Let us consider two kinds of field theories associated with the exceptional structures. We can use the group G (taken to be an exceptional group although this might also be extended to exceptional supergroups) as the global group of a nonlinear σ-model (first introduced in connection with the internal symmetry group $O(4)$).[26] If H is a maximal subgroup invariant under an involution, G/H is a symmetric space. If H leaves the element η invariant and U, an element of G, is decomposed as

$$U = VW, \quad (W \in H, V \in G/H),$$

with

$$\eta V \eta = V^\dagger, \quad \eta W \eta = W, \quad \eta^2 = E = \text{unit matrix},$$

then, the quantity

$$N = U\eta U^\dagger = V\eta V^\dagger$$

is invariant under local gauge transformations

$$U \to UW(x)$$

with respect to the subgroup H. Hence a Lagrangian $\mathcal{L}$ constructed out of N and $\partial_\mu N$, which is invariant under the global group G, will be gauge invariant under its subgroup H, which is also called the isotropy group of G/H.

On the other hand, we can directly construct a Lagrangian that is gauge-invariant under the local group H. The first method results in embedding a gauge theory of the group H in a nonlinear σ-model based on the global group G. The vector bosons of the theory are expressed in terms of the representative V of the coset G/H.

Let us give an example for the symplectic group $G = Sp(n + 1)$, $H = Sp(1) \times Sp(n)$. For η we choose

$$\eta = \begin{pmatrix} I & 0 \\ 0 & -I \end{pmatrix} \quad (I = n \times n \text{ unit matrix}),$$

and we denote the $(n + 1) \times (n + 1)$ unit matrix η^2 by E. Then if we parametrize V by the quaternionic $n \times 1$ column u and represent W by the block-diagonal matrix constructed from a quaternion r and an $n \times n$ quaternionic matrix R,

$$W = \begin{pmatrix} r & 0 \\ 0 & R \end{pmatrix}, \quad (|r| = 1, RR^\dagger = I),$$

we can write for the generalized stereographic projection

$$V = \gamma^{-1}(u) \begin{pmatrix} 1 & -u^\dagger \\ u & \Lambda(u) \end{pmatrix} \quad (\gamma(u) = \sqrt{1 + u^\dagger u}, \; \Lambda = \gamma (I + uu^\dagger)^{-1/2}).$$

Then $N = N(u)$, and the manifestly $Sp(n + 1)$ invariant Lagrangian

$$\mathcal{L} = \mathrm{Sc} \left\{ \frac{E + N}{2} (\partial_\mu N \, \partial_\nu N \, \partial_\mu N \, \partial_\nu N - \partial_\mu N \, \partial_\mu N \, \partial_\nu N \, \partial_\nu N) \right\}$$

is proportional to the Yang–Mills Lagrangian $\mathrm{Sc}(F_{\mu\nu} F_{\mu\nu})$ where

$$F_{\mu\nu} = \partial_\mu a_\nu - \partial_\nu a_\mu + [a_\mu, a_\nu],$$

$$a_\mu = \tfrac{1}{2} i \tau \cdot a_\mu = \tfrac{1}{2} \gamma^{-2}(u) (u^\dagger \partial_\mu u - (\partial_\mu u^\dagger) u),$$

and Sc denotes the scalar part of a quaternion. Hence $\mathcal{L}$ has local invariance under $SU(2) \sim Sp(1)$ and global invariance under $Sp(n + 1)$. When $u = u(x)$ are definite rational functions of x one obtains the instanton solutions to the Yang–Mills gauge theory.[27] The Kähler–Einstein structure also arises naturally.[28] If we put

$$\gamma^{-2}(u) \Lambda (u) = \Gamma^{1/2}, \quad \varphi_\mu = \Gamma^{1/2} \partial_\mu u,$$

then we obtain

$$F_{\mu\nu} = \varphi_\mu^\dagger \varphi_\nu - \varphi_\nu^\dagger \varphi_\mu = 2 \, \mathrm{Vec} (\varphi_\mu^\dagger \varphi_\nu).$$

Defining a metric $g_{\mu\nu}$ by

$$g_{\mu\nu} = \varphi_\mu^\dagger \varphi_\nu + \varphi_\nu^\dagger \varphi_\mu = 2 \, \mathrm{Sc}(\varphi_\mu^\dagger \varphi_\nu),$$

we can introduce the Kähler metric

$$K_{\mu\nu} = \varphi_\mu^\dagger \varphi_\nu = \tfrac{1}{2}(g_{\mu\nu} + F_{\mu\nu}) = (\partial_\mu u^\dagger) \Gamma (\partial_\nu u),$$

where Γ is a Kähler metric in the n-dimensional space. Instanton solutions are those for which $g_{\mu\nu}$ is conformally flat and $F_{\mu\nu}$ is self-dual or anti-self-dual.

This structure can now be generalized to the exceptional case. Let $G = E_6$, $H = SO(10) \times SO(2)$. The 16-dimensional coset space G/H is then represented by a spinor representation of $SO(10)$. A nonlinear σ-model based on E_6 will admit $SO(10)$ as a gauge group. This is equivalent to the grand unified $SO(10)$ theory proposed independently by Fritzsch and Minkowski[29] and by Georgi.[30] The $SO(10)$ spinor splits into the 1-, $\bar{5}$-, and 10-dimensional representations of the $SU(5)$ group of Georgi and Glashow[31] with $\hat{d}_R = \tfrac{1}{2}(1 + \gamma_5)d^c$, ν_L^e and e_L in ($\bar{5}$) and $u_L, d_L, \hat{u}_R, \hat{d}_R$ in (10). In this case the 16-dimensional representation of $SO(10)$ represents one generation of basic fermions, including the lepton doublet (v_L^e, e_L) and the quark doublet (u,d). To accommodate the known particles three generations are needed, namely (v_L^μ, μ_L; c, s) and (v_L^τ, τ_L; t,b). Thus, a hitherto unobserved top quark with charge $-2/3$ is predicted. Each generation lies in a 16-dimensional symmetric space that is also Hermitian (see Helgason[32]). Such a space is also called a positivity domain and it is a generalization of the Hilbert space $SU(n + 1)/U(n)$.

We may ask if there are other exceptional compact Hermitian symmetric spaces. In fact there is only one more,[33] that is the space $E_7/E_6 \times SO(2)$ of complex dimension 27. It corresponds to the 27-dimensional representation of E_6. Then, a gauge theory based on the local E_6 group would be embedded in a σ-model based on the global group E_7. If we decompose the 27-dimensional representation of E_6 with respect to the subgroup $SO(10)$ we obtain $27 = 16 + 10 + 1$.

Identifying (16) with the (e, v^e; u,d) generation, (10) can accommodate the colored quark b (b_L, $\hat{b}_R$) with charge $-1/3$ and the heavy leptons (τ_L, v_L^e), ($\hat{\tau}_R$, $\hat{N}_R$) where $\hat{N}_R$ is a heavy neutral lepton arising in a weak doublet that includes τ_R. The $SO(10)$ singlet can represent another neutral left-handed lepton. With 27-dimensional generations we only need two to accommodate all known basic fermions. Then a sixth quark with charge $-1/3$ (b' or h) is predicted along with a new charged heavy lepton (τ' or M), so that the second generation includes the leptons (μ, v^μ, M, v^M), the quarks (c, s, h) and some additional neutral leptons.

A gauge theory based on local E_6 was proposed earlier.[34] In this new topless six-quark theory a spontaneous symmetry breakdown can be induced by considering a 78-dimensional Higgs field that develops a superheavy mass of the order of 10^{16} GeV as in the $SU(5)$ theory. For the smaller mass breakings one needs a 27-dimensional and a 351-dimensional Higgs multiplet.[35] It is possible that the 351-Higgs multiplet is not elementary but is dynamically generated. Then E_6 is broken down to $SU(3) \times SU(2) \times U(1)$ or $SU(3) \times SU(2) \times SU(2) \times U(1)$. The unrenormalized Weinberg parameter in this theory is $3/8$.[36] The renormalized value is around 0.2 if the weak and electromagnetic subgroup is the Weinberg–Salam $SU(2) \times U(1)$ or $SU(2) \times SU(2) \times U(1)$ with the extra $SU(2)$ being associated with new neutral weak gauge bosons that may be responsible for the decays of the b and h quarks.

For the moment both grand unified theories based on $SO(10)$ and E_6 are compatible with experiment. We have seen that in both cases a generation corresponds to a point in an internal charge space that has a geometrical interpretation. The 16-dimensional representation of $SO(10)$ and the 27-dimensional representation of E_7 are uniquely selected for the fermion states because these give respectively the only exceptional Hermitian compact symmetric spaces $E_6/SO(10) \times SO(2)$ and $E_7/E_6 \times SO(2)$ with isotropy groups $SO(10) \times SO(2)$ and $E_6 \times SO(2)$.

It would be interesting to investigate if the Hermitian symmetric spaces, which are phenomenologically so successful in describing the geometry of the internal degrees of freedom, have a generalization to cosets of supergroups with respect to their maximal subsupergroups. This might lead to a deeper understanding of the occurrence of the group E_7 in 0(8) extended supergravity.

To conclude this section I would also like to draw attention to a generalization of symmetric spaces proposed by Wolf.[37] They are homogeneous spaces with irreducible isotropy groups acting linearly in the space. Again, if using Einstein's uniqueness principle, we single out the exceptional spaces, they are

$$G_2/SU(3) = S^6, \quad F_4/SU(3) \times SU(3), \quad E_6/SU(3) \times SU(3) \times SU(3),$$
$$E_7/SU(3) \times SU(6), \quad E_8/SU(9), \text{ and } E_8/SU(3) \times E_6.$$

Once more we note the remarkable role played by the color group $SU(3)$. The larger spaces might be useful if the color group is extended to hypercolor as proposed by Susskind[38] for dynamically broken gauge theories.

5. Concluding Remarks

The unification of fields will not be complete without gravity. In the schemes we have reviewed, $E_6/SO(10) \times SO(2)$, $E_7/E_6 \times SO(2)$, there are two main problems. One is the need for several generations of fermions (three for the $SO(10)$ gauge group and two for the E_6 gauge group). The generation number can only be predicted if the corresponding Hermitian symmetric spaces are embedded in higher spaces associated with a larger group or supergroup. An explanation could be sought in the larger structure required for unification with gravity. The other problem is the occurrence of the superheavy mass scale M that is needed for several different reasons: 1) The quasi-stability of the proton: in grand unified theories based on $SU(5)$, $SO(10)$, or E_6, decays of the nucleon into a meson and an antilepton like $p \to e^+ + \pi^0, p \to \bar{\nu}^e + \pi^+, n \to e^+ + \pi^-$, $n \to \bar{\nu}^e + \pi^0$ are allowed through the interchange of superheavy gauge bosons between two of the three constituent quarks of the nucleon.[39] A proton lifetime of around 10^{31} years results if $M \sim 10^{16}$ GeV. 2) The separation of the fine structure constant from the strong $\alpha_s = g_s^2/4\pi$ at energies of a few GeVs: the ratio α_s/α depends on M logarithmically, the value just stated bringing α_s from the initial value of the order of α to a value in the interval 0.2–0.3 as required by experiment.[40] 3) The renormalized value of the Weinberg parameter $\sin^2\theta_w$: this value is changed from the bare value 3/8 at superheavy energies to 0.2 in agreement with experiment.[41] 4) Explanation of the cosmological charge conjugation asymmetry, which is of the order of 10^{-8}, in terms of CP violating decay of superheavy bosons in the early history of the universe[42]: as shown in Professor Weinberg's chapter in this volume, again a mass of order M is required to explain the observed asymmetry. What is the origin of M? Again M being of the order of αM_p, where M_p is the Planck length that arises in quantum gravity, it is natural to seek the explanation within a further unification of grand unified theories with gravity.

Thus, we are led to consider possible scenarios for the further unification of gauge fields with gravity. One way is to unify gauge fields with supergravity.[43] Geometrization of such a superunified theory requires the extension of space-time to superspace. Then elementary gauge bosons associated with the internal 0(8) group can be introduced and, as shown by Julia and Cremmer, a local $SU(8)$ group associated with composite gauge bosons (that include the weak bosons) results from the scheme, which further yields a noncompact form of E_7 global symmetry.[44]

Another possibility would be the unification of the conformally invariant Weyl theory with Lagrangian $g R_{\mu\nu\alpha\beta} R^{\mu\nu\alpha\beta}$ with the grand unified gauge theory. Such a new gravitational Lagrangian, as also proposed by C. N. Yang,[45] would be formally very similar to the Yang-Mills Lagrangian. The resulting equations are of fourth order in the metric, hence unacceptable as a conventional causal theory. However, Belavin and Burlankov[46] have shown that the self dual solutions

$$**R^{\mu\nu\alpha\beta} = \tfrac{1}{4} g \, \varepsilon^{\mu\nu\rho\sigma} \varepsilon^{\alpha\beta\gamma\delta} R_{\rho\sigma\gamma\delta}$$

of this equation obey Einstein's equations with a cosmological term. Such solutions yield a finite Euler–Poincaré index associated with the integrand $g^{**}RR$. Then the Weyl action also equals the index and the causal solutions of the Weyl–Yang theory become finite action solutions (generalized instantons) that also satisfy Einstein's equation. This geometry can now be extended either in the Einstein–Kähler fashion by making the metric Hermitian, or in the Kaluza–Klein fashion by increasing the number of dimensions, or by combining both methods. If we follow the Kaluza–Klein path and introduce a generalized vielbein of the form

$$E^{\alpha}{}_a = \begin{pmatrix} h^{\alpha}_{\mu} & L^{k}_{\nu} \\ 0 & K^{\lambda}_{m} \end{pmatrix}$$

where h^{α}_{μ}, the square roots of the metric, are associated with the space-time vierbein and K^{λ}_{m} with the internal frame, then the volume element in the large space factorizes in terms of the space-time volume element and the volume element of the internal space.[47] Thus the requirement of finite action simultaneously reduces Weyl's theory to Einstein's in the space-time sector and constrains the pure Yang–Mills part of the theory to the instanton sector. That sector is described by an internal symmetric space invariant under the isotropy group. This is essentially the phenomenon of dimensional reduction,[48] which is seen to follow from the finite action requirements of the part of the action associated with the extra dimensions. The curvatures of the symmetric spaces and the cosmological constant of space-time sector can provide the renormalization lengths needed for the quantized unified field theory.

It has been my purpose to show that Einstein's attempts for enlarging differential geometry to include electromagnetism still remain valid for the unification of gravitation with grand unified gauge theories that extend Maxwell's theory. The Einstein–Kähler structures have enriched and extended Riemannian geometries and are essential in giving a geometrical interpretation to the instanton solutions of gauge theories. Furthermore, Einstein's program of topological quantization serves to classify classical solutions that dominate the contribution of field configurations to the action in the Feynman path integral formulation of quantum field theory. Finally, the topological quantization of the action in Kaluza–Klein type theories with factorized volume element, provides a basis for the dimensional reduction of gauge theories formulated in the larger space as well as the extraction of Einstein's theory from Weyl's theory.

In view of the neo-Einsteinian possibilities for achieving a unified quantum field theory in the future, Einstein's "failure" should be reassessed as the beginning of a new level of geometrization of modern relativistic quantum physics.

Acknowledgment

Research for this paper was supported in part by the U. S. Department of Energy under Contract No. EY-76-C-02-3075; Yale Preprint No. YTP79-07.

NOTES

1. M. Gell-Mann, talk at the 1977 Washington Meeting of the American Physical Society.
2. E. Cremmer and B. Julia, Ecole Normale Supérieure preprint (1979).
3. See for instance S. Helgason, *Differential Geometry and Symmetric Spaces* (Academic Press, New York, 1962).
4. R. Moufang, Abh. Math. Sem. Univ. Hamburg **9**, 207 (1933); P. Jordan, J. von Neumann, and E. P. Wigner, Ann. Math. **35**, 29 (1934); C. Chevalley and R. D. Schafer, Proc. Nat. Acad. Sci. U.S. **36**, 137 (1950).
5. See for instance R. D. Schafer, introduction to *Non-Associative Algebras* (Academic Press, New York, 1966).
6. For an explicit demonstration see M. Günaydin, C. Piron, and H. Ruegg, Comm. Math. Phys. **61**, 69 (1978).
7. H. Freudenthal, Proc. Colloq. Utrecht, p. 145 (1962). Also see F. Gürsey, in *International Symposium on Mathematical Problems in Theoretical Physics*, edited by H. Araki (Springer Verlag, Berlin, 1975), p. 189.
8. V. G. Kac, Adv. Math. **26**, 8 (1977).
9. I. Bars and M. Günaydin, J. Math. Phys. **20**, 1977 (1979).
10. A. Einstein, Preuss. Akad. Wiss. Berlin, Sitzber. 143 (1917).
11. W. DeSitter, Ned. Akad. Wet. **20**, 229 (1917).
12. See J. A. Wolf, *Spaces of Constant Curvature* (Publish or Perish, Berkeley, 1977).
13. A. Einstein, Preuss. Akad. Wiss. Berlin, Ber. **5**, 32, 76, 137 (1923).
14. See for instance L. P. Eisenhart, *Non Riemannian Geometry* (American Mathematical Society Publication, New York, 1927) and E. Schrödinger, *Space-Time Structure* (Cambridge University Press, Cambridge, England, 1950).
15. A. Einstein, Preuss. Akad. Wiss. Berlin Ber. **17**, 217 (1928).
16. S. Deser and B. Zumino, Phys. Lett. **62B**, 335 (1976).
17. A. Einstein, Preuss. Akad. Wiss. Berlin Ber. **22**, 414 (1925); A. Einstein, Ann. Math. **47**, 731 (1946); Rev. Mod. Phys. **20**, 35 (1948).
18. E. Kähler, Hamb. Abh. **9**, 173 (1933).
19. See for instance J. Morrow and K. Kodaira, *Complex Manifolds* (Holt, Rinehart and Winston, New York, 1971); A. Weil, Introduction a l'étude des Variétés Kählériennes, *Actualités Sci. Indust.* (Hermann, Paris, 1951).
20. See for instance S. Ishihara, J. Diff. Geom. **9**, 483 (1974).
21. A. M. Perelomov, Commun. Math. Phys. **63**, 237 (1978); F. Gürsey and C. Tze, to be published in Ann. Phys., (N.Y.).
22. See chapter 3 of S. Kobayashi, *Transformation Groups in Differential Geometry* (Springer-Verlag, New York, 1972).
23. S. S. Chern, *Complex Manifolds without Potential Theory* (D. Van Nostrand, New York, 1967); E. J. Flaherty, *Hermitian and Kählerian Geometry in Relativity*, Lectures in Physics No. 46 (Springer-Verlag, New York, 1976).
24. M. Günaydin, J. Math. Phys. **17**, 1875 (1976); F. Gürsey, in *Group Theoretical Methods in Physics*, edited by R. T. Sharp and B. Kolman (1976), p. 213.
25. M. Günaydin and F. Gürsey, Phys. Rev. **D9**, 3387 (1974).
26. F. Gürsey, Nuovo Cimento **16**, 230 (1960).
27. M. F. Atiyah et al., Phys. Lett. **65A**, 185 (1978); V. G. Drinfeld and Yu. I. Manin, Comm. Math. Phys. **63**, 177 (1978); N. Christ, E. Weinberg, and N. Stanton, Phys. Rev. **D18**, 2013 (1978).
28. Gürsey and Tze, op. cit. in n. 21.
29. H. Fritzsch and P. Minkowski, Ann. Phys. (N.Y.) **93**, 193 (1974).
30. H. Georgi, in *Proceedings of the A. P. S. Williamsburg Meeting*, edited by R. Carlson (1974).
31. H. Georgi and S. L. Glashow, Phys. Rev. Lett. **32**, 433 (1974).
32. Helgason, op. cit. in n. 3.
33. Ibid.
34. F. Gürsey, P. Ramond, and P. Sikivie, Phys. Rev. **D12**, 2155 (1975), and Phys. Lett. **60B**, 177 (1976).

35. F. Gürsey in *Second Workshop on Current Problems in High Energy Particle Theory,* edited by G. Domokos and S. Kövesi-Domokos (Johns Hopkins University, Baltimore, 1978), p. 3; Y. Achiman and B. Stech, Phys. Lett. **77B,** 389 (1978).
36. Gürsey, Ramond, and Sikivie, op. cit. in n. 34.
37. H. Wolf, Acta Mathematica **120,** 59 (1968).
38. L. Susskind, SLAC Report No. SLAC-PUB-2142 (1978); S. Dimopoulos and L. Susskind, Nucl. Phys. **B155,** 237 (1979).
39. J. Pati and A. Salam, Phys. Rev. **D8,** 1240 (1973); H. Georgi, H. R. Quinn, and S. Weinberg, Phys. Rev. Lett. **33,** 451 (1974).
40. Georgi, Quinn, and Weinberg, op. cit. in n. 39.
41. Ibid.
42. M. Yoshimura, Phys. Rev. Lett. **41,** 281 (1978); L. Dimopoulos and L. Susskind, Phys. Rev. **D18,** 4500 (1978); S. Weinberg, Phys. Rev. Lett. **42,** 850 (1979).
43. Gell-Mann, op. cit. in n. 1.
44. Cremmer and Julia, op. cit. in n. 2.
45. N. Yang, Phys. Rev. Lett. **33,** 445 (1974).
46. A. A. Belavin and D. E. Burlankov, Phys. Lett. **58A,** 7 (1976).
47. Cremmer and Julia, op. cit. in n. 2.
48. J. Scherk and J. Schwarz, Nucl. Phys. **B81,** 118 (1974); E. Cremmer and J. Scherk, Nucl. Phys. **B108,** 409 (1976).

Discussion

Following the Paper by F. GÜRSEY

Anon.: I want to ask whether all the recent theories are not science fiction.

Gürsey: This is for a science fiction writer to answer. As to the relation of modern theories to reality, only time can tell.

J. A. de Azcarraga (Valencia U.): I would like to ask you a question, a rather general one, concerning the first part of your talk. You have emphasized the importance of fiber bundles in the description of physics phenomena, as Professor Yang did. Usually in these descriptions fields are described as cross-sections of trivial fiber bundles, in the sense that if you wish to include inner symmetries in it, the fiber bundle is the direct product of Minkowski space and the inner space. But the question I would like to ask you is the following: Taking into account the difficulties of mixing inner and space-time symmetries, do you think that this difficulty could be traced to the fact that the bundle should not be trivial, in the same sense that for instance the description of Dirac's monopole, which has a string of singularities, becomes nonsingular in Yang's version, which takes the form of a nontrivial fiber bundle? Could you comment on that, or could, perhaps, Professor Yang if he is still here?

Gürsey: Locally, we have a direct product.

Azcarraga: That is precisely the point, could the *global* structure of the space be important?

Gürsey: If you embed the fiber bundle in a simple group, as I have tried to show, the fiber bundle comes from a coset decomposition. Now consider the base manifold. The base can also be regarded as a coset space. Take the conformal group and choose as a maximal subgroup the Poincaré group times the dilations, which defines the Weyl group. Then the coset space of the conformal group with respect to the Weyl group describes the space-time manifold. This is the coset space. $SO(4,2)/W$, where W is the Weyl group.

If you also have an internal symmetry group G^I, which acts in an internal symmetric space $S = G^I/H$, both the conformal group and G^I act on the product of the space-time manifold with S. Now embed this whole structure in a single group G^T, which admits as subgroups both the conformal group and the internal symmetry group G^I. We can now take the coset of G^T with respect to both the conformal group C and the internal symmetry group G^I. Then G^T acts on the manifold $G^T/(C \times G^I) = M$ or $G^T/(W \times H) = m$. The generators of G^T that do not belong to $C \times G^I$ will now act nontrivially on m and change the fibering of the base. This phenomenon occurs each time you can embed the local direct product structure in a larger simple group. If the bundle is an exceptional one, the embedding becomes more difficult because the exceptional groups cannot be embedded in higher groups having the same general structure. Suppose you take E_6 as the internal group and you take the conformal group as the group of space-time. Then, there is no simple group G^T such that E_6 or F_4 times the conformal group is a maximal subgroup. In that case one is forced to have at each point a direct product in the fiber bundle, and one cannot put together space-time and the internal space in a single base manifold. But if you take the conformal group, and for instance $SU(n)$ as the internal group, then one can embed $SU(2,2) \times SU(n)$ into $SU(n+2,2)$ and slice it, and get a nontrivial bundle. When the internal symmetry group is exceptional, this cannot be done. This might be one reason why at each point we have always the direct product of internal space and space-time. That is actually one of the points that led me to the consideration of exceptional structures. So, it is a very deep point that you have raised.

PART II: GENERAL RELATIVITY IN THE LARGE

This part selects two topics in "classical" general relativity that have a bearing on our main theme. Bekenstein's chapter reviews the discovery of gravitational entropy and black hole thermodynamics. This series of developments (influenced by the thinking of J. A. Wheeler), starting in 1970, has revealed an unsuspected, extremely rich structure in the intersection of quantum theory, gravitation, and statistical physics. Basically, it is quantum tunneling that allows for the decay of black holes, and so on, and it is astonishing that these quantum effects should at the same time make gravity add its "classical" contribution to entropy. Perhaps the only other case in which we deal with a direct quantum-originated component of entropy is the CP violation, as discussed by S. Weinberg in Part VI. Indeed, yet another effect, baryon nonconservation, which is thought to arise through black hole physics (see S. W. Hawking's paper in the Princeton volume) has also been derived from the interplay of the CP violation and a grand unification group, as discussed by Weinberg. The reader should be warned that in this editor's view, grand unification as promulgated at this stage is more speculative than black hole physics. Returning to our basic unification scheme — the physics of strong gravitational fields seems to contain extremely valuable clues.

Rosen's chapter starts by rederiving the equations demonstrating Mach's principle for a system of particles, first derived by Einstein but criticized by Davidson. Rosen's results agree with Einstein's. They describe the increment of inertial mass acquired by a particle when ponderable masses are gathered in its neighborhood, and so on. The inertial interaction potential goes like R, with the distant masses making the main contribution. The case of an expanding universe is somewhat unclear.

Rosen then opens the discussion of the correct interpretation of Mach's principle. Considering that the gravitational field — gravitons — carries energy, it would seem as if we are facing a tautology. This is exemplified in the various anti-Mach universes, in which geometry is fixed by a gravitational field (geometry itself), with no matter present. In recent years, the discussion has become less academic. This is because we can also account for the boundary conditions. The homogeneity of the 3 °K background radiation tells us that the boundary conditions are those of a relatively smooth matter distribution and fit the homogeneity of inertia. The issue also relates to the question of an open or closed universe and to the availability of enough matter (including

---

Yuval Ne'eman (ed.), To Fulfill A Vision: Jerusalem Einstein Centennial Symposium on Gauge Theories and Unification of Physical Forces
ISBN 0-201-05289-X

Copyright © 1981 by Addison-Wesley Publishing Company, Inc., Advanced Book Program.
All rights reserved. No part of this publication may be reproduced, stored in a retrieval system, or transmitted, in any form or by any means, electronic, mechanical photocopying, recording, or otherwise, without the prior permission of the publisher.

gravitons) to account for the whole of inertia. This is discussed in the papers by M. Rees, P. J. E. Peebles, G. B. Field, W. L. W. Sargent, D. Sciama, and C. W. Misner in the Princeton Einstein Centennial volume. Note that the observation of an anisotropy in the background radiation provides the type of information about our "true" motion, which was sought after by Michelson and Morley. However, that motion is not absolute, it refers to a convenient cosmological frame of reference, that of the background radiation.

Session Chairman's Remarks

J. GEHENIAU, U. Libre de Bruxelles

Before asking the speaker to give his lecture, I shall read an address from the Académie Royale des Sciences, des Lettres et des Beaux Arts de Belgique, to the Israel Academy of Science and Humanities.

"Avec Einstein disparait une de plus vives lumières de tout les temps." Ainsi s'exprimaient nos confrères de la Classe des Sciences au lendemain de la disparition du grand savant. À l'hommage qu'ils rendaient de la sorte a un Membre Associé de l'Académie Royale de Belgique, (élu en 1931) se joignait la fierté d'avoir vu leur pays recevoir sa visite à de nombreuses reprises: en 1911, 1913, 1927, 1930, années de quatre Conseils de Physique Solvay, auxquels il participa a Bruxelles, et encore, en cette douloureuse année de 1933, pour ne citer que les moments les plus importants. Albert Einstein fut present en Belgique, ou il noua des relations fructueuses tant avec les milieus scientifiques, qu'avec la famille royale. Nous prendrons part a l'hommage universel qui est rendu à l'illustre savant a l'occasion du centenaire de sa naissance en presentant en Mai 1979 une exposition qui aura pour thème "Einstein et la Belgique," et en publiant la correspondance encore inedite qu'échangèrent entre 1929 et 1932 Albert Einstein et Elie Cartan, autre associé de notre Académie. Nous formulons les voeux les plus chaleureux pour la reussite du Symposium du Centenaire que vous organisez à Jérusalem.

Bruxelles, le 14 Mars, 1979

Signé: Le Sécrétaire Perpétuel
Maurice Leroy
et le President
Louis Davin

This correspondence between Elie Cartan and Einstein will be published very soon in French and in German, but we also have the English translation of all the letters.

4. GRAVITATION, THE QUANTUM, AND STATISTICAL PHYSICS
Jacob D. Bekenstein

Einstein and the Unity of Physics

Of the many disciplines to which Einstein made fertile contributions, quantum theory, statistical physics, and gravitational theory drew much of his attention. To each he provided central ideas: the very concept of the quantum of radiant energy, the existence of stimulated emission, Bose–Einstein statistics, the theory of statistical fluctuations, the idea of gravitation as geometry of space-time, to mention just a few. To Einstein physics was unity, with understanding in one branch going hand in hand with understanding in another. Indeed, since Einstein and to a large extent because of Einstein, statistical physics is indivisible from quantum physics. Einstein must have pondered on the unification of the quantum with gravitation. In his famous discussion of the gravitational red shift we glimpse some of his thoughts.[1] But he did not leave us a unification. He must also have wondered about connections between gravitation and thermal physics, but again he left us no unification.

The dream of joining gravitation and the quantum has never been forgotten. Programs for unification were proposed over the decades.[2] But difficulties of concept and technique stood in the way, and no real unification in terms of physical concepts emerged from those efforts. Likewise, no deep unity of thermodynamics and gravitation was disclosed by wide-ranging investigations[3] if one excepts the gravitational red shift of temperature. Only in the 1970s did signs appear of an impending three-way synthesis of gravitation, quantum theory, and thermal physics. It is not my purpose here to describe it; it does not yet exist except as fragments. What I want to do is to recall how the clues were uncovered, to point out how deeply are the new developments rooted in Einstein's heritage, and to give a suggestion of the richness of new phenomena to be expected from the synthesis when it is completed. For my own contribution to the subject I draw on some early publications, although a number of results appear here for the first time.

The Dawn of the New Developments

The clues turned up in black hole physics.[4] The black hole made its appearance as a solution of Einstein's general relativity, representing a region of space-time causally invisible from the rest of the world. Then black hole solutions came to be regarded as representing the gravitational fields of completely collapsed objects, be they stars or galactic nuclei. We cannot say how Einstein would have taken to such interpretations. But at least in one respect he should have been pleased. The black hole solutions more than any others representing physical systems (save perhaps cosmological solutions) draw their very meaning from Einstein's conception of gravitation as geometry. Today we know that black holes are not confined to general relativity. A variety of other metric theories of gravitation also have black holes, and surprisingly, the same black holes as general relativity.[5] Thus in the black hole one has an entity embodying the geometrization of gravitation but not necessarily subject to all the other principles of general relativity.

For a long time black holes were regarded as inert absorbers of anything that approached them — hence their name, coined by Professor John Wheeler in the late 1960s. The 1970s brought a fresh recognition that black holes are not so passive, starting with Penrose's discovery[6] that energy can be drawn out of a rotating (Kerr) black hole by exploiting the curious properties of its ergosphere. Hard on the heels of this came Christodoulou's analysis[7] of the efficiency of the Penrose processes, leading to the surprising conclusion that the most efficient extraction corresponds to a reversible change of the black hole, whereas less efficient processes lead to an irreversible increase of the "irreducible mass" of the black hole. The thermodynamic flavor of this result was widely noted. The irreversibility uncovered by Christodoulou for Kerr black holes was codified by Hawking in a theorem that verified a hunch of Penrose and Floyd, namely, that "the surface area A of the horizon (boundary) of a black hole cannot decrease, and generally increases in a dynamical process."[8] Christodoulou's squared irreducible mass turned out to be proportional to A. For the Kerr black hole in units with $G = c = 1$,

$$A = 4\pi\{[M + (M^2 - Q^2 - L^2/M^2)^{1/2}]^2 + L^2/M^2\} , \qquad (4.1)$$

where M, Q, and L are the black hole's mass, charge, and angular momentum.

Parallel to these developments came the realization by Zel'dovich and, independently, by Misner that a Kerr black hole can superradiate,[9] that is, amplify waves scattered off it, when they satisfy

$$\omega < m \Omega , \qquad (4.2)$$

where ω and m are the angular frequency and azimuthal "quantum" number of the wave, respectively, and Ω is the rotational frequency of the hole. The energy for amplification comes from the rotation. Interestingly enough, the existence of superradiance when (4.2) holds can be inferred directly from Hawking's theorem.[10] The analysis when pushed to its logical extreme reveals that quantum effects must eventually violate the theorem, which is strictly classical.[11] It was widely realized that clasical superradiance must reflect stimulated emission by the black hole at the quantum level. We have then the first link between gravitation in its black hole guise and the quantum world.

Area and Entropy

The irreversibility in the growth of black hole surface area[12] suggested to me to interpret some monotonically increasing function of A as black hole entropy, since entropy also likes to increase irreversibly.[13] Various questions then presented themselves: What function? How to go from dimensions of area to dimensions of entropy? What is black hole entropy physically? What role, if any, does it play in the second law?

The first question was easy. The natural requirement that black hole entropy be additive for a system of black holes suggested that black hole entropy S_{bh} be taken as proportional to A.[14] The constant of proportionality was a separate problem. I remember telling Professor Wheeler, my thesis adviser, on one occasion that I thought A represented entropy, but that it was not clear to me how much entropy corresponded to a unit area. He replied that there could be no option but to ascribe a unit of entropy (one k) to something of the order of the square of the fundamental length $(\hbar G/c^3)^{1/2}$, which for many other good reasons is known as the Planck-Wheeler length. Later developments have amply justified the wisdom of the choice[15]:

$$S_{bh} = \eta k \frac{A}{(\hbar G/c^3)^{1/2}}, \qquad (4.3)$$

where η is a number of order unity to be determined separately.

This formula is actually a fitting embodiment of the threefold synthesis to be. It not only involves the constants of quantum theory $\hbar$, thermodynamics k, and relativistic gravitation c, G, but it relates a purely thermodynamic notion — entropy — to a gravitational one — black hole horizon area. And the relation is meaningful only because the quantum of action $\hbar$ is *not* negligible ($\hbar \neq 0$). (From here on I shall work in units where $k = c = G = 1$).

In contemporary science entropy is synonymous with missing information. It was thus natural to inquire what missing information is reflected in S_{bh}. The principle colloquially known as "Black holes have no hair"[16] enunciated by Wheeler in the late 1960s, and by then firmly established, suggested the answer. According to it M, Q, and L are a complete set of parameters for a stationary black hole so far as its exterior properties are concerned. But behind each triplet M, L, Q is a multitude of possibilities for internal black hole configurations — different structures, different compositions, even different microscopic states. (A similar comment could be made about a nonstationary black hole for which some other external parameters will exist.) S_{bh} was the natural candidate for quantifying our ignorance as to the precise internal configuration of the black hole. In the tradition of statistical physics I adopted $\exp(S_{bh})$ as the number of distinct internal configurations compatible with the given black hole exterior.[17] This interpretation has proved a valuable guide, and I will here mention one instance.

The importance of entropy in physics is due to its central role in the second law of thermodynamics, probably the law with the widest applicability we know. Now, black holes seemed to transcend the second law, as Wheeler was fond of pointing out. A black hole could act as a sink of entropy, and its external descriptors M, Q, and L clearly could not reveal how much entropy had fallen in. The external observer unable to peer into the black hole could never be sure that the second law was not being violated. Black hole entropy made possible a generalization of the second law for systems containing black holes that led out of this impasse: "The sum of the black hole entropy and the ordinary thermal entropy outside black holes cannot decrease."[18] It is to be noted that this generalized second law (GSL) is intrinsically a quantum law, being meaningless in the classical limit $\hbar \to 0$ because $\hbar$ occurs in a denominator.

A variety of *Gedanken* experiments showed the generalized second law (GSL) to be respected when a quantity of entropy goes down a black hole — the growth of S_{bh} being more than sufficient to compensate for the decrease in exterior thermal entropy.[19] The appearance of $\hbar$ in the definition of S_{bh} proved essential to secure this behavior. Thus the GSL invested S_{bh} with an almost phenomenological significance.

But problems remained. Seriously interpreting S_{bh} as entropy meant that one should take the associated temperature for a Kerr black hole,[20]

$$T_{bh} = (\partial M/\partial S_{bh})_{L,Q} = \frac{\hbar D}{32\pi\eta M[M - \tfrac{1}{2} Q^2/M + D]} , \qquad (4.4)$$

$$D = (M^2 - Q^2 - L^2/M^2)^{1/2} , \qquad (4.5)$$

just as seriously. But what is the physical significance of such temperature? Further, if one imagined a black hole immersed in a blackbody radiation environment at a temperature lower than T_{bh}, the inflow of radiation entropy exceeded the growth of S_{bh}, and the GSL seemed to be violated. Attempts to resolve this paradox[21] were complicated and inelegant. A third problem was the apparent asymmetry of nature with respect to the GSL. The law only required that black hole entropy plus exterior entropy increase. Examples existed where both entropies increased, or where an increase in S_{bh} outweighed a decrease in exterior entropy. But why, as Remo Ruffini asked during my thesis exam, can't an example be provided where a decrease in S_{bh} is outweighed by an increase in exterior thermal entropy? The answer I gave then, the trivial one, is that Hawking's area theorem forbids a decrease in S_{bh}. As time went on it became clear, as I mentioned, that the area theorem could be violated by quantum effects, so my answer was clearly an oversimplified one.

It turned out that all the three difficulties I mentioned had one root: the existence of a phenomenon nobody had the imagination or courage to foresee in 1972, spontaneous emission by a black hole. The key discovery was that of Hawking.[22] But the subject had a prehistory providing at once resemblance and contrast.

Black Hole Radiance

Zel'dovich in discussing his argument for the existence of superradiance[23] used a typical Einsteinian line of reasoning to surmise that the same quantum radiance that, by stimulated emission, manifests itself as classical superradiance must also be manifested as spontaneous emission in modes capable of superradiance. Unruh proved this to be the case by second-quantizing a scalar field in the background geometry of an eternal Kerr black hole and examining the expectation value of the Poynting vector operator.[24] Interestingly, he also demonstrated that spontaneous emission of neutrinos occurs in modes satisfying (4.2), although the Pauli principle forbids stimulated emission (and hence superradiance) for the neutrino.

But none of these developments prepared anybody for the surprise occasioned by Hawking's discovery.[25] In Unruh's approach the nonrotating uncharged (Schwarzschild) black hole does not radiate. Hawking studied a second-quantized scalar field in the background geometry of a "star" collapsing to form a Schwarzschild black hole. By considering the expectation value of the number operator, he found that quanta are emitted during the collapse, which is hardly surprising. But at late times, as the static black hole forms, the outgoing radiation is not extinguished but attains a standard form independent of details of the collapse. This *is* surprising. And to top off the surprise, Hawking discovered that the radiation spectrum is thermal. The mean number of quanta emitted in a mode of energy ε ($\varepsilon = \hbar\omega$) for which the classical black hole absorption coefficient (absorptivity) is Γ, is given by

$$\bar{n}_{sp} = \Gamma(e^{\varepsilon/T_0} - 1)^{-1} \ , \qquad (4.6)$$

$$T_0 = \hbar/8\pi M \ . \qquad (4.7)$$

Thus the black hole radiates just as a body at temperature T_0 endowed with the same absorptivity as the hole!

Hawking also considered the effects of rotation and charge of the hole, and of the nature of the quanta, and concluded that if the quanta are bosons (possibly charged) then[26]

$$\bar{n}_{sp} = \Gamma(e^x - 1)^{-1} \ , \qquad (4.8)$$

$$x \equiv (\varepsilon - \Omega\hbar m - \Phi e)T_0^{-1} \ , \qquad (4.9)$$

where m is the azimuthal "quantum" number of the mode, e its charge, Ω the hole's angular velocity, Φ its electrical potential, and T_0 a temperature identical to the T_{bh} defined by black hole thermodynamics [Eq. (4.4)] with the choice $\eta = 1/4$. Again the radiation is like that from a hot, rotating, and charged body. By integrating the thermodynamic relation between temperature and energy, Hawking arrived back at formula (4.3) for the entropy of the black hole with the choice $\eta = 1/4$.

Hawking's remarkable discovery, apart from fixing the constant η, cleared up the difficulties I mentioned. First of all it left no doubt as to the identity of the T_{bh} with the T_0 of the Hawking process. The operational physical meaning of T_{bh} became clear: it is the radiation temperature. Hawking's result showed that a source of entropy had been overlooked in the problem of a black hole in a colder radiation bath: the entropy of the emitted radiation. Hawking noted that by analogy with a hot body in a colder radiation bath, one expects this extra entropy to suffice to make the GSL work. Finally, Hawking's work provided an example in which a decrease in S_{bh} is outweighed by an increase in thermal entropy — the missing case of the GSL. The energy loss to radiation causes a decrease in area, thus violating the classical area theorem. Yet, as Hawking noted, the analogy with a hot body suggests that the radiation entropy will more than compensate for the decrease in S_{bh}. Both of Hawking's hunches were upheld by detailed statistical calculations[27] of the radiation entropy that took into account the nonblackbody spectrum (the fact that $\Gamma \neq 1$) of black hole radiation.

The importance of Hawking's work can never be stressed enough. Before it the formal analogy between black holes and thermodynamics was recognized by a number of people[28] apart from myself. But the view that the connection between black holes and thermodynamics is a deep one via the quantum of action, which is the crux of black hole thermodynamics, was regarded with deep skepticism by these workers and nearly everybody else. The Hawking radiation illustrated vividly the correctness of black hole thermodynamics and brought it to maturity by clearing up the last paradoxes. It provided a test case that the GSL passed successfully, a test case that could not have been conceived at the time the law was formulated. And by way of the shock it caused, Hawking's discovery triggered a vast amount of work that is leading to a deeper understanding of the quantum processes in curved space-time, and even to an understanding of quantum gravitation. First among these results were works of Wald, Boulware, Hartle and Hawking, Gerlach, and others that verified Hawking's result for Schwarzschild black holes by diverse approaches.[29] Gerlach's work is seldom appreciated. But with its emphasis on zero-point energy, excitation of radiation oscillators by the space-time geometry, and entropy as the logarithm of the number of emitted modes, it is the sort of analysis that Einstein, with his reservations about standard quantum theory, would have found most convincing.

Statistical Black Hole Physics

There can now be little doubt that quantum theory, gravitation, and thermodynamics are deeply interrelated. Most of the work going on today to clarify the precise relation uses quantum field theory as its principal tool. This work is bedeviled by the usual problems of field theory in flat space-time (divergences, questions of renormalization), as well as by problems intrinsic to curved space-time (nonuniqueness of the vacuum, the very meaning of the particle concept). I want to take here a different, less traveled path in which one relies on statistical arguments to study quantum-thermal properties of the gravitational field in the context of black holes. This approach is technically much simpler than the field-theoretic one and nevertheless can obtain results of great generality at a relatively small cost in terms of effort. It does have the disadvantage of not being able to deal with certain questions that the field-theoretic method can attack. I want to use the statistical approach to investigate the characteristics of stimulated emission by a black hole and to study the fluctuations of black hole mass about its mean. (But I will not have the opportunity to discuss the interesting work of Candelas and Sciama on the relation between quantum fluctuations and dissipation in irreversible black hole thermodynamics,[30] a subject closely related to Einstein's pioneering work on fluctuation-dissipation theorems.) Whenever feasible I shall make contact with corresponding results from the field-theoretic approach.

The proofs of the validity of the GSL in the face of the Hawking process that I mentioned posed the problem of how to determine the entropy of the Hawking radiation from the known mean number of quanta emitted per mode, and that alone.[31] Since the radiation is not blackbody no ready recipe existed. The method I found useful is Jaynes's information-theoretic approach to statistical physics, which I now want to describe, as it is crucial to the investigation of stimulated emission by black holes.[32]

For a system that can be in each of its states with a certain probability P_i for state i, one defines the measure of ignorance as to the actual state by

$$S = -\sum_i P_i \ln P_i, \qquad \text{where} \sum_i P_i = 1 . \qquad (4.10)$$

S is called the entropy. According to Jaynes the least biased probability distribution for a given situation is that which maximizes S subject to all known information. This is not to say one assumes equilibrium; the situation might be a nonequilibrium one. The maximization of S is just the formal way of admitting that our information about the system is limited. If the information included as constraints in the maximization is actually the maximum *knowable* information, one expects the least biased probability distribution to coincide with the physical one, and the corresponding S to be the physical entropy.

For black hole radiation the natural states are those labeled by the occupation numbers for the various modes.[33] The known information is the normalization of probability, and the mean value [Eq. (4.8)] for the mean number of outgoing quanta in each mode. Maximization of S by the method of lagrange multipliers gave for the probability of spontaneous emission of n quanta in a given mode[34]

$$p_{sp}(n) = (1 - e^{-\beta})e^{-\beta n} , \qquad (4.11)$$

$$e^{-\beta} \equiv \Gamma(e^x - 1 + \Gamma)^{-1} . \qquad (4.12)$$

At about the same time by independent field-theoretic calculations Parker, Hawking, and Wald each also derived this result for the Schwarzschild case.[35] Hawking and Wald each took the fact that $\Gamma \neq 1$ into account. It thus became clear that the intuitive notions as to which information can be regarded as known a priori were on the mark. It also became clear that the information-theoretic approach is capable of yielding results of a generality difficult to equal with quantum field theory, and this with comparatively little labor.

Response of a Black Hole to Incident Radiation

The next logical step was to determine the probability distribution $p_0(n)$ that a black hole immersed in a thermal radiation bath at temperature T sends out n quanta in a given mode. The known information apart from normalization is the mean number $\bar{n}_0$ of quanta outgoing in the mode, the sum of the contribution [Eq. (4.8)] from spontaneous emission and the contribution from reflection by the hole of a fraction $1 - \Gamma$ of the incident mean number of quanta (a Planckian one)

$$\bar{n}_0 = \Gamma(e^x - 1)^{-1} + (1 - \Gamma)(e^y - 1)^{-1} , \qquad (4.13)$$

where $y = \varepsilon/T$. Maximization of S yields the deceptively simple distribution[36]

$$p_0(n) = (1 - e^{-\gamma})e^{-\gamma n} , \qquad (4.14)$$

$$e^\gamma - 1 = (\bar{n}_0)^{-1} . \qquad (4.15)$$

This expression is physically obscure because it depends on properties both of the black hole and of the radiation bath. In the effort to isolate the black hole contribution and in using it to investigate the Einstein A and B coefficients, I benefited from the cooperation of my student A. Meisels.

The probability distribution of the incident blackbody radiation modes is of Boltzmann form

$$p_{bb}(n) = (1 - e^{-y})e^{-yn} \ . \tag{4.16}$$

We can thus express $p_0(m)$ as

$$p_0(m) = \sum_{n=0}^{\infty} p(m|n) (1 - e^{-y})e^{-yn} \ , \tag{4.17}$$

where $p(m|n)$ is the probability that the black hole returns m quanta outward when exactly n are incident in the given mode. We already know one element of $p(m|n)$, namely

$$p(m|0) = p_{sp}(m) = (1 - e^{-\beta})e^{-\beta m} \ . \tag{4.18}$$

We notice $p(m|0)$ is independent of the exterior radiation; it is a natural assumption that all $p(m|n)$ are independent of T. Then one can isolate $p(m|n)$ by expanding $p_0(m)$ in Eq. (4.14) in powers of e^{-y} and identifying the coefficients with those of the series in (4.17). The result is[37]

$$p(m|n) = \frac{(e^x - 1)e^{xn}\Gamma^{m+n}}{(e^x - 1 + \Gamma)^{n+m+1}} \sum_{k=0}^{\min(n,m)} \frac{(m+n-k)!}{k!(n-k)!(m-k)!} X^k \ ,$$
$$X = 2\Gamma^{-2}(1 - \Gamma)(\cosh x - 1) - 1 \ . \tag{4.19}$$

It is easy to see that (4.18) is a special case of (4.19). For the case $Q = 0$, $L = 0$ the general formula (4.19) has been confirmed by a direct field-theoretic calculation of Panangaden and Wald.[38] Again the economy of the information-theoretic approach is evident by comparison. The $p(m|n)$ obeys the interesting symmetry relation

$$e^{-xn}p(m|n) = e^{-xm}p(n|m) \ . \tag{4.20}$$

As noted by Hartle and Hawking[39] who first derived the $n = 0$ case of this relation for the Schwarzschild black hole by the method of Feynman path integrals, the relation guarantees that detailed balancing can come about between black hole and surrounding thermal radiation. We recall that Einstein, starting from the assumption that detailed balancing holds, inferred the existence of stimulated emission by atoms. Let us take Eq. (4.20) as proof that we are on the right track, and go on to explore the role of stimulated emission in black hole physics.

The $p(m|n)$ looks complicated because it is a composite distribution containing contributions from spontaneous emission, scattering by the effective potential barrier about the hole, and stimulated emission. One can convince oneself that there is stimulated emission by recognizing that spontaneous emission and scattering by themselves cannot reproduce the form of $p(m|n)$. Consider first the ordinary (elastic) scattering. If $1 - \Gamma_0$ is the probability that an incident quantum is scattered in the same mode, then the probability that k will be scattered and $n - k$ absorbed when precisely n are incident is[40]

$$p_{sc}(k|n) = \frac{n!}{k!(n-k)!} \Gamma_0^{n-k} (1 - \Gamma_0)^k \ . \tag{4.21}$$

The combinatorial factor is just the number of ways to partition n incident quanta into a class of k indistinguishable scattered quanta, and a class of $n - k$ indistinguishable absorbed quanta (Bose-Einstein statistics).

The composite probability for spontaneous emission or scattering of m quanta when n are incident is

$$p_{sp+sc}(m|n) = \sum_{k=0}^{\min(m,n)} p_{sc}(k|n) \, p_{sp}(m - k) \ . \tag{4.22}$$

The upper limit on the sum reflects the fact that if n quanta are incident, no more than n can be scattered. It is easy to do the sum for $m \geq n$; it is just a binomial sum then. The result is

$$p_{sp+sc}(m|n) = (\Gamma_0 + (1 - \Gamma_0)e^{-\beta})^n (1 - e^{-\beta})e^{-\beta m} \ , \tag{4.23}$$

which does not even resemble $p(m|n)$. Hence the black hole "emission" is not just by spontaneous emission and scattering. This all sounds familiar. Faced by a similar paradox in atomic physics Einstein realized that there has to be stimulated emission. We cannot but follow his steps and go on to isolate the probability distribution for stimulated emission from a black hole.

A word of perspective is appropriate here. That there should be stimulated emission from black holes is no new idea. It was widely recognized after the early work on superradiance. But it was believed that stimulated emission can occur only in modes capable of superradiance [condition (4.2) for uncharged quanta], and this view seemed to be supported by field-theoretic work.[41] Yet we find evidence for stimulated emission in *all* modes; our arguments in this section are also valid for modes that cannot superradiate. Thus, for example, a Schwarzschild black hole should be capable of stimulated emission even though it cannot superradiate in any mode. This may sound strange, but it is not without precedent. Every atomic transition is capable of stimulated emission, but this expresses itself as amplification at the classical level only if the atoms are in a special environment such as that in a laser, or that in an interstellar molecular cloud that we detect as an astrophysical maser.

The Probability Distribution for Stimulated Emission

By considering special cases of $p(m|n)$ I was able to guess the form of the probability for stimulated emission of j quanta when exactly n are incident. In the general case it is

$$p_{st}(j|m) = \frac{(n + j - 1)!}{(n - 1)! j!}(1 - e^{-\beta})^n \, e^{-\beta j} \ . \tag{4.24}$$

To verify this we first compute the composite probability that k quanta go out by spontaneous or stimulated emission when n are incident:

$$p_{sp+st}(k|n) = \sum_{j=0}^{k} p_{sp}(k - j) \, p_{st}(j|n) . \tag{4.25}$$

In view of the identity

$$\sum_{j=0}^{k} \frac{(n + j - 1)!}{(n - 1)! j!} = \frac{(n + k)!}{n! k!} , \tag{4.26}$$

which is easily verified by mathematical induction on k, we get

$$p_{sp+st}(k|n) = \frac{(n + k)!}{n! k!} (1 - e^{-\beta})^{n+1} e^{-\beta k} . \tag{4.27}$$

We now put together the overall emission probability:

$$p_{sp+st+sc}(m|n) = \sum_{k=0}^{\min(m,n)} p_{sc}(k|n) \, p_{sp+st}(m - k|n) , \tag{4.28}$$

where the limit on the summation index reminds us that no more than n quanta can be scattered. Comparison of (4.28) with the expression (4.19) shows $p_{sp+st+sc}$ to be identical to $p(m|n)$ if we recall the definition (4.12) of β, and if we make the identification

$$\Gamma = \Gamma_0 (1 - e^{-x}) . \tag{4.29}$$

Thus the black hole sends out quanta by three independent processes with distributions given by (4.11), (4.21), and (4.24). (It will be shown presently that the process called here stimulated emission is in all respects like atomic stimulated emission.) It has been shown in the foregoing that the (classical) absorptivity Γ is not just the complement of the quantum-scattering coefficient $1 - \Gamma_0$, but contains an extra negative contribution $-\Gamma_0 e^{-x}$, the existence of which was surmised before the calculations were done.[42] This contribution is a reflection of stimulated emission that acts to reduce the effective absorptivity. As long as $x > 0$ ($\omega > m\Omega$ for uncharged quanta) stimulated emission is overpowered by absorption and $\Gamma > 0$. But when $x < 0$ ($\omega < m\Omega$) and stimulated emission wins out, Γ turns negative, which implies superradiance. We clearly have here a purely thermodynamic derivation of superradiance.[43] Also, since $\Gamma_0 < 1$, Eq. (4.29) represents a thermodynamic upper bound on Γ for nonsuperradiant modes. Thus, if in some complex calculation we yield to temptation and make the approximation $\Gamma = 1$, we will be flying in the face of thermodynamics.

The Einstein Coefficients for a Black Hole

The Einstein A and B coefficients can be defined for a black hole if attention is paid to one point. In an atomic transition the probability for stimulated emission of a quantum is $B_\downarrow n$, where $B_\downarrow$ is the corresponding Einstein coefficient and n the number of quanta incident in the given mode. Since only one quantum can be emitted, $B_\downarrow n$ is also the mean number of quanta emitted. But a black hole can emit any number of quanta per mode; emission of just one is nothing special. Thus it is appropriate to define $B_\downarrow$ and the other coefficients from mean numbers, not from probabilities.

This being clear, let me calculate the mean number emitted by stimulated emission,

$$\bar{k}_{st} = \sum_{k=1}^{\infty} k\, p_{st}(k|n) \;, \tag{4.30}$$

where n is the number incident. Substituting Eq. (4.24), shifting the summation index to $k+1$, and setting $\xi = e^{-\beta}$, we can put (4.30) in the following form:

$$\bar{k}_{st} = (1-\xi)^n \xi^{2-n} \frac{\partial}{\partial \xi}\left[\xi^n \sum_{k=0}^{\infty} \frac{n(n+1)\ldots(n+k-1)}{k!} \xi^k\right] \;. \tag{4.31}$$

We recognize the sum as the Taylor expansion of $(1-\xi)^{-n}$. Carrying out the differentiation and substituting the definition of $e^{-\beta}$, we get

$$\bar{k}_{st} = \Gamma(e^x - 1)^{-1}\, n \;. \tag{4.32}$$

Comparing this with (4.6) we see that the mean number of quanta emitted by spontaneous *and* stimulated emission is proportional to $n+1$, just as in atomic physics.[44] This is perhaps the best reason for referring to the process whose distribution is (4.24) as stimulated emission. We can now identify two of the Einstein coefficients[45]

$$A = B_\downarrow = \Gamma(e^x - 1)^{-1} \;, \tag{4.33}$$

where A, the coefficient of spontaneous emission, is what multiplies the 1 in the mean number of emitted quanta.[46] The Einstein absorption coefficient $B_\uparrow$ is simply the complement of the quantum-scattering coefficient $1 - \Gamma_0$, namely Γ_0. By (4.29) we get[47]

$$B_\uparrow = \Gamma(1 - e^{-x})^{-1} \;. \tag{4.34}$$

In (4.33)–(4.34) we have the Einstein coefficients for a given mode of radiation.

Note the relation

$$B_\downarrow = B_\uparrow e^{-x} \;. \tag{4.35}$$

Although it looks rather undistinguished, it corresponds to the Einstein relation for an atomic transition,

$$g_U B_\downarrow = g_L B_\uparrow \; , \tag{4.36}$$

where g_U and g_L are the degeneracy factors of the upper and lower levels of the transition that emits the mode in question. This can be seen by writing the black hole version of the first law of thermodynamics including the electrical and rotational work[48]:

$$T_{bh} dS_{bh} = dM - \Omega dL - \Phi dQ \; . \tag{4.37}$$

For the change of black hole parameters due to *emission* of one quantum, one can identify $dM = -\varepsilon$, $dQ = -e$, $dL = -\hbar m$. It follows from (4.9) that

$$x = -dS_{bh} = S_{bh}(M,Q,L) - S_{bh}(M - \varepsilon, Q - e, L - m\hbar) \; , \tag{4.38}$$

so that (4.35) takes the form

$$e^{S_{bh}(M,Q,L)} B_\downarrow = e^{S_{bh}(M-\varepsilon,Q-e,L-m\hbar)} B_\uparrow \; . \tag{4.39}$$

Recall that the exponential of S_{bh} is to be interpreted as the number of distinct internal configurations or states of the black hole corresponding to the external state described by M, Q, L. Thus this exponential is really a degeneracy factor, and it then becomes clear that the content of (4.39) is the same as that of (4.36). Therefore, the Einstein A and B coefficients for a black hole obey the Einstein relations first uncovered in atomic physics. This result above all gives a vivid confirmation of the reality of internal black hole configurations and their relation to S_{bh}. It does not tell us what the configurations are like. For this one must invoke a specific model of the radiation process, as in the work of Gerlach.[49]

The Fundamental Transition Probability

The analogy just discussed between black hole emission and atomic transitions begs for further inquiry. One observation may prove helpful in trying to understand it. Let us bear in mind the limitations of the analogy. An atom emits one quantum per mode at most; a black hole can emit any number. So a slight difference in viewpoint is needed. I propose to visualize a black hole transition resulting in the emission of a given mode as a sequence of jumps down a ladder of levels equally spaced in mass, charge, and angular momentum, each jump giving one quantum for the mode in question. The levels with definite M, Q, L are to be regarded as Kerr solutions of Einstein's equations. The black hole itself is to be thought of as a system whose spectrum is made up of all allowed Kerr solutions. (It helps conceptually to think of the spectrum as discrete.) The black hole, then, is the analog of the atom. The Kerr solution is the analog of the atomic level. The uniform spacing of the levels makes possible multiquantum emission in the black hole case. The uniform spacing is to be taken in the sense that over a range of energy small compared to the mean black hole mass, the spacing is nearly uniform (like levels $n = 100, 101, 102, 103$ of hydrogen).

Together with this physical picture one can give an intuitively appealing, if simplistic, interpretation of the probability distributions for spontaneous and stimulated emission. The crux of this interpretation is the concept of a fundamental probability for the black hole to jump down one level in the chain associated with emission in a given mode. We call it p. It must depend on the mode, but we shall assume it is identical for spontaneous and stimulated scattering. The probability for spontaneous emission of n quanta in the given mode must equal the probability of n successive jumps, p^n, times the probability no further jump takes place, $1 - p$. This is of the same form as our earlier result (4.11) with the identification $p = e^{-\beta}$.

Now consider stimulated emission due to n incident quanta. The probability that the black hole makes exactly j_i jumps down due to the influence of the ith incident quantum is $(1 - e^{-\beta})e^{-\beta j_i}$. The incident quanta are assumed to act independently, so we multiply these factors to get $(1 - e^{-\beta})^n e^{-\beta j}$, where j is the sum of all the j_i. We must also multiply this factor by the number of ways that j indistinguishable emitted quanta can be partitioned into n classes, one for each of n indistinguishable incident quanta, in order to take care of Bose–Einstein statistics. We get exactly the expression (4.24) for the probability of stimulated emission of j quanta due to n incident ones.

Thus our simple viewpoint provides a unified treatment of spontaneous and stimulated emission in terms of a single transition probability $e^{-\beta}$. From this point of view the fundamental quantity is $e^{-\beta}$, not Γ. (Scattering is described by its own scattering probability $1 - \Gamma_0$.) Only a deeper theory will be able to ascertain the value of the approach presented here, but it is certainly a suggestive one.

Black Hole Mass Distribution

We are accustomed to regard a black hole in a stationary state as having definite M, Q, and L. But the Hawking radiation denies this comfortable idea. Since the black hole emits quanta according to a probability distribution, the change in the hole's mass (and charge and angular momentum) is not known sharply, and neither can the M, Q, and L after emission. A radiating black hole must thus be described by a statistical distribution over M, Q, and L with certain dispersions. Yet all theoretical work indicates that it makes sense to think of M, Q, and L as well defined. Hence the distribution must be a sharp one about mean values $\overline{M}$, $\overline{Q}$, $\overline{L}$ that will play the role of effective mass, charge, and angular momentum of the hole.

It is clearly of interest to find out what the distribution looks like. This is possible by way of a slight reformulation of Wheeler's dictum "a black hole has no hair"[50] to conform with the fact that M, Q, and L are not known precisely. There is little doubt that the principle must be held to stipulate that $\overline{M}$, $\overline{Q}$, and $\overline{L}$ are the only parameters of the probability distribution that determine external properties of the hole. Let us then deduce the form of the distribution. To keep things simple I shall only consider the case with $\overline{Q} = \overline{L} = 0$ and concentrate on the (marginal) distribution for mass alone.

We designate the distribution by $P(M,\overline{M})$ and regard it as continuous for convenience. It must satisfy

$$\int_0^\infty P(M,\overline{M})dM = 1 , \qquad (4.40)$$

$$\int_0^\infty MP(M,\overline{M})dM = \overline{M} . \qquad (4.41)$$

Let the black hole radiate over a short time and emit a representative sample of quanta in all modes (including different species of particles) according to the distributions $p_{sp}(n)$ like (4.11) and its fermion analog.[51] The new distribution of the hole's mass after emission shall be given by the composite distribution

$$\widetilde{P} = \sum_{\{n\}} P(M + \sum_i n_i \varepsilon_i, \overline{M}) \prod_i p_{sp}(n_i) , \qquad (4.42)$$

where the product is over all modes i emitted, and the sum over $\{n\}$ means a sum over all possible sets of occupation numbers n_i. In (4.42) we take the probability that the hole originally has an excess energy $\sum_i n_i \varepsilon_i$ over M, multiply it by the probability (the $\prod p_{sp}$) of one possible event by which it may radiate just this excess, and sum over all possible radiation events, and over all possible excess energies. Now let us expand P in (4.42) in powers of $\sum_i n_i \varepsilon_i$ to second order. We get

$$\widetilde{P} = P(M,\overline{M}) + \frac{\partial}{\partial M} P(M,\overline{M}) \cdot \sum_i \overline{n}_i \varepsilon_i + \tfrac{1}{2} \frac{\partial^2}{\partial M^2} P(M,\overline{M}) \cdot \sum_{ij} \overline{n_i n_j} \varepsilon_i \varepsilon_j , \qquad (4.43)$$

where bars indicate averaging with respect to the distribution p_{sp}. Since $\overline{n_i n_j} = \overline{n}_i \overline{n}_j$ for $i \neq j$, (4.43) can be put in the more convenient form [also correct to $O(n_i^2 \varepsilon_i^2)$]

$$\widetilde{P} = P(M - \Delta\overline{M}, \overline{M}) + \tfrac{1}{2} \frac{\partial^2}{\partial M^2} P(M,\overline{M}) \cdot s^2 , \qquad (4.44)$$

where

$$\Delta\overline{M} = -\sum_i \overline{n}_i \varepsilon_i , \qquad (4.45)$$

$$s^2 = \sum_i (\overline{n_i^2} - \overline{n}_i^2)\varepsilon_i^2 . \qquad (4.46)$$

Clearly $\Delta\overline{M}$ is the change in the mean mass of the hole, while s^2 is the variance of the emitted energy, which must equal the change in the variance

$$\sigma^2 \equiv \int_0^\infty (M - \overline{M})^2 P(M,\overline{M})dM , \qquad (4.47)$$

of M. By assumption σ^2 can only depend on $\overline{M}$, so we can write $s^2 = (d\sigma^2/d\overline{M})\Delta\overline{M}$ to first order. Also by assumption $\widetilde{P}$ can only be $P(M, \overline{M} + \Delta\overline{M})$ since $\overline{M}$ determines the distribution entirely. Making the identifications in (4.44) and expanding to first order in $\Delta\overline{M}$, we get the equation

$$\frac{\partial P}{\partial \overline{M}} = - \frac{\partial P}{\partial M} + \tfrac{1}{2} \frac{\partial^2 P}{\partial M^2} \frac{d\sigma^2}{d\overline{M}} , \qquad (4.48)$$

which determines $P(M,\overline{M})$. It is easy to verify that

$$P(M,\overline{M}) = (2\pi\sigma^2)^{-1/2} \exp\left[-\tfrac{1}{2}(M - \overline{M})^2 \sigma^{-2}\right] \qquad (4.49)$$

is a solution satisfying conditions (4.40), (4.41), and (4.47) (provided $\overline{M} \gg \sigma$ so the integrals can be extended to $-\infty$). In fact all our physical experience would have made us expect such a Gaussian distribution, so (4.49) must be *the* solution to our question.

The Width Paradox

The real physical question now is what is the width of the Gaussian, σ. We know that

$$s^2 = \frac{d\sigma^2}{d\overline{M}} \Delta\overline{M} , \qquad (4.50)$$

so let us calculate s^2 in (4.46). Making use of $p_{sp}(n)$ in (4.11) and of its fermion analog[52] we get

$$s^2 = \sum_i \varepsilon_i^2 e^{\beta_i} (e^{\beta_i} \mp 1)^{-2} = \sum_i \varepsilon_i^2 \Gamma_i (e^{x_i} \mp 1 \pm \Gamma_i) \times (e^{x_i} \mp 1)^{-2} , \qquad (4.51)$$

where upper signs are for boson modes and lower for fermion modes. To get the second equality we have used the definition of $e^{-\beta}$ [Eq. (4.12)] and its fermion analog.[53] By x_i we mean ε_i/T_{bh} (where $T_{bh} = \hbar/8\pi\overline{M}$).

In view of (4.8) and its fermion analog

$$\sum_i \varepsilon_i^2 \Gamma_i e^{x_i} (e^{x_i} \mp 1)^{-2} = - \partial/\partial(1/T_{bh}) \sum_i \varepsilon_i \Gamma_i (e^{x_i} \mp 1)^{-1} + \sum_i \varepsilon_i \Gamma_i' (e^{x_i} \mp 1)^{-1} \qquad (4.52)$$

where the prime stands for derivative with respect to $1/T_{bh} = 8\pi M/\hbar$. For massless quanta Γ can depend only on the dimensionless variable ε/T_{bh} for given angular momentum, so Γ_i' really stands for $\varepsilon_i (d\Gamma/dx)_i$. The first sum in the r.h.s. of (4.52) is just $-\Delta\overline{M}$ which is proportional to T_{bh}^4.[54] Thus the first term in the r.h.s. of (4.52) (call it H) is

$$H = - 4 T_{bh} \Delta\overline{M} = - \hbar \Delta\overline{M}/2\pi\overline{M} . \qquad (4.53)$$

By comparing Eqs. (4.50)–(4.53) we see that

$$\frac{d\sigma^2}{d\overline{M}} = - (\hbar/2\pi\overline{M}) \left\{ 1 + H^{-1} \sum_i \varepsilon_i^2 \left[(d\Gamma/dx)_i (e^{x_i} \mp 1)^{-1} \mp \Gamma_i (1 - \Gamma_i)(e^{x_i} \mp 1)^{-2} \right] \right\} \qquad (4.54)$$

In (4.50) s^2 is clearly positive and $\Delta\overline{M}$ negative. It follows that the quantity in curly brackets in (4.54) must be positive. Call it $1 + K$. We now argue that K is independent of $\overline{M}$ for $\overline{M} > 10^{17}$g, so that we can integrate (4.54).

For this mass range the hole is too cold to emit anything but massless particles,[55] and for these our earlier comments about Γ apply. Represent the H in (4.54) by the difference of the first and last sums in (4.52). The factors ε^2 can be reduced to x^2 by division of T_{bh}^2 which cancels out in the ratio in (4.54). When the sums in (4.54) are converted to integrals in the familiar manner, x enters as a dummy variable of integration; $\overline{M}$ has disappeared. Hence K is independent of $\overline{M}$. Let us integrate the equation (4.54) in the range $\overline{M} > 10^{17}$ g to get

$$\sigma^2 = \sigma_0^2 - [\hbar(1 + K)/2\pi] \ln \overline{M}/M_0 , \qquad (4.55)$$

where M_0 and σ_0^2 (together the constant of integration) are a reference mass and its corresponding variance, respectively. Note that σ depends on $\overline{M}$ very weakly.

There is a disturbing feature in (4.55). For a sufficiently large $\overline{M}$, σ vanishes — bigger masses are not meaningful in the framework of our treatment. Before we inquire into the significance of our finding let us get some idea as to how big the maximum mass may be. Let us take the reference mass M_0 as 10^{17} g; the exact value is not important because it appears in a logarithm. What shall we take for σ_0^2? We cannot take squared masses of particles for it; the σ^2 arises because of radiation of massless particles, so massive particles have nothing to do with it. The only thing left is $\hbar$ so we set $\sigma_0^2 = \alpha\hbar$ where α should be of order unity in the broad sense. The dispersion in M is thus of order $\hbar^{1/2}$ or 10^{-5} g. For the maximum mass we get

$$\overline{M}_{max} = 10^{17} g \cdot \exp\left[2\pi\alpha/(1 + K)\right] . \qquad (4.56)$$

If $\alpha \approx 1$, $\overline{M}_{max} \approx 5 \times 10^{19}$ g which is distressingly small. But if $\alpha \approx 10$, $\overline{M}_{max} \approx 10^{11} M_\odot$, which is as big a mass as has ever been discussed, even for astrophysical black holes. Thus the maximum mass does not pose a "practical" problem, only one of principle.

But the problem is a deep one. The existence of a maximum mean mass for black holes described by a distribution in mass should mean that above this limit the mass is known precisely. But then the radiation process cannot be operative as it inexorably leads to broadening of the distribution. Yet no study of Hawking's process has revealed a dependence of the *character* of the process on the scale of mass. The paradox is clear.

What are we doing wrong? Must we give up the straightforward approximation in (4.43) that led us to the Gaussian distribution that has served us so well in other contexts? Or is it that mean mass is not the sole descriptor of an uncharged nonrotating black hole? Or could it be that the field-theoretic calculations of the Hawking process are leaving out something crucial, and the process really turns off at some large mass? The answer is not clear today. But just as the paradox posed by black hole temperature in 1972 was only unraveled by the discovery of black hole radiance in 1974, so we may hope that the width paradox will be cleared up by some upcoming breakthrough in our understanding of the interplay of gravitation, the quantum, and statistical physics that first revealed itself in black holes.

Acknowledgment

I want to acknowledge the cooperation of A. Meisels in part of the work described here.

NOTES

1. A. Einstein, Ann. d. Physik **35** (1911).
2. P. G. Bergmann, Helv. Phys. Acta Suppl. **4**, 79 (1956); P. A. M. Dirac, Phys. Rev. **114**, 924 (1959); R. Arnowitt, S. Deser, and C. W. Misner in *Gravitation: An Introduction to Current Research,* edited by L. Witten (Wiley, New York, 1962); J. A. Wheeler in *Relativity, Groups and Topology,* edited by B. S. DeWitt and C. DeWitt (Gordon and Breach, New York, 1964); B. S. DeWitt, Phys. Rev. **160**, 1133, and **162**, 1195, 1239 (1967).
3. R. C. Tolman, *Relativity, Thermodynamics, and Cosmology* (Oxford University Press, London, 1934).
4. For the elements see C. Misner, K. Thorne, and J. Wheeler, *Gravitation* (Freeman, San Francisco, 1973).
5. S. W. Hawking, Commun. Math. Phys. **25**, 167 (1972); J. D. Bekenstein and A. Meisels, Phys. Rev. **D18**, 4 (1978).
6. R. Penrose, Riv. Nuovo Cimento **1**, 252 (1969).
7. D. Christodoulou, Phys. Rev. Letters **25**, 1596 (1970); see also D. Christodoulou and R. Ruffini, Phys. Rev. **D4**, 3552 (1971).
8. S. W. Hawking, Phys. Rev. Letters **26**, 1344 (1971); R. Penrose and R. M. Floyd, Nature **229**, 177 (1971).
9. Ya. B. Zel'dovich, Zh. Eksp. Teor. Fiz. Pis'ma **14**, 270 (1971), translated in Sov. Phys. JETP Letters **14**, 180 (1971); C. W. Misner, Bull. Amer. Phys. Soc. **17**, 472 (1972).
10. J. D. Bekenstein, Phys. Rev. **D7**, 949 (1973); A. A. Starobinski and S. M. Churilov, Zh. Eksp. Teor. Fiz. **65**, 3 (1973), translated in Sov. Phys. JETP **38**, 1 (1974).
11. Bekenstein, op. cit. in n. 10.
12. Christodoulou, op. cit. in no. 7; Christodoulou and Ruffini, ibid.; Hawking, op. cit. in n. 8; Penrose and Floyd, ibid.
13. J. D. Bekenstein, Ph.D. thesis, Princeton University, 1972.
14. Bekenstein, ibid., and Phys. Rev. **D7**, 2333 (1973).
15. Bekenstein, ibid., and Lett. Nuovo Cimento **4**, 737 (1972).
16. See op. cit. in n. 4.
17. Bekenstein, op. cit. in n. 13.
18. Bekenstein, op. cit. in nn. 13, 14, 15.
19. Bekenstein, ibid., and Phys. Rev. **D12**, 3292 (1974).
20. Bekenstein, op. cit. in n. 14.
21. Bekenstein, op. cit. in n. 13, and 1974 op. cit. in n. 19.
22. S. W. Hawking, Nature **248**, 30 (1974).
23. Op. cit. in n. 9
24. W. Unruh, Phys. Rev. **D10**, 3194 (1974). See also L. Ford, Phys. Rev. **D12**, 2963 (1975).
25. Hawking, op. cit. in n. 22, and Commun. Math. Phys. **43**, 199 (1975).
26. 1975 op. cit. in n. 25.
27. J. D. Bekenstein, Phys. Rev. **D12**, 3077 (1975).
28. B. Carter, Nature Phys. Sci. **238**, 71 (1972); B. Carter and also J. Bardeen in *Black Holes,* edited by B. DeWitt and C. M. DeWitt (Gordon and Breach, New York, 1973); J. M. Bardeen, B. Carter, and S. W. Hawking, Commun. Math. Phys. **31**, 161 (1973).
29. R. M. Wald, Commun. Math. Phys. **45**, 9 (1975); D. G. Boulware, Phys. Rev. **D13**, 2169 (1976); J. B. Hartle and S. W. Hawking, Phys. Rev. **D13**, 2188 (1976); U. Gerlach, Phys. Rev. **D14**, 1479 (1976).
30. P. Candelas and D. W. Sciama, Phys. Rev. Letters **38**, 1372 (1977).

31. Bekenstein, op. cit. in n. 27.
32. E. T. Jaynes, Phys. Rev. **106,** 620 (1957).
33. Strictly speaking, we should define S from the density matrix [E. T. Jaynes, Phys. Rev. **108,** 171 (1957)]. Definition (4.10) amounts to neglecting correlations between modes, a procedure justified by the quantum-field-theoretic results.
34. Bekenstein, op. cit. in n. 27.
35. L. Parker, Phys. Rev. **D12,** 1519 (1975); S. W. Hawking, Phys. Rev. **D13,** 191 (1976); Wald, op. cit. in n. 29.
36. Bekenstein, op. cit. in n. 27.
37. J. D. Bekenstein and A. Meisels, Phys. Rev. **D15,** 2775 (1977).
38. P. Panangaden and R. M. Wald, Phys. Rev. **D16,** 929 (1977).
39. Op. cit. in n. 29.
40. One may also expect stimulated scattering, but for the problem at hand it is not physically distinct from stimulated emission, and so let us treat them together under the label stimulated emission.
41. R. M. Wald, Phys. Rev. **D13,** 3176 (1976).
42. Bekenstein, op. cit. in n. 27.
43. For a very different derivation of superradiance from thermodynamics for a rotating body see Ya. B. Zel'dovich, Zh. Eksp. Teor. Fiz. **62,** 2076 (1972), translated in Sov. Phys. JETP **35,** 1085 (1972).
44. A. Meisels and I, in our previous treatment, op. cit. in n. 37, *assumed* this result; here its proof is given.
45. Bekenstein and Meisels, op. cit. in n. 37.
46. Ordinarily the ration $A/B_\downarrow$ involves a phase space factor proportional to ω^3. By treating each mode separately we avoid this complication.
47. Bekenstein and Meisels, op. cit. in n. 37.
48. Bekenstein, 1972 op. cit. in n. 13.
49. Gerlach, op. cit. in n. 29.
50. Wheeler in op. cit. in n. 4.
51. Bekenstein, op. cit. in n. 27.
52. Ibid.
53. Ibid.
54. This is seen going over to the dimensionless variable x, and converting the sum to an integral over x. Recall that the density of states goes like ε^2. The result is valid for massless quanta, which is all we are interested in.
55. D. Page, Phys. Rev. **D13,** 198 (1976).

Discussion

Following the Paper by J. BEKENSTEIN

Chairman, J. A. WHEELER, U. of Texas

Wheeler: This was not just a nice and an interesting paper but an exciting paper. We should remember that our speaker is a very modest person. He said nothing at all about his own part in bringing about the discovery of the Hawking radiation. However, one has only to refer the contribution of Brandon Carter to the splendid Les Houches book, edited by the DeWitts, on Black Hole Physics, to read of the dismay that both Carter and Hawking felt when they read the arguments of Bekenstein, the original arguments that said in effect: "Not only does the surface area of a black hole look like entropy, it is entropy. Not only does the surface gravity of a black hole look like temperature, it is temperature." These physical arguments of Bekenstein forced us into a new world. Especially interesting in the method of analysis of Bekenstein is his combination of physical reasoning and Talmudic scholarship, if I can use that phrase, where one goes over and over again the various points in the argument to come to new relations. Professor Sciama, I hope, will comment also because he is much more of an expert in this subject than I am, but I remember so much the concern that Einstein expressed, in conversations over the years, about why *initial* conditions should appear in physics and not *final* conditions! Where does the asymmetry come from? In ordinary radiation processes, this question is a microscopic kind of question, but when one deals with black holes, with a whole Niagara Falls of matter falling into (and not a Niagara Falls of matter pouring out from) a concentration of mass, one sees friction, one sees one-sidedness in time coming in on the most enormous scale we have ever seen anywhere in physics. Therefore one is led once more to ask: what is the relation between this one-sidedness in time and the cosmological issues that trouble us all? There is a wonderful feature about the great paradox with which Professor Bekenstein has ended. It directs our attention to the realm of very large masses. Therefore it leads us to ask if the solution of the beautiful problem he presented to us may not in the end come back to cosmological issues.

D. Sciama (U. of Texas and Oxford U.): I would like to add a historical footnote to what John Wheeler has said, because I remember very well that one of the most outspoken critics of our speaker's important contributions to black hole's thermodynamics was Steven Hawking. He simply could not believe in 1973 that the entropy was a real entropy and he just thought it was a vague mechanical analogy. That makes it all the more remarkable that he made the discovery that you have been talking about, and even then he was not trying to do it to obtain the effect! That's psychologically very interesting. I can recall that he told me he felt it was about time for him to study the combination of quantum theory and gravitation, and I remember saying to myself: "Aha, he has found a problem too difficult for him now," and then, a year later, to his own surprise, he came out with the black hole radiance, and of course came entirely into Bekenstein's own camp.

I. Segal (MIT): I am wondering about the foundations. Usually with the physical system we associate a system of observables. These observables then have various states and we know what the entropy of a state is, the expectation of a density matrix, and so on. I think it would make things much clearer if you could specify the system of observables for a black hole and derive their entropy in that way.

Bekenstein: I think the main point is that the black hole entropy that I mentioned is to be taken, if you want, in a symbolic sense as empirical entropy. One does not know why the area is the entropy. One cannot calculate it by taking the trace of the density matrix of something. To date there is no understanding of how to get that black hole entropy to be proportional to the area by deep fundamental methods. Take it as an empirical result, even though it is not even an experiment but only a *Gedanken* experiment.

Sciama: I would like to come back to that issue because Hawking now has a calculation from quantum gravity in which the entropy of the black hole does come out to be the area. You must have seen that. Would you comment on it?

Bekenstein: I have seen that groundbreaking work of Hawking and Gibbons using Euclidean quantum gravity. Actually, I think there is an advantage in starting a new branch of physics directly from empirical facts or *Gedanken* experiments, and not from abstract theoretical ideas. Later you can try to understand theoretically what is happening. But I think the advantage of the approach I described is in its being nearly phenomenological. If you had experiments you would say it was phenomological. Of course, you only have *Gedanken* experiments.

Ne'eman (Tel Aviv U.): Can I ask a question? These last paradoxes you were describing: is it conceivable that they might be resolved when the quantum aspect is completed on the gravitational field itself? It would involve the same kind of masses also.

Bekenstein: It is of course possible. I know the scale of mass could appear in a problem. The problem is very true. Actually, you put your finger on what could be wrong in those three questions that I asked here. But I feel one of these three issues must be basic.

Ne'eman: Another point that may be a coincidence: 10^{11} solar masses is roughly where quasars start to appear.

Bekenstein: Right.

G. Horwitz (Hebrew U.): I think that there is a problem about what you said that is also common to Hawking's description. It somewhat ignores the thermodynamic complexities, namely, to be very precise, you have black holes — that is, a nonclosed system in some surroundings. You then treat this Hawking black hole as though one could simply superpose the black hole plus the surroundings. I think there is a problem with the precise definition of your system in terms of what kind of ensemble you are dealing with. In other words, the question that I would ask about your paradox, is: Can you properly treat the black hole independently from the radiation field that surrounds it or should you be regarding the total combination?

Bekenstein: I think that just to try to come through the back door with such simple statistical arguments whose power has been proved throughout the history of physics, is to learn something without doing the full theory, and that is essentially what I was trying to show you here in a brief time.

Note added in proof: Further work in cooperation with U. Ben-Yaacov has revealed that the paradox can be disposed of, if in the Hawking radiation formula Γ depends on the instantaneous mass M while T_0 depends, as earlier assumed, only on $\overline{M}$. It is also important to treat fluctuations of the angular momentum alongside those of mass.

5. MACH'S PRINCIPLE AND THE GENERAL RELATIVITY THEORY
Nathan Rosen

In order to consider the role of Mach's principle in the general theory of relativity, let us begin with what Einstein wrote about this principle in *The Meaning of Relativity*.[1] In the introduction to his discussion of the general theory of relativity, referring to the preferred status assigned to inertial systems both in Newtonian mechanics and in the special relativity theory, Einstein wrote:

> We can think of no cause for this preference for definite states of motion to all others, according to our previous considerations, either in the perceptible bodies or in the concept of motion; on the contrary, it must be regarded as an independent property of the space-time continuum. The principle of inertia, in particular, seems to compel us to ascribe physically objective properties to the space-time continuum, . . . having a physical effect, but not itself influenced by physical conditions.
>
> As long as the principle of inertia is regarded as the keystone of physics, this standpoint is certainly the only one which is justified. But there are two serious criticisms of the ordinary conception. In the first place, it is contrary to the mode of thinking in science to conceive of a thing (the space-time continuum) which acts itself but which cannot be acted upon. This is the reason why E. Mach was led to make the attempt to eliminate space as an active cause in the system of mechanics. According to him, a material particle does not move in unaccelerated motion relatively to space, but relatively to the center of all the other masses in the universe. . . . In order to develop this idea within the limits of the modern theory of action through a medium, the properties of the space-time continuum which determine inertia must be regarded as field properties of space, analogous to the electromagnetic field. . . . In the second place, classical mechanics indicates a limitation which directly demands an extension of the principle of relativity to spaces of reference which are not in uniform motion relatively to each other.

After having developed the ideas and the formalism of general relativity, Einstein wrote:

> If the universe were quasi-Euclidian, then Mach was wholly wrong in his thought that inertia, as well as gravitation, depends upon a kind of mutual action between bodies. For in this case, with a suitably selected system of coordinates, the $g_{\mu\nu}$ would be constant at infinity, as they are in the special theory of relativity, while within finite regions the $g_{\mu\nu}$ would differ from these constant values by small amounts only . . . as a result of the influence of the masses in finite regions. The physical properties of space would not then be wholly independent, that is,

uninfluenced by matter, but in the main they would be, and only in small measure conditioned by matter. Such a dualistic conception is even in itself not satisfactory; there are, however, some important physical arguments against it, which we shall consider.

The hypothesis that the universe is infinite and Euclidean at infinity, is, from the relativistic point of view, a complicated hypothesis. In the language of the general theory of relativity it demands that the Riemann tensor of the fourth rank R_{iklm} shall vanish at infinity, which furnishes twenty independent conditions, while only ten curvature components $R_{\mu\nu}$ enter into the laws of the gravitational field. . . .

But in the second place, the theory of relativity makes it appear probable that Mach was on the right road in his thought that inertia depends upon a mutual action of matter. For we shall show in the following that, according to our equations, inert masses do act upon each other in the sense of the relativity of inertia, even if only very feebly. What is to be expected along the line of Mach's thought?

1. The inertia of a body must increase when ponderable masses are piled up in its neighborhood.
2. A body must experience an accelerating force when neighboring masses are accelerated, and, in fact, the force must be in the same direction as the acceleration.
3. A rotating hollow body must generate inside of itself a "Coriolis field," which deflects moving bodies in the sense of the rotation, and a radial centrifugal field as well.

We shall now show that these three effects, which are to be expected in accordance with Mach's ideas, are actually present according to our theory, although their magnitude is so small that confirmation of them by laboratory experiments is not to be thought of.

Einstein then carried out a calculation in which he obtained the metric tensor for a given distribution of matter by solving his gravitational field equations in the weak-field approximation, and then wrote down the equations of motion for a particle in a form that showed the presence of the three effects listed by him.

Einstein's calculations were criticized by Davidson,[2] who proceeded to give his version of the calculations. It turns out, however, that Davidson's calculations are also open to criticism. Let us therefore do the calculations again, and let us try to do them in a completely consistent way.

Let us consider the metric for the case of a number of particles, or point masses, moving slowly in a nearly Galilean coordinate system, $(x^0, x^1, x^2, x^3) = (t, x, y, z)$, so that $v/c \ll 1$, where v is the speed of a typical particle and c the speed of light. Using conventional units, let us take $1/c$ as a small parameter, following Einstein, Infeld and Hoffmann,[3] and let us expand the metric tensor $g_{\mu\nu}$ in powers of this quantity, keeping terms up to order $1/c^4$. Let us solve the Einstein field equations for empty space,

$$R_{\mu\nu} = 0 , \qquad (5.1)$$

together with the de Donder coordinate condition

$$[(-g)^{1/2} g^{\mu\nu}]_{,\nu} = 0 , \qquad (5.2)$$

by successive approximations. We take as a boundary condition that the metric has the Galilean form at infinity, and we look for a solution having suitable singularities at the particle positions. One finds, with Latin indices denoting space components,

$$g_{00} = 1 + \frac{2}{c^2}\phi + \frac{1}{c^4}(\psi + 2\phi^2), \tag{5.3}$$

$$g_{0k} = \frac{1}{c^3}A_k, \tag{5.4}$$

$$g_{kl} = -\delta_{kl}\left(\frac{1}{c^2} - \frac{2}{c^4}\phi\right), \tag{5.5}$$

where

$$\phi = -G\sum\frac{m_a}{r_a}, \tag{5.6}$$

$$A_k = 4G\sum\frac{m_a \dot{y}_a^k}{r_a}, \tag{5.7}$$

$$\psi = G\sum\frac{m_a}{r_a}(x^k - y_a^k)\ddot{y}_a^k + G\sum\frac{m_a}{r_a^3}[(x^k - y_a^k)\dot{y}_a^k]^2$$

$$- 4G\sum\frac{m_a v_a^2}{r_a} + 2G^2\sum{}'\frac{m_a m_b}{r_a r_{ab}}. \tag{5.8}$$

The space coordinates of a field point are written x^k, those of a particle, y^k. Particles are labeled by indices a and b, with m_a the mass of a particle, r_a the (Euclidean) distance from the field point to a particle, r_{ab} the distance between two particles. Summations are carried out on particle indices (a prime indicating $a \neq b$) and on repeated coordinate indices. A dot denotes a time derivative, and G is the gravitational constant.

Let us now consider the equations of motion of a test particle, that is, the equations of the geodesic

$$\frac{d^2x^\mu}{ds^2} + \left\{\begin{matrix}\mu\\ \alpha\beta\end{matrix}\right\}\frac{dx^\alpha}{ds}\frac{dx^\beta}{ds} = 0. \tag{5.9}$$

Taking $\mu = k$ and using the foregoing components of the metric, one finds that, to order $1/c^2$,

$$\frac{d^2x^k}{ds^2} + \phi_{,k}\left(\frac{dt}{ds}\right)^2 + \frac{1}{c^2}\left(\tfrac{1}{2}\psi_{,k} - A_{k,0} + 4\phi\phi_{,k}\right)\left(\frac{dt}{ds}\right)^2$$

$$+ \frac{1}{c^2}(A_{q,k} - A_{k,q} - 2\phi_{,0}\delta_{kq})\frac{dt}{ds}\frac{dx^q}{ds} \qquad (5.10)$$

$$+ \frac{1}{c^2}(\delta_{pq}\phi_{,k} - 2\delta_{kp}\phi_{,q})\frac{dx^p}{ds}\frac{dx^q}{ds} = 0.$$

To the same order, the relation,

$$ds^2 = g_{\mu\nu}dx^\mu dx^\nu, \qquad (5.11)$$

gives

$$\left(\frac{dt}{ds}\right)^2 = 1 + \frac{1}{c^2}(v^2 - 2\phi), \qquad (5.12)$$

with $v^2 = \dot{x}^q\dot{x}^q$. Substituting (5.12) into (5.10) gives, to order $1/c^2$,

$$\frac{d^2x^k}{ds^2} = -\left(1 + \frac{2v^2}{c^2}\right)\phi_{,k} + \frac{1}{c^2}[A_{k,0} - \frac{1}{2}\psi_{,k} - 2\phi\phi_{,k} + (A_{k,q} - A_{q,k})\dot{x}^q$$

$$+ 2\phi_{,0}\dot{x}^k + 2\phi_{,q}\dot{x}^q\dot{x}^k]. \qquad (5.13)$$

If one multiplies both sides by $m_0(1 - 2\phi/c^2)$, where m_0 is a constant, one finds that the equation can be rewritten in the form

$$\frac{d}{ds}\left[m_0\left(1 - \frac{2\phi}{c^2}\right)\frac{dx^k}{ds}\right] = -m_0\left[\left(1 + \frac{2v^2}{c^2}\right)\phi + \frac{1}{2c^2}\psi\right]_{,k}$$

$$+ \frac{m_0}{c^2}[(A_{k,q} - A_{q,k})\dot{x}^q + A_{k,0}]. \qquad (5.14)$$

Let us write this as

$$\frac{dP^k}{ds} = F^k, \qquad (5.15)$$

where F^k is the (''special-relativity'') force acting on the particle, and P^k is the momentum vector,

$$P^k = m_0\left(1 - \frac{2\phi}{c^2}\right)\frac{dx^k}{ds}. \qquad (5.16)$$

To interpret this, we note that (5.11) can be written

$$ds^2 = dT^2 - \frac{1}{c^2} dX^k dX^k ,\qquad(5.17)$$

where, to second order,

$$dT = \left(1 + \frac{\phi}{c^2}\right) dt ,\qquad(5.18)$$

$$dX^k = \left(1 - \frac{\phi}{c^2}\right) dx^k .\qquad(5.19)$$

Let us call these the metric differentials, since dT is the time interval given by a clock (at rest in the coordinate system) corresponding to dt, and dX^k is the distance as determined with a measuring rod corresponding to dx^k. The metric velocity corresponding to dx^k/ds is then dX^k/ds. If we write

$$P^k = m \frac{dX^k}{ds} ,\qquad(5.20)$$

then the mass m is given, to second order, by

$$m = m_0 \left(1 - \frac{\phi}{c^2}\right) .\qquad(5.21)$$

It is amusing to note that this agrees with the mass obtained by Einstein and not with that of Davidson.

From (5.17) one can also write, to second order,

$$ds^2 = \left(1 - \frac{v^2}{c^2}\right) dT^2 ,\qquad(5.22)$$

so that (5.20) can be written

$$P^k = m^* \frac{dX^k}{dT} ,\qquad(5.23)$$

where, as in the special relativity theory,

$$m^* = \frac{m}{\left(1 - \frac{v^2}{c^2}\right)^{1/2}} ,\qquad(5.24)$$

and can be written

$$m^* = m_0\left(1 - \frac{\phi}{c^2} + \frac{1}{2}\frac{v^2}{c^2} + \ldots\right). \tag{5.24a}$$

If we multiply (5.14) by ds/dT, using (5.22), we get

$$\frac{dP^k}{dT} = f^k, \tag{5.25}$$

where f^k is the "Newtonian" force, given to second order by

$$\begin{aligned} f^k = &-m_0\left[\left(1 + \frac{3}{2}\frac{v^2}{c^2}\right)\phi + \frac{\psi}{2c^2}\right]_{,k} \\ &+ \frac{m_0}{c^2}\left[(A_{k,q} - A_{q,k})\dot{x}^q + A_{k,0}\right]. \end{aligned} \tag{5.26}$$

We see that the qualitative behavior of f^k is the same as that of F^k, the right side of (5.14),

$$\begin{aligned} F^k = &-m_0\left[\left(1 + \frac{2v^2}{c^2}\right)\phi + \frac{1}{2c^2}\psi\right]_{,k} \\ &+ \frac{m_0}{c^2}[(A_{k,q} - A_{q,k})\dot{x}^q + A_{k,0}]. \end{aligned} \tag{5.27}$$

From Eq. (5.21) we see, as Einstein pointed out, that bringing up a mass to the vicinity of our particle increases its inertial mass.

If we suppose that all the masses have the same acceleration, $\ddot{y}^k$, and that they are distributed with spherical symmetry about our particle, it is found that in F^k or f^k the terms in $\psi_{,k}$ and $A_{k,0}$ involving accelerations contribute a term that can be written

$$F_i^k = \frac{11Gm_0}{3c^2}\ddot{y}^k\sum\frac{m_a}{r_a}. \tag{5.28}$$

We see that this force is in the direction of the acceleration, as Einstein expected it to be. One can regard it as an inertial force acting on the particle if one considers the latter as being in a reference frame moving with an acceleration $-\ddot{y}^k$ with respect to the masses.

Similarly one finds that, for a spherically symmetric distribution of (distant) masses rotating about an axis through the origin with constant angular velocity $\vec{\omega}$, the terms involving $A_{k,q} - A_{q,k}$ give a contribution to the force that one can write

$$\mathbf{F}_c = \frac{8Gm_0}{3c^2} \left(\sum \frac{m_a}{r_a} \right) \vec{\omega} \times \mathbf{v} , \qquad (5.29)$$

with $\mathbf{v}$ the particle velocity. This can be regarded as a kind of Coriolis force if one thinks of the frame of reference as rotating with angular velocity $-\vec{\omega}$ with respect to the masses.

If one takes this axis of rotation to be the Z axis, one also finds that terms in ψ contribute a force of the form

$$F_r^k = \frac{Gm_0}{5c^2} \left(\sum \frac{m_a}{r_a} \right) \omega^2 x^k \qquad (k = 1,2) , \qquad (5.30)$$

$$F_r^3 = - \frac{2Gm_0}{5c^2} \left(\sum \frac{m_a}{r_a} \right) \omega^2 z . \qquad (5.31)$$

[In Eqs. (5.29)–(5.31) it is assumed that the particle is so near to the origin that one can consider r_a as the distance from the origin to particle a.] We see that (5.30) has a resemblance to the centrifugal force in a rotating reference frame. Equation (5.31) is more difficult to understand from this point of view, but Einstein, in a footnote, pointed out that the centrifugal force must be inseparably connected with the existence of the Coriolis force in the case of a coordinate system rotating uniformly relatively to an inertial system.

The expressions (5.29)–(5.31) are similar to those obtained by Thirring[4] for the case of a particle inside a rotating spherical shell, but some of the numerical coefficients are different.

Einstein concluded that the foregoing effects, although small for masses in our neighborhood, are in the right direction and they suggest that, if we took into account all the masses in the universe and carried out exact calculations on the basis of general relativity, we would confirm Mach's principle.

In this connection it should be noted that if, as a rough calculation, we try to apply equations such as (5.6), (5.21), (5.28), and (5.29) to the whole universe on the assumption that the masses are distributed with a uniform density, then we find that most of the contribution to the inertial properties of a particle come from the distant masses. For if we consider the matter in a thin spherical shell of thickness ΔR and radius R with center near the particle, then the mass is proportional to R^2 and the contribution to $\Sigma m_a/r_a$ is proportional to R.

Let us go back to (5.21), giving the same mass m as Einstein obtained. Einstein's result was criticized by Brans[5] as being coordinate dependent and therefore as not describing a real effect. According to Brans one should use a coordinate system in which the velocity of light is equal to one. Since this is what was essentially done above [cdT and dX^k correspond to such coordinates according to (5.17)], it appears that, in spite of his criticism, the relation (5.21) does have a physical significance.

It is possible to generalize, in a certain sense, the preceding procedure. Suppose we have a physical situation and a coordinate system for which

$$ds^2 = \Phi^2 dt^2 - \Psi^2(dx^2 + dy^2 + dz^2) \ , \tag{5.32}$$

with Φ and Ψ given functions of the coordinates. The equations of motion can be written (for $k = 1,2,3$)

$$\frac{d}{ds}\left(\Psi^2 \frac{dx^k}{ds}\right) + \left[\Phi\Phi_{,k}\left(\frac{dt}{ds}\right)^2 - \Psi\Psi_{,k}\frac{dx^q}{ds}\frac{dx^q}{ds}\right] = 0 \ . \tag{5.33}$$

Here again we take metric coordinate differentials,

$$dT = \Phi dt, \qquad dX^k = \Psi dx^k \ , \tag{5.34}$$

and corresponding metric velocity components, so that

$$\left(\frac{dT}{ds}\right)^2 = 1 + \frac{dX^q}{ds}\frac{dX^q}{ds} \ . \tag{5.35}$$

One can write (5.33) in the form

$$\frac{d}{ds}\left(\Psi \frac{dX^k}{ds}\right) = -\frac{\Phi_{,k}}{\Phi} + \left(\frac{\Psi_{,k}}{\Psi} - \frac{\Phi_{,k}}{\Phi}\right)\frac{dX^q}{ds}\frac{dX^q}{ds} \ . \tag{5.36}$$

If we multiply through by m_0, we can take as the particle momentum

$$P^k = m \frac{dX^k}{ds} \ , \tag{5.37}$$

with the inertial mass

$$m = m_0 \Psi \ , \tag{5.38}$$

depending on the gravitational field described by Ψ.

As a special case, consider an isotropic model of the universe, and suppose that we are working in a small enough region of space so that the curvature can be neglected. Then we can take coordinates so that (5.32) holds with $\Phi = 1$, $\Psi = \Psi(t)$. In this case (5.36) becomes

$$\frac{d}{ds}\left(\Psi \frac{dX^k}{ds}\right) = 0 , \qquad (5.39)$$

so that, by (5.37) and (5.38), we have

$$P^k = \text{const.} \qquad (5.40)$$

This is as it should be. Since our space is isotropic and hence homogeneous, we can expect conservation of momentum from symmetry considerations.

We see from (5.38) that in our expanding universe ($d\Psi/dt > 0$) the inertial mass increases with time. This can be regarded as a manifestation of Mach's principle. One finds[6] that in this case the passive gravitational mass also increases according to (5.38), while the active gravitational mass remains constant in time. It follows that if we take as the unit of mass the inertial mass of a given body, then the active gravitational mass will decrease in time in proportion to $1/\Psi$. This is equivalent to taking the gravitational constant as decreasing with time: that is,

$$G = \frac{G_0}{\Psi} . \qquad (5.41)$$

Can one carry out further generalizations? In an arbitrary coordinate system with a metric $g_{\mu\nu}$ one can write the equations of motion in the form of the equations of the geodesic, as given by (5.9). In this form one sees no sign of a mass depending on the gravitational field. On the other hand, one can also write the equations in the form

$$\frac{d}{ds}\left(g_{\mu\alpha}\frac{dx^\alpha}{ds}\right) - \tfrac{1}{2} g_{\alpha\beta,\mu}\frac{dx^\alpha}{ds}\frac{dx^\beta}{ds} = 0 , \qquad (5.42)$$

and if one multiplies through by the mass constant m_0 one has

$$\frac{dP_\mu}{ds} = \tfrac{1}{2} m_{\alpha\beta,\mu}\frac{dx^\alpha}{ds}\frac{dx^\beta}{ds} , \qquad (5.43)$$

with the momentum four-vector

$$P_\mu = m_{\mu\alpha}\frac{dx^\alpha}{ds} , \qquad (5.44)$$

and the mass tensor

$$m_{\mu\nu} = m_0 g_{\mu\nu} \,. \tag{5.45}$$

The latter is, of course, a highly coordinate-dependent quantity, and the only invariant associated with it is the constant m_0. Does this tensor have any physical significance? It can be argued that, as long as the observer works in the given coordinate system, it has physical significance in the same way that inertial forces in this system can have physical significance.

To understand the situation more clearly let us consider the effect of a nongravitational force. Suppose that our particle has a charge e and that an electromagnetic field is present. Corresponding to (5.9) we can write the equations of motion as

$$m_0 \left[\frac{d^2 x^\mu}{ds^2} + \begin{Bmatrix} \mu \\ \alpha\beta \end{Bmatrix} \frac{dx^\alpha}{ds} \frac{dx^\beta}{ds} \right] = F^\mu_{(e)} \,, \tag{5.46}$$

where $F^\mu_{(e)}$ is the Lorentz force

$$F^\mu_{(e)} = e \frac{dx^\alpha}{ds} F_\alpha{}^\mu \,, \tag{5.47}$$

with $F_\alpha{}^\mu$ the electromagnetic field tensor. Here too there is no indication of a variable mass. On the other hand, we can write, corresponding to (5.42),

$$\frac{d}{ds}\left(m_0 g_{\mu\alpha} \frac{dx^\alpha}{ds} \right) - \tfrac{1}{2} m_0 g_{\alpha\beta,\mu} \frac{dx^\alpha}{ds} \frac{dx^\beta}{ds} = g_{\mu\nu} F^\nu_{(e)} \,, \tag{5.48}$$

and here it is plausible to introduce the mass tensor (5.45).

One can now ask whether it is possible to observe the anisotropy of the mass tensor by means of a laboratory experiment based on (5.48). In practice one may encounter two difficulties:

1. In the weak-field case one is likely to be using a coordinate system corresponding to the metric tensor given by (5.3)–(5.5), the space part of which is isotropic.
2. The metric tensor components may be practically constant over the spatial region under consideration. In that case one has, practically,

$$g_{\alpha\beta,\mu} = \begin{Bmatrix} \mu \\ \alpha\beta \end{Bmatrix} = 0 \,, \tag{5.49}$$

and the metric tensor effectively cancels out of the equations.

It is therefore understandable that Hughes et al. and Drever obtained negative results in their searches for mass anisotropy.[7]

Let us return for a moment to Eq. (5.42), and let us consider it in a frame of reference in which the particle is always at rest. In this case we have $dx^k/ds = 0$, so that (5.42) becomes

$$\frac{d}{ds}\left(g_{\mu 0}\frac{dx^0}{ds}\right) - \tfrac{1}{2}g_{00,\mu}\left(\frac{dx^0}{ds}\right)^2 = 0 , \qquad (5.50)$$

while (5.11) gives

$$\frac{dx^0}{ds} = (g_{00})^{-1/2} . \qquad (5.51)$$

Using (5.51) to eliminate s from (5.50), one gets for $\mu = k$

$$A_{k,0} - \Phi_{,k} = 0 , \qquad (5.52)$$

provided one writes

$$g_{00} = \Phi^2 , \qquad g_{k0} = \Phi A_k . \qquad (5.53)$$

If we think of Φ and A_k as made up of contributions from matter in the vicinity of the particle, describing the gravitational field, and contributions from distant matter, describing the inertial forces, then (5.52) represents the condition for their equilibrium.

Let us now consider the broad question of the role of Mach's principle in the general theory of relativity. Much has been written on this subject, as one can see, for example, from the survey article by Reinhardt.[8] Hence the present discussion will be brief.

Since the properties of space-time appear to determine the inertial properties of a body, one can say that Mach's principle asserts that matter determines the properties of space-time. In the Einstein field equations

$$G_{\mu\nu} = -8\pi T_{\mu\nu} , \qquad (5.54)$$

the Einstein tensor $G_{\mu\nu}$ is a function of $g_{\mu\nu}$ and its first and second derivatives, and $T_{\mu\nu}$ is the energy-momentum density tensor of the matter, so that the matter is the source of the $g_{\mu\nu}$ field. This appears to be in accordance with Mach's principle, except for one difficulty: the solution of the set of partial differential equations (5.54) for $g_{\mu\nu}$ is determined not only by the form of $T_{\mu\nu}$, but also by the boundary conditions.

The influence of the boundary conditions, in addition to that of the matter, is in conflict with Mach's principle. In the foregoing example of a system of particles, the boundary condition taken is that $g_{\mu\nu}$ is Galilean at infinity, where no matter is present. This gives the solution (5.3)–(5.5) for the metric tensor, and from this we obtain the mass of the test particle (5.21). We see that at infinity, where ϕ vanishes, the mass is m_0, although there is no interaction there between the particle and other matter.

Einstein, who was strongly attracted by Mach's ideas, tried to remove this difficulty by looking for a model of the universe without any boundary, that is, a closed model. Since at that time (1917) it was not known that the universe is expanding, he proposed a static closed isotropic model filled with matter of uniform density and pressure.[9] However, in order to be able to get a solution of the field equations corresponding to this model he was forced to add the cosmological term $\Lambda g_{\mu\nu}$ to the left side of (5.54).

Shortly thereafter de Sitter[10] found another solution of the field equations (with the cosmological term) also describing a static closed model of the universe that differed from Einstein's in that the matter density and pressure vanished. This was a source of disappointment to Einstein, since one had here a case of a space-time without boundaries in which there was a preferred frame of reference without any matter to single it out.

Actually, de Sitter's model should not be a cause for surprise. The Λ term in the equations can be interpreted as describing the presence of an energy density and a stress throughout the universe, and one can say that these give space-time its properties. If one wants the matter to be the sole determiner of the space-time properties, one should omit the cosmological term. Later on, when it was found that the universe is expanding and not static, so that it is possible to get solutions of the field equations for closed models without this term, Einstein was in favor of omitting it. However, not everyone agreed with him.

Subsequently Gödel[11] found a solution of the Einstein field equations such that, for the same form of the matter energy-momentum density tensor as in the Einstein model, the metric tensor was quite different, with the matter rotating relatively to the compass of inertia. This solution can be characterized by the fact that the congruence of the matter world lines has nonvanishing rotation. Gödel's universe was not closed, and it had some unphysical features. However, Oszvath and Schücking[12] succeeded in obtaining a more satisfactory model that was closed. The existence of such solutions of the field equations was interpreted by them[13] as indicating that Mach's principle is not fulfilled in the general relativity theory. On the other hand, Hönl and Dehnen[14] explained the discrepancy between such solutions and Mach's principle as being due to the presence of a gravitational wave circulating about the universe and contributing to its energy. According to them, Mach's principle should serve to single out physically admissible solutions of Einstein's equations. According to Wheeler[15] Mach's principle serves to select the boundary conditions that physically significant solutions of the Einstein field equations must satisfy.

In discussing Mach's principle one should not overlook the fact that physics has progressed considerably since Mach's time and that it now has data and concepts that were unknown to him. Let us recall that Mach was greatly impressed by the fact that the rate of rotation of the earth as determined by means of a Foucault pendulum agrees (to the accuracy of the measurements) with that determined from observations on the fixed stars. He concluded that there was a causal connection between the two. However, we now know that, in addition to the fixed stars — it is better to talk about the galaxies — there also exist dark matter and other invisible objects such as neutrinos in the universe, and these should also be taken into account in our looking for such a causal connection.

The fact is that nature is more complicated than Mach realized. Recently Sciama, Waylen, and Gilman and others put the Einstein field equations into the form of integral equations.[16] Gilman[17] proposed using the latter in order to select physically admissible solutions conforming to Mach's principle, these being space-times that are entirely source-generated. However, it is not certain that this approach is satisfactory. As Higbie[18] pointed out, the situation is complicated by the finite propagation velocity of gravitation and the existence of independent field excitations of the geometry, that is, modes of oscillation of the free gravitational field, or — from the standpoint of quantum theory — the existence of gravitons. Because of the nonlinearity of the Einstein equations, the gravitational field can be its own source. This was brought out by the work of Brill and Hartle and of Isaacson in which it was shown that it is possible to solve the field equations, when short-wavelength gravitational waves are present, by treating the energy of these waves as the source of a slowly varying gravitational field.[19] One can imagine a universe in which a considerable fraction of the energy is in the form of gravitational waves, so that the geometry differs appreciably from what one would expect if only matter were present. An extreme example of this is the Taub model of the universe which is closed and contains no matter, only gravitational radiation.[20] Wheeler pointed out the difficulty encountered in such a case: Mach's principle requires that the energy determine the geometry; but the gravitational radiation represents both energy and geometry.[21]

One gets the impression that, as time goes on, the discussions of Mach's principle in the general relativity theory are becoming more and more subtle and sophisticated, and one wonders if one should go on in this direction. One can argue that, if one believes in Mach's principle, it would be better to go back to the simple and clear-cut position taken by Einstein and to assume that the universe is closed. In that case there are no difficulties with Mach's principle. Since there is no boundary, there is no influence from the outside. Hence whatever happens in the universe must be due to whatever is in it — matter, radiation, gravitation, and so on — and this can be regarded as a realization of Mach's principle, suitably formulated. To be sure, one can argue that the question of the closure of the universe has to be settled by means of observations and is not something that one can decide on a priori. However, the conclusions to be drawn from observations depend on theoretical considerations. Is there a cosmological term in the Einstein field equations? Are there perhaps other fields in the universe that are unimportant for small-scale phenomena but become important on a cosmic scale and bring about closure? Just as nature is more complicated than Mach realized, it may be more complicated than we realize as well.

NOTES

1. A. Einstein, *The Meaning of Relativity* (Princeton University Press, Princeton, N. J., 1946).
2. W. Davidson, Monthly Notices, Roy. Astron. Soc. **117**, 212 (1957).
3. A. Einstein, L. Infeld, and B. Hoffmann, Ann. Math. **39**, 65 (1938).
4. H. Thirring, Phys. Z. **22**, 29 (1921).
5. C. Brans, Phys. Rev. **125**, 2194 (1962).
6. N. Rosen, Ann. Phys. **35**, 426 (1965).
7. V. W. Hughes, H. G. Robinson, and V. Beltran-Lopez, Phys. Rev. Lett. **4**, 342 (1960); R. W. P. Drever, Phil. Mag. **6**, 683 (1961).
8. M. Reinhardt, Z. Naturforsch. **28a**, 529 (1973).
9. A. Einstein, S. Berlin. Akad. d. Wiss., p. 142 (1917).
10. W. de Sitter, Proc. Akad. Wetensch. Amsterdam **19**, 1217 (1917).
11. K. Gödel, Rev. Mod. Phys. **21**, 447 (1949).
12. I. Oszvath and E. L. Schücking, Ann. Phys. **55**, 166 (1969).
13. I. Oszvath and E. L. Schücking, Nature **193**, 1168 (1962).

14. H. Hönl and H. Dehnen, Z. Phys. **171,** 178 (1963).
15. J. A. Wheeler, *Gravitation and Relativity,* edited by Chiu and Hoffmann (W. A. Benjamin, 1964), ch. 15.
16. D. W. Sciama, P. C. Waylen, and R. C. Gilman, Phys. Rev. **187,** 1762 (1969); B. L. Al'tshuler, Zh. Eksp. i Teor. Fiz. **51,** 1143 (1966), translated in Sov. Phys. JETP **24,** 766 (1967); D. Lynden-Bell, Monthly Notices Roy. Astron. Soc. **135,** 413 (1967).
17. R. C. Gilman, Phys. Rev. **D2,** 1400 (1970).
18. J. H. Higbie, Gen. Rel. Grav. **3,** 101 (1972).
19. D. R. Brill and J. B. Hartle, Phys. Rev. **135,** B271 (1964); R. A. Isaacson, Phys. Rev. **166,** 1272 (1968).
20. A. H. Taub, Ann. Math. **53,** 472 (1951).
21. Wheeler, op. cit. in n. 15.

Discussion

Following the Paper by N. ROSEN

Chairman, J. A. WHEELER, U. of Texas

Wheeler: I comment only to agree with the mathematical description that you gave in the beginning, based on asymptotically flat space. One immediately sees by the method that you outlined that the influence of one mass on another should be thought of in terms of the point of view that you sketched in the latter part of your paper. We understand Mach's principle as part of the initial value equation of general relativity, which is an equation that has a unique solution. Moreover, a solution always exists under suitable and very simple conditions. This formulation of Mach's principle in terms of the initial value equation of general relativity, instead of being in contrast with your first treatment, is, I would think, indeed in conformity with it and is a part of it. I ought to add that I had the great privilege of asking Einstein to give a relativity seminar. This was the last talk he ever gave in his life, about a year before he died. In that, he discussed how he came to general relativity, what general relativity meant to him, what quantum theory meant. However, he also remarked, as you have already, that the boundary condition of closure is the one that really makes things simple. I wish also to remark that the day before yesterday in Haifa, our colleagues at the Technion gave a special celebration in honor of the seventieth birthday of Professor Rosen, and I wonder if we, representing a still larger community, might have the privilege of associating ourselves with their expression of felicitations to Professor Rosen?

Rosen: Thank you very much.

G. Tauber (Tel Aviv U.): This sounds like an anticlimax after Professor Wheeler's felicitation, which we all share, but I am going back to physics. I have a small comment about the cosmological case. If you were to use, for example, a conformally flat line element, then you have not only the advantage of having a simple line element, but as Infeld and Schild quite some time ago have shown, it will include all cosmological Friedmann type line elements without the need of assuming that the curvature is small locally.

Rosen: If you mean that I could use it for the expression that I used for the mass, I agree; this could be an advantage.

A. Kastler (Paris): May I make again a historical remark on Einstein's papers on gravitation? We know that he developed his complete mathematical theory of gravitation, using non-Euclidean geometry and tensor analysis in the *Annalen der Physik* in 1916. But before he did this, in the years 1911 and 1912 he wrote three papers in *Annalen der Physik,* starting from the equivalence principle within a uniformly accelerated system, and a system with a static field of gravitation. By very simple schoolboy mathematics he showed the behavior of most physical quantities in the gravitational field, the behavior of time and clocks, the behavior of frequencies, of periodic motion, the behavior of energy, the behavior of the velocity of light, the behavior of the electromagnetic quantities and even the behavior of the thermodynamic quantities. He did not discuss in these three papers the behavior of mass in a gravitational field. But he did this in a strange paper, published in a strange place. This paper is cited in Cornelius Lanczos's book on the Einstein Decades. It was published in a journal edited in Berlin, *The Quarterly Journal of Legal Medicine and Public Health (Vierteljahrsschrift für Gerichtliche Medizin und Öffentliches Sanitatswesen).* Of course the paper of Einstein has nothing to do with legal medicine nor with public health. The title of his paper is a question: "Is there a gravitational effect which is an analog of electromagnetic induction?" (Gibt es eine Gravitationswirkung, die der elektromagnetischen Induktion analog ist?). It is in this paper where Einstein for the first time shows how mass varies in a gravitational field, as a function of the gravitational potential. As he shows, it is very misleading to state, taking the famous equation: $E = Mc^2$, that mass is the same as energy, and energy is the same as mass. This is only true in the special theory of relativity, which starts from the postulate of the universal constancy of the velocity of light. But this is no more true in Einstein's theory of gravitation where he shows that the velocity of light is a function of the gravitational potential. He proves that the velocity of light varies just in the same way as the energy. So, if you take the relation $M = E/c^2$ you see that mass does not vary like energy, but like E^{-1}. Einstein discusses in this paper Mach's principle as a cosmological consequence. He shows that we must take into account the whole mass of the universe and especially the very distant masses to account for the Mach force producing the inertia of a body. These conclusions remain valid in Einstein's complete mathematical theory as established later on. See for example: Albert Einstein, *The Meaning of Relativity,* 5th ed., Princeton University Press, Princeton, N. J., 1955, p. 102.

PART III: GAUGE THEORIES OF GRAVITY AND THEIR GENERALIZATIONS

From a gauge point of view, gravity is the gauge theory of the parallel-transport modification of the Poincaré group. The introduction to this volume can be read to introduce the subject, together with the arguments for the gauge approach presented in my chapter in the Princeton volume. Once this is assumed, the possibility of developing a "larger" theory (one with additional fields coupled to additional currents) is determined by possible embeddings of the Poincaré group. See Fig. III.1.

In the figure are sketched three possible embeddings: A, the affine group $GA(4,R)$; C, the conformal group $SU(2,2) = \overline{SO}(4,2)$; S, supersymmetry or the super-Poincaré group. S intersects

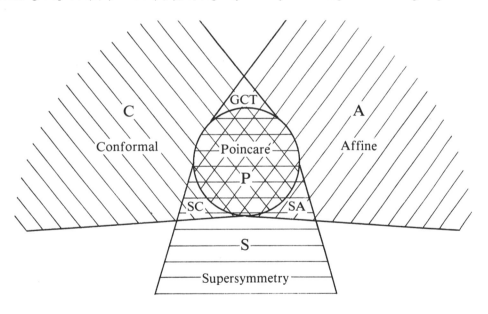

Fig. III.1. Embeddings of the Poincaré group.

C and A, generating SC, the superconformal group $SU(2,2/N)$ and SA, the superaffine infinite group $gGA(4,R)$. By Ogievetsky's theorem, the intersection of C and A generates the general covariance group.

Freedman's chapter deals with S, and the chapter by van Nieuwenhuizen in the Princeton volume relates to SC. The Ne'eman chapter deals with P, A and SA.

Freedman presents an excellent review of supergravity, including a sketch of the advances in renormalization. The removal of paradoxes relating to spin 3/2 fields is mentioned in the description of supergravity itself. One should add on the credit side of the sheet the proof of positive definiteness of the gravitational field energy. The discussion of renormalization also indicates what are the next hurdles to overcome.

Unification comes in through the possibility of extending supersymmetry by an $SO(N)$. These theories are described — including the $SO(2)$ theory, which comes closest to Einstein's goal of a unification of gravity and electromagnetism. The difficulty relates to the appearance of a large cosmological term. It is possible that this will prove useful for renormalization, in the light of Hawking's realization of the Wheeler "foam" (related to the quantization of space-time).

$SO(8)$ was regarded as the maximal possible $SO(N)$ model, because for $N > 8$ one gets more than one $J = 2$ field, and fields with $J > 2$. The limit is not one of principle, and there may exist models that overcome these difficulties. $SO(8)$ was thought on the other hand to be too small a group to provide for both the weak-electromagnetic gauge group and color $SU(3)$, the strong gauge group (setting aside flavors). The recent Cremmer-Julia-Scherk model appears to provide room for $SU(8)$ and other groups larger than $SO(8)$.

The Ne'eman chapter presents the case for quadratic Lagrangians in both torsion and curvature, as recently developed by Hehl, Ne'eman, Nitsch, and von d. Heyde. Such Lagrangians provide for a confined strong (?) gravity with a dimensionless coupling, besides having the usual long-range Einstein-Newton gravity. If the group is extended to $GA(4,R)$, the confined component contains additional features resembling the strong interactions. The superaffine group has an infinite superalgebra. One senses a possible connection to strings, lumps, or bags.

A recent analysis[2] throws new light on the quantization of models of the type discussed by Ne'eman. Specifically, allowing for two independent propagating fields e_μ^a and $\Gamma_\mu^{cd.}$ and without requiring power-counting renormalizability one finds that the following models are unitary (no ghosts or tachyons): Einstein's and the Einstein-Cartan model and four specific models of the type $R + R^2$, with 2, 5, 5, 6 parameters, respectively, and 0^-, $2^- + 1^-$, $1^+ + 0^-$, $0^+ + 0^-$ propagating massive fields in addition to the massless 2^+ graviton. The specific model described by Ne'eman (note 9 in Chapter 7) is found to contain a 1^+ ghost and double poles. This is somehow trivial, since in its present state the theory is not locally Lorentz-invariant; one has to impose vanishing curvature as a constraint to reproduce the teleparallelism limit that makes it equivalent to Einstein's theory up to 5th ppn. This implies the need for adding something like a spontaneous-breakdown mechanism, to destroy local Lorentz invariance. If this can be done, it might well make the theory unitary.

The models by C. N. Yang and by E. E. Fairchild (see note 1 of Chapter 7) are also not unitary, with dipole ghosts at $k^2 = 0$ and a 2^- tachyon, respectively.

NOTES

1. E. Sezgin and P. van Nieuwenhuizen, New Ghost-Free Gravity Lagrangians with Propagating Torsion, State University of New York at Stony Brook report ITP-SB-79-97.

Session Chairman's Remarks

SYBREN R. DE GROOT

I have been asked to offer a few words on behalf of the Royal Netherlands Academy of Arts and Sciences. I think it is apt to recall first that Albert Einstein was a foreign member of our Academy from the year 1920. Furthermore I would like to remind you that he was tied by links of profound friendship to Hendrik Antoon Lorentz and to Paul Ehrenfest. Lorentz had discovered Einstein's genius at an early stage, and they discussed frequently scientific matters of common interest. With Ehrenfest Einstein entertained a lasting relationship, which involved, besides questions of theoretical physics, matters of personal and public relevance, such as music and politics.

Einstein wrote a short note on Lorentz in German in 1953, on the occasion of an exhibition in Leyden in commemoration of Lorentz and Kamerlingh Onnes. An English version of this text was published in 1957 in a booklet, edited by Lorentz's daughter, Mrs. G. L. de Haas-Lorentz. Einstein speaks characteristically of Lorentz's "act of intellectual liberation" (*erlösende Tat*) in putting forward his microscopic theory of field and matter, with only *one* electric and *one* magnetic field, seated in empty space and created by atomistic electric charges.

Einstein wrote in "Paul Ehrenfest in Memoriam" about Ehrenfest's faculty "to strip a theory of its mathematical accoutrements until the simple basic idea emerged with clarity."

In 1911 Einstein was offered a chair in Utrecht. In the same year, when Lorentz had decided to go to Haarlem, he tried to convince Einstein to be his successor in Leyden. But Einstein had just accepted a chair in Zürich, and Lorentz then chose Ehrenfest in his place. After Einstein had accepted to move to Berlin in 1913, he visited, in March 1914, Ehrenfest and Lorentz, because, as Martin J. Klein says in his brilliant and sensitive book on Ehrenfest: "The one place where Einstein did look for a real and understanding interest in his work was Leyden." After that, still in 1914, Einstein started working in Berlin, and visited Leyden again in 1916. He was nominated in 1919 as "special professor" at the University of Leyden, and took the chair in 1920, in the same year therefore as his nomination at the Royal Netherlands Academy.

The session of today is devoted to gauge theories of gravity and their generalization. It thus involves Einstein's dream of many years: the unification of the fundamental interactions, which in his time — with its incomplete knowledge of the forces acting between particles — was hard to realize. Now enormous progress has been made to unify the four forces. The authors of the two chapters that follow have contributed essentially to various unifying points of view.

6. UNIFICATION IN SUPERGRAVITY

Daniel Z. Freedman

Supergravity theories are extensions of general relativity that incorporate a fundamental symmetry between bosons (integer spin particles) and fermions (half-integer spin particles). These theories bring gravitation and particle physics closer together and promise great unification of the fundamental laws of nature. At present there is neither experimental support for supergravity nor direct contradiction, and the subject is very much a case of free invention or free theoretical speculation, of the type discussed by Einstein in his later writing. Yet the fundamental ideas of supergravity are related to an elegant and fruitful tradition in theoretical physics, and they may well turn out to be correct.

The fundamental ideas of supergravity involve concepts associated with Einstein:

1. *The symmetry of physical law* plays a central role, and the global symmetry of special relativity and the local symmetry of general relativity are both significantly extended. Global supersymmetry is the only known invariance compatible with quantum field theory that unifies spin and internal symmetry. It does this by relating bosons and fermions, the two broad classes of particles found in nature. Local supersymmetry — that is, supergravity itself — is a fermionic gauge principle; the gauge field is the spin 3/2 Rarita–Schwinger field. In supergravity theories the classical concept of general covariance is closely linked with the quantum ideas of spin and statistics.

2. The second important idea concerns *geometry and physics*. Einstein's fundamental insights concerning space-time geometry are extended, because the appropriate geometrical setting for supergravity appears to be superspace manifolds parametrized by Bose and Fermi coordinates.

3. Finally we have the hope for a *unified field theory* — that is, a few simple equations following from a single principle that contain all the fundamental laws of physics. In supergravity the graviton can be unified with very restrictive combinations of lower spin particles in irreducible representations of supersymmetry algebras. The field theories of these systems raise hope for the eventual unification of gravity with the weak, electromagnetic, and strong interactions. Much further progress will be required before we can say that these theories are correct. If they do succeed, it will be because of the union of quantum mechanical and geometrical ideas in their formulation.

Yuval Ne'eman (ed.), To Fulfill A Vision: Jerusalem Einstein Centennial Symposium on Gauge Theories and Unification of Physical Forces
ISBN 0-201-05289-X

Copyright © 1981 by Addison-Wesley Publishing Company, Inc., Advanced Book Program.
All rights reserved. No part of this publication may be reproduced, stored in a retrieval system, or transmitted, in any form or by any means, electronic, mechanical photocopying, recording, or otherwise, without the prior permission of the publisher.

Basics of Global Supersymmetry

The essential idea of supersymmetry is to construct field theories with conserved spinor charges Q_α, where α is a Dirac index with values 1, 2, 3, 4. This is in contradistinction to the vector and antisymmetric tensor charges, P^a and M^{ab} of translations and Lorentz transformations, and the Lorentz scalar charges T^i of internal symmetry. The three types of charges combine in algebraic systems called *graded Lie algebras*, with structure relations of the form

$$\{F, F\} = B,$$

$$[B, F] = F,$$

$$[B, B] = B,$$

where B and F are generic labels for boson and fermion operators, respectively. Anticommutators are required for fermion operators in agreement with the physical connection of spin and statistics. Previous no-go theorems concerning the unification of space-time and internal symmetries are bypassed in supersymmetry because of the essential use of more general algebraic systems than traditional Lie algebras.

Supersymmetry was first found in quantum field theory by Golfand and Likhtman.[1] There was further work by Volkov and Akulov, but the subject really became alive after its rediscovery and systematic study by Wess and Zumino, who were then joined by many others.[2]

Several graded Lie algebras have been used in theoretical physics. The fundamental Poincaré supersymmetry algebra involves the Poincaré group generators P^a and M^{ab} and a single spinor charge Q_α which is self-conjugate, that is, subject to the Majorana condition, which is a Lorentz invariant constraint, $Q = C\bar{Q}^T$, involving the charge conjugation matrix. Majorana spinors are more fundamental because they involve half as many degrees of freedom (that is, four real components rather than four complex). They are equivalent to the two component complex Weyl spinors used by relativists in the Penrose–Newman formalism. The structure relations are

$$\{Q_\alpha, \bar{Q}_\beta\} = (\gamma^a)_{\alpha\beta} P_a, \qquad \{\gamma^a, \gamma^b\} = 2\eta^{ab},$$

$$[M^{ab}, Q_\alpha] = -i(\sigma^{ab})_{\alpha\beta} Q_\beta, \qquad \sigma^{ab} = \tfrac{1}{4}[\gamma^a, \gamma^b],$$

$$[P^a, Q_\alpha] = 0,$$

together with standard Poincaré commutators for $[M,M]$, $[M,P]$, $[P,P]$. The first relation is the most important, and it suggests the terminology that the supersymmetry charges are the "square root of translations." One might even say that supersymmetry is a deeper invariance underlying the Poincaré group, because translations in space can be built up from repeated supersymmetry transformations. Clearly, supersymmetry involves the structure of space-time, and it is this connection that is fully developed in supergravity. The second equation simply states that Q transforms as a spinor, and the third states that the spinor charges are conserved and translation-invariant.

It is now very easy to see that supersymmetry implies a relation between particles of different spin. One simply applies the spinor charge Q_α to a particle state $|\mathbf{p}, s\rangle$ of definite momentum and helicity:

$$Q_a | \mathbf{p}, s \rangle = a | \mathbf{p}, s + \tfrac{1}{2} \rangle + b | \mathbf{p}, s - \tfrac{1}{2} \rangle .$$

Because $[P, Q] = 0$, the right side is a superposition of particles of the same momentum and energy — therefore the same mass. Addition of angular momentum implies that these particles have helicities $s \pm \tfrac{1}{2}$. Supersymmetry transformations connect states which differ by one-half unit of spin. They therefore relate bosons and fermions!

So far we have used only the existence of a conserved spinor charge. The basic structure relation $\{Q, \overline{Q}\} = \gamma \cdot P$ has not entered, but it, too, can be motivated very simply. It depends on the fact that Bose fields $b(x)$ have dimension 1 (second derivative kinetic actions), while Fermi fields have dimension 3/2 (first derivative kinetic actions). Global supersymmetry transformations of the fields involve constant spinor parameters ε_a treated as infinitesimals. Schematically the Bose field variation is

$$\delta_s b(x) = \bar{\varepsilon} f(x) .$$

One can regard ε as an infinitesimal "angle" in a hypothetical vector space with Bose and Fermi axes, so that under a supersymmetry transformation the Bose field rotates and acquires a small admixture of the Fermi field. The equation also implies that the dimension of the parameter ε is $-\tfrac{1}{2}$. In the companion transformation

$$\delta_s f(x) = \partial b(x) \varepsilon + m b(x) \varepsilon ,$$

one must insert a derivative or dimensional parameter such as mass m in order to maintain proper dimensions. The derivative must be present because massless free field kinetic Lagrangians are invariant and do not contain dimensional parameters. Thus supersymmetry necessarily involves displacement in space-time. More precisely, the commutator of two supersymmetry transformations gives, on both $b(x)$ and $f(x)$

$$[\delta_s(\varepsilon_1), \delta_s(\varepsilon_2)] \begin{array}{c} b(x) \\ \text{or} \\ f(x) \end{array} = i\bar{\varepsilon}_1 \gamma^\mu \varepsilon_2 \partial_\mu \begin{array}{c} b(x) \\ \text{or} \\ f(x) \end{array} ,$$

that is, a translation in space-time.

A third fundamental aspect of supersymmetry can also be seen at a simple level. Fermion creation and annihilation operators anticommute

$$f f' + f' f = 0 .$$

Under a supersymmetry rotation, $f \to f + b\varepsilon$ (f acquires a small amount of the boson operator). In order to maintain anticommutativity the transformation parameters must satisfy the following equation:

$$\varepsilon_a \varepsilon_\beta + \varepsilon_\beta \varepsilon_a = 0 ,$$

The transformation parameters cannot be ordinary numbers; they must be anticommuting numbers or Grassmann variables. Such a number system was introduced into physics long ago by Schwinger as an adjunct technique in fermion field theories, and mathematicians have systematized the rules of calculation for such quantities. In supersymmetry and supergravity the new number system is essential to the formulation of the invariance. This reflects again the connection between spin and statistics.

Irreducible Representations of Supersymmetry

We have seen that supersymmetry transformations connect bosons and fermions. The irreducible representations of the graded Poincaré algebra tell us more precisely what sets of Bose and Fermi particles constitute closed multiplets under these transformations. These multiplets are the building blocks of supersymmetric field theories.

The techniques for finding irreducible representations were given by Salam and Strathdee.[3] They involve the use of Wigner's method of induced representations, and an isomorphism of the "little algebra" of spinor charges with mathematical Clifford algebras. For the fundamental Poincaré supersymmetry algebra under discussion now, the results are extremely simple.

The basic irreducible representations consist of massless boson-fermion doublets of adjacent spin, $(s, s - \frac{1}{2})$, and involve particles of all allowed spin values; that is, $s = \frac{1}{2}, 1, 3/2 \ldots$. There are also representations involving massive particles. They are usually larger and we will not discuss them here. Global supersymmetric field theories of the $(\frac{1}{2}, 0)$ and $(1, \frac{1}{2})$ doublets have been studied extensively in the literature, and the $(2, 3/2)$ representation is the basis of the simplest supergravity theory.

A Global Supersymmetric Field Theory

The most elegant global theory is the supersymmetric Yang–Mills theory.[4] This is the field theory of the $(1, \frac{1}{2})$ representation of supersymmetry in direct product with an arbitrary internal symmetry gauge group G. The fields are the gauge potentials $A_\mu^i(x)$ and Majorana spinors $\chi^i(x)$ both in the adjoint representation of G. One forms field strengths and covariant derivatives in the usual way:

$$F_{\mu\nu}^i = \partial_\mu A_\nu^i - \partial_\nu A_\mu^i + g f^{ijk} A_\mu^j A_\nu^k ,$$

$$(D_\mu \chi)^i = \partial_\mu \chi^i + g f^{ijk} A_\mu^j \chi^k .$$

The remarkable fact is that the minimal coupling Lagrangian of this system

$$\mathcal{L} = -\tfrac{1}{4} F_{\mu\nu}^i F^{\mu\nu i} + \tfrac{1}{2} \bar\chi^i \gamma^\mu (D_\mu \chi)^i$$

is supersymmetric. $\mathcal{L}$ changes by a total derivative under the transformations

$$\delta A_\mu^{\ i} = \frac{i}{\sqrt{2}} \bar{\varepsilon} \gamma_\mu \chi^i \ ,$$

$$\delta \chi^i = \frac{1}{\sqrt{2}} \sigma^{\mu\nu} F_{\mu\nu}^{\ i} \varepsilon \ ,$$

which illustrate the Bose–Fermi character of supersymmetry rotations. The proof of invariance is an interesting blackboard exercise.

The consequences of the new symmetry are of two kinds. First, there are relations among boson and fermion scattering amplitudes.[5] For helicity amplitudes two examples are:

$$<1,1|\ S\ |1,1> = <\tfrac{1}{2},\tfrac{1}{2}|\ S\ |\tfrac{1}{2},\tfrac{1}{2}> \ ,$$

$$<1,\tfrac{1}{2}|\ S\ |1,\tfrac{1}{2}> = \sqrt{\frac{-u}{s}} <1,1|\ S\ |1,1> \ ,$$

where "1" denotes a vector boson of helicity 1 and "½" denotes a fermion of helicity ½, and where u and s are Mandelstam variables.

Second, there are relations among the renormalization constants[6] of the theory, for example $Z_A = Z_\chi$ (for a particular gauge choice). Thus supersymmetry restricts the number of independent divergent quantities. Indeed this was one of the motivations for supergravity. It was hoped that the new spinor invariance would ameliorate the serious problem of nonrenormalizability in quantum gravity.

Superspace

To paraphrase Voltaire: If superspace did not exist, it would be necessary to invent it. It is one of those inevitable concepts. More seriously, superspace is probably the correct geometrical setting for supersymmetry and supergravity, and it is appropriate to discuss it as an extension of Einstein's geometrical ideas about space-time.

Superspace was introduced by Volkov and Akulov and by Salam and Strathdee as the quotient space—graded Poincaré/Lorentz group.[7] It is a manifold parametrized by four Bose coordinates x^a and four Fermi coordinates θ^a. The latter are Grassmann variables. Supersymmetry transformations are motions in superspace involving the parameter ε^a:

$$\delta_s \theta^a = \varepsilon^a \ ,$$

$$\delta_s x^a = \frac{i}{2} \bar{\theta} \gamma^a \varepsilon \ .$$

The commutator of two transformations is

$$[\delta_s(\varepsilon_1), \delta_s(\varepsilon_2)] \theta^a = 0 ,$$

$$[\delta_s(\varepsilon_1), \delta_s(\varepsilon_2)] x^a = i\bar{\varepsilon}_1 \gamma^a \varepsilon_2 ,$$

that is, a translation in the ordinary space-time part of the manifold. Thus we recover in the geometric context of motions in superspace the key algebraic property of supersymmetry discussed earlier.

Superfields $\Phi(x,\theta)$ can be defined on the manifold. They transform as

$$\delta_s \Phi(x,\theta) = \bar{\varepsilon} \left[\frac{\partial}{\partial \bar{\theta}} - \frac{i}{2} \gamma^a \theta \frac{\partial}{\partial x^a} \right] \Phi(x,\theta) .$$

To make contact with ordinary field theory a superfield is expanded in powers of θ^a as

$$\Phi(x,\theta) = A(x) + \bar{\theta}\psi(x) + \tfrac{1}{4}\bar{\theta}\theta F(x) + \tfrac{1}{4}\bar{\theta}i\gamma_5\theta G(x) +$$

$$\tfrac{1}{4}\bar{\theta}\gamma^\mu\gamma_5\theta A_\mu(x) + \tfrac{1}{4}\bar{\theta}\theta\bar{\theta}\chi(x) + \tfrac{1}{32}(\bar{\theta}\theta)(\bar{\theta}\theta)D .$$

The components are ordinary scalar, vector, or spinor fields, and the expansion terminates at fourth order because of anticommutativity. Any product of five or more θ_a vanishes!

The supersymmetric Yang–Mills theory discussed earlier was first formulated in terms of $\Phi(x,\theta)$. Roughly speaking, the physical fields are A_μ and χ, while A, ψ, F, G are generalized gauge excitations and D is an auxiliary field that can be eliminated from the Lagrangian. Such auxiliary fields are important because they simplify the structure of the more complicated supersymmetric theories.

Even the rigid superspace of global supersymmetry has a nontrivial differential geometry. The 1-forms

$$dx^a + \frac{i}{2}\bar{\theta}\gamma^a d\theta ,$$

$$d\theta^a$$

are invariant, and the conjugate derivatives

$$D_a = \frac{\partial}{\partial x^a} ,$$

$$D_a = \frac{\partial}{\partial \theta^a} + \frac{i}{2}\bar{\theta}\gamma^a \frac{\partial}{\partial x^a}$$

are covariant under supersymmetry transformations. However, these covariant derivatives do not commute, since

$$[D_\alpha, D_\beta] = i(\gamma^\mu)_{\alpha\beta} D_a ,$$

which shows that the manifold has torsion.

Einstein discovered general relativity by introducing Riemannian geometry on the rigid Minkowski space-time of special relativity. Therefore it was natural to try to formulate supergravity by introducing further geometrical structures on superspace. With coordinates x^a, θ^α jointly denoted by Z^M, one can have a vielbein $V^A{}_M$, metric R_{MN}, connection $\omega_M{}^{AB}$, curvatures $R_{MN}{}^{AB}$, and so on. Several such formulations preceded the type of supergravity theory that I have been involved in and will discuss shortly. They include work by Volkov and Soroka, Arnowitt and Nath, and Wess and Zumino.[8] The difficulty with these early formulations was that physical aspects such as the particle spectrum and questions of ghosts and consistency could not be clarified because of the large number of nonphysical components of superfields.

Recently, great progress has been made in the superspace formulation of supergravity, and I refer you to papers of Arnowitt and Nath; Wess and Zumino; Ne'eman and Regge; Brink, Gell-Mann, Ramond, and Schwarz; Siegel; Gates and Shapiro; Howe and Tucker; Ogievetsky and Sokachev; Romau, Ferber, and Freund; Taylor; and others.[9] The length of this list of authors bears testimony to the inevitability of the concept of superspace.

Supergravity

The approach P. van Nieuwenhuizen, S. Ferrara, and I took[10] to supergravity early in 1976 was to strip the formalism to the barest physical essentials. We wanted to combine supersymmetry with general relativity. Since only general covariant concepts are permitted in a gravitational theory, global supersymmetry was insufficient. A constant spinor parameter ε_α is not a covariant notion, since a spinor acquires space-time dependence after a local Lorentz transformation. We needed a local or gauge form of supersymmetry, where the parameters $\varepsilon_\alpha(x)$ are arbitrary functions of space-time.

We assumed for reasons of simplicity that the particle content of the desired theory would be an irreducible representation of the algebra that contained the spin 2 graviton. The choices were the (2, 3/2) or (5/2, 2) doublets. The first choice is clearly preferred because a spin 3/2 particle is described by a vector-spinor field $\psi_{\mu\alpha}(x)$. It was natural to assume that this would be the gauge field of supersymmetry, because a gauge field in mathematical physics always has one more vector index than the parameters of the transformation to be gauged. However, this line of thinking could be successful only if the spin 3/2 field was a gauge field.

A free-field theory for spin 3/2 was proposed in 1941 by Rarita and Schwinger, who concentrated on the massive case. For $m = 0$ an equivalent form of the Lagrangian is

$$\mathcal{L}_{3/2} = -\tfrac{1}{2} \varepsilon^{\lambda\rho\mu\nu} \overline{\psi}_\lambda \gamma_5 \gamma_\mu \partial_\nu \psi_\rho .$$

This Lagrangian is clearly invariant under the change

$$\delta\psi_\rho = \partial_\rho \varepsilon(x) ,$$

where $\varepsilon(x)$ is an arbitrary spinor function. This is an obvious analog of the electromagnetic gauge transformation. There is one sentence at the end of the Rarita–Schwinger paper where the gauge invariance of the massless spin 3/2 field is noted (in a somewhat more complicated way). This free-field invariance became one of the cornerstones of supergravity.

Starting from these observations, we constructed the supergravity theory in the second-order gravitational formalism. The independent fields are the vierbein $V_{a\mu}(x)$, which describes the spin 2 graviton, and the Rarita–Schwinger field $\psi_\mu(x)$, which describes its spin 3/2 partner, which many people now call the gravitino. The theory contained the minimal coupling terms of the second-order formalism, plus complicated contact terms involving ψ_μ.

An important simplification was then found.[11] The contact terms could be removed by using the first-order Cartan–Palatini formalism, where the independent fields are $V_{a\nu}(x)$, $\psi_\mu(x)$, and the spin connection $\omega_{\mu ab}(x)$. This formalism is natural from the viewpoint of gauging the graded Poincaré group. $V_{a\mu}$, ψ_μ, and $\omega_{\mu ab}$ can be regarded as gauge fields for the generators P^a, Q_a, and M^{ab}. The spin connection is a nonpropagating auxiliary field. When eliminated, it gives back the contact terms of the equivalent second-order form.

The simplest way to present the complete theory now is midway between the first-order and the second-order form and is appropriately called the 1.5-order formalism.[12] The independent fields are $V_{a\mu}$ and ψ_μ, but one retains the grouping of terms suggested by the first-order formalism, with $\omega_{\mu ab}(V,\psi)$ dependent on V and ψ according to the Cartan–Palatini equations. The Lagrangian is then a minimal coupling of spin 2 and spin 3/2:

$$\mathcal{L} = -\frac{1}{4\kappa^2} (\det V) V^{a\mu} V^{b\nu} R_{\mu\nu ab} - \tfrac{1}{2} \varepsilon^{\lambda\rho\mu\nu} \overline{\psi}_\lambda \gamma_5 \gamma_\mu \mathcal{D}_\nu \psi_\rho .$$

The first term is the standard gravitational Lagrangian with coupling constant $\kappa^2 = 4\pi G$ and curvature tensor

$$R_{\mu\nu ab} = \partial_\mu \omega_{\nu ab} - \partial_\nu \omega_{\mu ab} + \omega_{\mu a}{}^c \omega_{\nu cb} - \omega_{\nu a}{}^c \omega_{\mu cb} .$$

The second term is just the Rarita–Schwinger Lagrangian taken over into curved space-time with local Lorentz covariant derivative:

$$\mathcal{D}_\nu \psi_\rho = (\partial_\nu + \tfrac{1}{2} \omega_{\nu ab} \sigma^{ab}) \psi_\rho .$$

The spin connection is

$$\omega_{\mu ab} = \tfrac{1}{2} [V_a^\nu (\partial_\mu V_{b\nu} - \partial_\nu V_{b\mu}) + V_a^\rho V_b^\sigma (\partial_\rho V_{c\sigma}) V^c_\mu$$
$$+ i\kappa^2 \overline{\psi}_\mu \gamma_a \psi_b - \tfrac{1}{2} i\kappa^2 \overline{\psi}_a \gamma_\mu \psi_b] - [a \leftrightarrow b] .$$

and has both first derivative terms analogous to those of standard Christoffel symbol and terms bilinear in ψ_μ due to torsion.

The theory is locally supersymmetric and thus invariant under transformation rules involving an arbitrary anticommuting Majorana spin function $\varepsilon_a(x)$. They are

$$\delta V_{a\mu}(x) = i\kappa\bar{\varepsilon}(x)\gamma_a\psi_\mu(x) ,$$

$$\delta\psi_\mu(x) = \kappa^{-1}\mathcal{D}_\mu\varepsilon(x) .$$

The last term is just the covariant derivative of the gauge parameter. Since this is the typical gauge field transformation law (true in Yang–Mills theory and ordinary gravitation), it confirms that $\psi_\mu(x)$ is the gauge field of supersymmetry transformations.

The proof of invariance involves an intimate mixture of Dirac matrix algebra, Fierz rearrangement of spinors, and basic geometrical identities such as the Ricci identity $[\mathcal{D}_\mu,\mathcal{D}_\nu]\varepsilon = \frac{1}{2} R_{\mu\nu ab}\sigma^{ab}\varepsilon$ and the cyclicity property $\varepsilon^{\mu\nu\rho\sigma}R_{\mu\nu\rho\tau} = 0$ (second-order form). If you review the proof and see how these elements fit together, you will be convinced that the union of spin 2 and spin 3/2 is a marriage made in heaven.

It is worth noting that all previous attempts to write interacting spin 3/2 field theories led to inconsistencies such as propagation faster than light or ghosts. These problems are overcome in supergravity exactly because the interactions respect a fermionic gauge principle. Whether or not the unification of the actual laws of physics suggested by supergravity is ever realized, this achievement in mathematical physics shall remain. Spin 3/2 field theories can be mathematically respectable!

Unified $SO(N)$ Models

March 1979 is the third anniversary of supergravity. No one can say that its one-hundredth anniversary will be worth noting, but supergravity has now become a very active field of research, and many locally supersymmetric field theories have been constructed and their properties explored. There is time to discuss only one class of theory called the *unified SO(N) models*. The new element introduced in these theories is internal symmetry, and it is here that supergravity comes closest to Einstein's vision of unified field theory.

The $SO(N)$ theories are the gauge theories of extended supersymmetry algebras involving N spinor charges Q_α^i with spinor index α and internal symmetry index i. The Q_α^i transform as an N-dimensional vector under the group $SO(N)$. The fundamental structure relation is now

$$\{Q_\alpha^i, \bar{Q}_\beta^j\} = \delta^{ij}(\gamma^a)_{\alpha\beta}P_a ,$$

which involves a Kronecker δ^{ij} in internal symmetry indices as well as the previous $\gamma \cdot P$.

The extended supersymmetry algebras are larger systems and their irreducible representations are also larger than the previous. In general, they are towers of several different spins, beginning with some maximum helicity value s_{max} and ending with minimum value $s_{min} = s_{max} - \frac{1}{2}N$. The particles at each level transform in definite representations of $SO(N)$. The unified theories are based on massless irreducible representations with the following properties:

1. The highest-spin particle is $s_{max} = 2$. It is an $SO(N)$ singlet, and we therefore identify it as the graviton.
2. Lower-spin particles have a remarkable correlation between spin and internal symmetry, as shown in the table accompanying, which lists the number of particles of a given spin s in several of these theories.

	$N = 1$	$SO(2)$	$SO(3)$	$SO(4)$	$SO(8)$
$s = 2$	1	1	1	1	1
$s = 3/2$	1	2	3	4	8
$s = 1$		1	3	6	28
$s = 1/2$			1	4	56
$s = 0$				1 + 1	35 + 35'

The $N = 2$ theory contains 1 graviton, two Majorana gravitinos (equivalent to one charged spin 3/2 particle), and one vector particle, which, mathematically at least, could be the photon. In fact, this field theory could be the unification of gravitation and electromagnetism that eluded Einstein. For $N \geq 3$ particles of spin 1/2 and spin 0 enter, so that fundamental spinors can be unified in principle with the vector particles that mediate forces between them and with scalar fields that can lead to spontaneous breakdown of symmetry. The $N = 8$ theory is the largest that can be constructed in this framework. It contains one graviton, eight gravitinos, twenty-eight spin 1, fifty-six spin ½, and thirty-five each of scalars and pseudoscalars.

The extraordinary feature of these theories is their high degree of unification. Supersymmetry transformations connect particles of different spin, and $SO(N)$ rotations relate particles of given spin. All particles are therefore unified with the graviton. As just described, this unification is algebraic — that is, associated with an irreducible representation of an algebraic system. It is also geometric. This is clear from the complete superspace formulations for $N = 1$, and superspace and other formalisms developed for higher N.

There are always $\frac{1}{2}N(N - 1)$ vector particles in each of these theories. Since this is the dimension of the adjoint representation of $SO(N)$, the internal symmetry can be gauged leading to a marriage between the gauge principles of local internal symmetry and local supersymmetry. The resulting theories depend on two parameters — the gravitational constant $\varkappa$ and the gauge charge e. Unfortunately, this is a troubled marriage, largely due to a required cosmological constant of the form $\Lambda \approx e^2/\varkappa^2$, whose natural magnitude is set by the Planck energy scale, that is, $\Lambda \approx 10^{38}(\text{GeV})^2$. This is a nonsensical value for macroscopic cosmology, but it may be required at the quantum level in the picture of space-time "foam" discussed long ago by Wheeler and recently given a mathematical basis by Hawking.[13] I hope that the marriage can be saved.

Phenomenology

At first glance the $SO(8)$ theory seems large enough to accommodate the elementary particles suggested by present experiments and even to allow for future expansion. However, the opposite is true. The $SO(8)$ theory is too small, essentially because the $SO(8)$ group is not big enough to contain the product subgroup $SU(3) \times SU(2) \times U(1)$ required in present ideas in particle physics to describe the conserved color quantum chromodynamics symmetry of strong interactions and the spontaneously broken flavor symmetry of weak and electromagnetic forces. Nevertheless, one can classify particles[14] of the theory with respect to $SU(3)_{color} \times U(1)_{e.m.}$, which is the largest subgroup of $SO(8)$ that seems physically relevant. The good feature of this classification is the prediction of four flavors of quarks with canonical charges, $u(2/3)$, $d(-1/3)$, $s(-1/3)$, and $c(2/3)$. The bad feature is that several elementary particles do not appear, such as the $W^{\pm}$ intermediate bosons of the weak interactions and the μ and T leptons. The framework of extended supergravity is very rigid; there is little freedom to adjust the internal symmetry group or the representations assigned to various particles. Ultimately I hope that this rigidity will be advantageous, but at present we must say that the phenomenology appears unnatural.

If field theories for supersymmetry algebras with $N \geq 9$ can be found, the phenomenology could be improved. Indeed, the present limitation to $N \leq 8$ may not be fundamental. The problem is that the smallest representations of supersymmetry algebras with $N \geq 9$ necessarily contain particles of spin $s \geq 5/2$, and it is part of the negative folklore of theoretical physics that consistent interacting high-spin field theories cannot be formulated. It is unwise to rely on folklore, especially in supergravity, where similar consistency problems for spin 3/2 were overcome.

Higher-Spin Gauge Fields

Recently the high-spin barrier was cracked for massless free fields. Integer spins have been treated by Fronsdal, half-integer spins by Fang and Fronsdal, with spin 5/2 case discussed independently by Behrends, van Holten, van Nieuwenhuizen, and de Wit.[15] Quite simple Lagrangians are given for each spin s that involve a single totally symmetric tensor of rank s for integer s (or a tensor spinor of rank $s - \frac{1}{2}$ for half-integers). The Lagrangians have gauge invariances that generalize the free spin 3/2 Rarita–Schwinger invariance and free spin 2 Pauli–Fierz invariance, which is the primitive form of general covariance. Spin 5/2 is described by a symmetric tensor-spinor $\psi_{\mu\nu}$ with gauge transformation $\delta\psi_{\mu\nu}(x) = \partial_{\mu}\varepsilon_{\nu}(x) + \partial_{\nu}\varepsilon_{\mu}(x)$ with the algebraic constraint $\gamma^{\mu}\varepsilon_{\mu}(x) = 0$ on the gauge parameter. Spin 3 is described by a third-rank totally symmetric tensor $\phi_{\mu\nu\rho}$ with gauge variation

$$\delta\phi_{\mu\nu\rho} = \partial_{\mu}\xi_{\nu\rho} + \partial_{\nu}\xi_{\rho\mu} + \partial_{\rho}\xi_{\mu\nu}$$

with constraint $\xi_{\mu}{}^{\mu} = 0$. The key new point is that algebraic constraints on the gauge parameters are sufficient and even necessary to guarantee a positive Hilbert space metric with two propagating helicity $\pm s$ modes and Lorentz invariant coupling to external sources. Construction of interacting field theories is being studied, although it is a difficult problem. Indeed, the analysis of Haag, Lopuszanski, and Sohnius suggests that it is not possible to extend these free-field gauge invariances to include interactions.[16] However, one must remember that the free spin 3/2 gauge invariance was one of the key ingredients of supergravity and that it took thirty-five years to find the correct interacting theory!

Recent Work on the $SO(8)$ Theory

I now want to discuss the recent work on the $SO(8)$ unified theory in a little more detail. The explicit form of the theory has recently been obtained by very interesting techniques. The basic idea is the dimensional reduction of Kaluza and Klein, who showed that a theory of gravity formulated in five dimensions leads to a theory of gravity and electromagnetism in four dimensions after suitable restrictions are placed on the five-dimensional manifold. As applied to supersymmetry and supergravity, the idea is that a simple supergravity (that is, no internal symmetry) in some dimension $4 + N$ will lead to extended $SO(N)$ invariant theories in four dimensions. The reason is that the tangent space group $SO(3 + N, 1)$ of the higher-dimensional theory reduces to $SO(3,1)_{Lorentz} \times SO(N)_{internal}$ if the fields $(x^0, x^1, x^2, x^3, y^1, y^2, \ldots, y^N)$ are restricted to be independent of y.

The first step in this program was to formulate the simple supergravity[17] in eleven dimensions, chosen because eleven is the largest dimension where the graded Poincaré group has a representation corresponding to maximum spin 2. The spinor gauge field of the theory is a vector-spinor $\psi_{\hat{\mu}\hat{\alpha}}(x,y)$ with an eleven-dimensional vector index and a thirty-two-dimensional spinor index. Corresponding to dimensional reduction there is a reduction of the group representation of $\psi_{\hat{\mu}\hat{\alpha}}(x,y)$ according to the following index decomposition:

$$\hat{\mu} = \mu \oplus i$$
$$\mu = 0,1,2,3 \qquad i = 1,2,\ldots,7$$
$$\text{space-time} \qquad \text{vector irrep. of } SO(7)$$

$$\hat{\alpha} = \alpha \otimes A$$
$$\alpha = 1,2,3,4 \qquad A = 1,\ldots,8$$
$$\text{Dirac spinor} \qquad \text{8-dim. spinor irrep. of } SO(7).$$

Thus the simple vector spinor field in eleven dimensions gives

$$\psi_{\mu\alpha}{}^A \quad \text{i.e., 8 spin 3/2 gravitino fields}$$

and

$$\psi_\alpha{}^{iA} \quad \text{i.e., } 7 \times 8 \text{ spin 1/2 fields}$$

in four dimensions.

By similar analysis of the boson fields one finds that the counting of the degrees of freedom of the four-dimensional theory is exactly that of the $N = 8$ model (this is actually expected on simpler grounds), but the manifest internal symmetry is only $SO(7)$.

Rearrangement of the Lagrangian so that the invariances of the $N = 8$ theory are manifest is a difficult problem. This was solved recently by Cremmer and Julia[18] and involves unsuspected new invariances. Roughly speaking, the group of the eleven-dimensional vierbein $V_{\hat{a}\mu}$, that is $SO(10,1) \times GL(11,R)$, contains an internal symmetry part $SO(7) \times SL(7,R)$ after reduction, and this can be extended to an $SU(8) \times E_7$ invariance. The $SU(8)$ invariance is local, and gauge potentials are introduced that are nonpropagating in the classical approximation and constrain physical scalars to the coset space $E_7/SU(8)$, which has the proper dimension, viz. $133 - 63 = 70$. Cremmer and Julia speculate that the $SU(8)$ gauge potentials are promoted to physical particles by radiative corrections and may account for some of the missing particles of the previous phenomenological analysis, such as the $W^{\pm}$. The work is very new and must be studied further. Clearly, the ideas involved are refreshing, and we may hope for new physics!

If the ideas of supersymmetry and supergravity are ever applicable to nature, it must be in a form in which the boson-fermion symmetry is spontaneously broken. There are few Bose-Fermi mass equalities among the fundamental particles. A super-Higgs mechanism[19] has been worked out for some supergravity theories, but until very recently, not for the unified models. Now, Scherk and Schwarz have shown that one can obtain breakdown of supersymmetry in the dimensional reduction procedure by allowing very particular dependence of the fields on the extra coordinates.[20] For example, in a simple model involving the reduction from four to three dimensions, the spin 3/2 field has the dependence

$$\psi_\mu(x^0, x^1, x^2, y) = e^{im\gamma_5 y} \psi_\mu(x^0, x^1, x^2) .$$

The y dependence drops out of the final Lagrangian because of chiral invariance, but it leaves mass terms for various particles including the gravitino. As applied to the $SO(8)$ theory, the authors argue that all eight supersymmetries are broken and that one finds eight massive spin 3/2 particles. The procedure avoids previous difficulties involving cosmological terms and badly behaved scalar field potentials.

Renormalizability in Quantum Gravity

In any quantum field theory the total amplitude is a sum of Feynman diagrams. For example, electron-electron scattering in quantum electrodynamics can be expressed as a sum of a tree diagram involving exchange of a photon and loop diagrams involving closed loops of virtual particles. The three diagram gives the contribution to scattering of the classical electromagnetic force, $F = e^2/r^2$, while the loop diagrams are the actual quantum corrections. The Feynman rules assign definite mathematical expression for each diagram. For a diagram with V vertices and L closed loops, one finds an expression

$$e^V \int d^4k_1 \ldots d^4k_L f(k_1, \ldots, k_L) ,$$

where f is a function of the independent loop momenta that is then integrated. Although such integrals frequently diverge at large momenta, the situation is controllable. The divergences can be absorbed in redefinition of an electron charge e and mass m and in the asymptotic field strength Z. This is the renormalization program.

We now consider electron-electron scattering in quantum gravity where gravitons replace photons. Again, the tree diagram describes effects of the classical gravitational force, which takes the familiar form $F = -\kappa^2 m^2/r^2$ in the nonrelativistic limit. Thus κ carries negative dimensions of energy. The loop diagram contribution is given by

$$\kappa^V \int d^4k_1 \ldots d^4k_L f'(k_1 \ldots k_L) k^V ,$$

where $f'(k_1 \ldots k_2)$ is similar to f (it has the same dimension and ultraviolet behavior). The overall dimension of the scattering amplitude is the same in quantum electrodynamics and quantum gravity. Therefore the extra V negative units of energy in the coupling constant dependence are compensated by V positive powers of loop momentum under the integrals. The ultraviolet behavior is much worse, and the renormalization program fails.

Traditional renormalization fails for dimensional reasons in quantum gravity, and this has led to several possible scenarios for a microscopic theory of gravitation. Within supergravity, there has been hope for a finite S-matrix in which the divergences cancel in directly observable scattering amplitudes but persist in off-shell Green's functions. Before supergravity, the status of this hope was as follows:

1. One-loop graviton-graviton scattering was shown to be finite[21] due to a special identity applicable in four-dimensional space-time,
2. One-loop amplitudes involving gravity plus matter fields of spin 0, ½, 1 are infinite,[22]
3. For two or more loops there are no explicit calculations of amplitudes, but there are candidate counterterms even for divergences of graviton-graviton scattering.[23]

In supergravity the situation is

1. In the eight unified theories all the one-loop scattering amplitudes are finite.[24] Cancelation occurs in lower-spin matter amplitudes because of the gravitino contributions that were not previously included. More physically cancelation of divergences is a direct consequence of the unification of the theories; that is, lower-spin amplitudes can be related to finite graviton-graviton amplitudes by combined supersymmetry and $SO(N)$ transformations. Arguments involving locally supersymmetric counterterms[25] have also been applied to the study of one-loop divergences.
2. Cancelation extends to the two-loop level in supergravity, because there are no locally supersymmetric counterterms of the appropriate dimension.[26]
3. For three or more loops, there are no concrete calculations, but there are serious indications that infinities persist.[27] Dangerous counterterms are shown rigorously[28] to exist for $N = 1$, using the formalism of a locally supersymmetric tensor calculus.[29] The three-loop counterterm seems to be allowed for $N = 2$.[30] The case $N = 8$ may be special, but nothing is really known.
4. Complete ultraviolet finiteness (both on- and off-shell) has been proved in the superspace theory of Arnowitt and Nath.[31] However, it is suspected that this result is due to the presence of negative metric ghost states — a physically unsatisfactory feature.

Conclusion

In this discussion of supergravity I have emphasized the physical aspects of the subject and have largely neglected the formal work that has been vital in establishing physical results. If your curiosity is aroused by what I have said, then I urge you to consult further references.

Supergravity has an elegant theoretical foundation: it is rooted in a gauge principle that unifies bosons and fermions and leads potentially to great unification of the fundamental laws of physics. At present we are faced with difficulties in finding realistic applications and in proving full renormalizability. I believe that these difficulties will be overcome, because I have great faith in gauge principles. The number of gauge principles allowed in theoretical physics is very limited. There is the spin 1 gauge principle of electromagnetism and Yang–Mills theory. There is the spin 2 gauge principle of general relativity. Both play an important role in nature. One can formulate as a theorem the statement that the only other gauge principle allowed in quantum field theory with interactions is the spin 3/2 gauge principle of local supersymmetry. I think there is a good chance that nature makes use of it also.

NOTES

1. Y. A. Golfand and E. P. Likhtman, JETP Lett. **13**, 452 (1971).
2. D. V. Volkov and V. P. Akulov, JETP Lett. **16**, 438 (1972); J. Wess and B. Zumino, Nucl. Phys. **B70**, 39 (1974).
3. A. Salam and J. Strathdee, Nucl. Phys. **B80**, 499 (1974).
4. A. Salam and J. Strathdee, Phys. Lett. **51B**, 353 (1974); S. Ferrara and B. Zumino, Nucl. Phys. **B79**, 413 (1974).
5. M. T. Grisaru and H. Pendleton, Nucl. Phys. **B124**, 81 (1977).
6. J. Iliopoulos and B. Zumino, Nucl. Phys. **B76**, 310 (1974).
7. Volkov and Akulov, op. cit. in n. 2; A. Salam and J. Strathdee, Nucl. Phys. **B76**, 477 (1974).
8. D. V. Volkov and V. A. Soroka, JETP Lett. **18**, 529 (1973); P. Nath and R. Arnowitt, Phys. Lett. **56B**, 177 (1975); see the lectures of B. Zumino in *Gauge Field Theories and Model Field Theory,* edited by R. Arnowitt and P. Nath (MIT Press, Cambridge, Mass., 1976).
9. R. Arnowitt and P. Nath, Phys. Rev. Lett. **42**, 138 (1979); J. Wess and B. Zumino, Phys. Lett. **74B**, 51 (1978); Y. Ne'eman and T. Regge, Phys. Lett. **74B**, 54 (1978); L. Brink et al., Phys. Lett. **74B**, 336 (1978); W. Siegel, Harvard University preprint 1978; J. Gates and J. Shapiro, MIT preprint 1978; P. S. Howe and R. W. Tucker, Phys. Lett. **80B**, 138 (1978); V. Ogievetsky and E. Sokatchev, Dubna preprint 1978; J. C. Romau, A. Ferber, and P. G. O. Freund, Nucl. Phys. **B126**, 429 (1977); J. G. Taylor, Phys. Lett. **79B**, 399 (1978).
10. D. Z. Freedman, P. van Nieuwenhuizen, and S. Ferrara, Phys. Rev. **D13**, 3214 (1976).
11. S. Deser and B. Zumino, Phys. Lett. **62B**, 335 (1976).
12. E. S. Fradkin and M. A. Vasiliev, Lebedev Institute preprint 1976; P. Townsend and P. van Nieuwenhuizen, Phys. Lett. **67B**, 439 (1977).
13. J. A. Wheeler, in *Relativity, Groups, and Topology*, edited by B. S. and C. M. DeWitt (Gordon and Breach, New York, 1964); S. W. Hawking, Nucl. Phys. **B144**, 349 (1978).
14. M. Gell-Mann, lecture at the Washington meeting of the American Physical Society, 1977, unpublished.
15. C. Fronsdal, Phys. Rev. **D18**, 3624 (1978); J. Fang and C. Fronsdal, Phys. Rev. **D18**, 3630 (1978); F. A. Behrends et al., Phys. Lett. **83B**, 188 (1979).
16. R. Haag, J. T. Lopuszanski, and M. Sohnius, Nucl. Phys. **B88**, 257 (1975).
17. E. Cremmer, B. Julia, and J. Scherk, Phys. Lett. **76B**, 61 (1978).
18. E. Cremmer and B. Julia, Phys. Lett. **80B**, 48 (1978), and Ecole Normale Superieure preprint, 1979.
19. A. Das, M. Fischler, and M. Roček, Phys. Lett. **69B**, 186 (1977); J. Polonyi, Budapest preprint KFK1-1977-93; E. Cremmer et al., Phys. Lett. **79B**, 231 (1978).

20. J. Scherk and J. H. Schwarz, Phys. Lett. **82B,** 60 (1979); Nucl. Phys. **B153,** 61 (1979).
21. G. 't Hooft and M. Veltman, Ann. Inst. H. Poincaré **20,** 69 (1974).
22. Ibid.; S. Deser and P. van Nieuwenhuizen, Phys. Rev. **D10,** 401, 410 (1974).
23. P. van Nieuwenhuizen and C. C. Wu, J. Math. Phys. **18,** 182 (1977).
24. M. Grisaru, J. Vermaseren, and P. van Nieuwenhuizen, Phys. Rev. Lett. **37,** 1662 (1976).
25. S. Deser, J. Kay, and K. Stelle, Phys. Rev. Lett. **38,** 527 (1977).
26. M. Grisaru, Phys. Lett. **66B,** 75 (1977); E. Tomboulis, Phys. Lett. **67B,** 75 (1977).
27. Deser, Kay, and Stelle, op. cit. in n. 25; S. Ferrara and P. van Nieuwenhuizen, Phys. Lett. **78B,** 573 (1978).
28. Ferrara and van Nieuwenhuizen, op. cit. in n. 27.
29. S. Ferrara and P. van Nieuwenhuizen, Phys. Lett. **76B,** 404 (1978); K. Stelle and P. C. West, Phys. Lett. **77B,** 376 (1978).
30. Deser, Kay, and Stelle, op. cit. in n. 25.
31. Arnowitt and Nath, Phys. Rev. Lett. **42,** 138 (1979).

Discussion

Following the Paper by D. Z. FREEDMAN

P. G. Freund (U. of Chicago): I would like to put a slightly different emphasis on a remark that you have made, relating to an idea of Einstein's. In unified theories of the type that he contemplated, or even of the type discussed here earlier, like the Weinberg–Salam theory, once you unify two kinds of force, you still put practically no constraints on the forms of matter that you can still add to the theory. Unlike these theories, in maximally extended supergravity, once you unify the forces (gravity plus weak interactions, color and so forth), you automatically also prescribe the forms of matter that are in the theory. Einstein is known to have always been worried that his equations have a right-hand side, the $T_{\mu\nu}$ of matter. He always wanted to eliminate this and I think that this suggestion of his is essentially realized in supergravity.

Freedman: Yes, I agree with you, but perhaps I should mention that because of reasons of time I had to omit the discussion of a class of supergravity theories in which one can add matter in a rather arbitrary way. However, like you, I thought the emphasis should be placed on unified theories.

Y. Ne'eman (Tel Aviv U.): I wanted to make a point about the graded Lie algebras and the supersymmetries mentioned in the beginning of your paper. In a recent theory that I am not discussing in this meeting, but at the coming International Colloquium on the Applications of Group Theory to Physics here, I found a way of using a graded Lie algebra and a supergroup without changing the spins when the statistics are changing. I am using the existence of those "ghosts" that Professor Freedman has been mentioning. This means that the group is really a symmetry of the effective quantum Lagrangian, connecting particles to ghosts whose spins by definition do not fit the statistics and vice versa.

7. GAUGED AND AFFINE QUANTUM GRAVITY

Yuval Ne'eman

1. Motivation: The Search for Quantum Gravity

There is a good and ever-improving fit between those observations and experiments that probe physical phenomena in the large and the corresponding predictions based upon Albert Einstein's highly aesthetic theory of the gravitational field. Without doubt this success story has contributed much to the motivation for celebrating Einstein's Centenary with great enthusiasm. There is unanimity among the physics community that his theory has provided a deeper and more precise understanding of gravitational and inertial forces than Newton's, plus the foundations for the new physical science of cosmology.[1] Most will admit that further modifications of the theory in the large are possible, even though they appear at first to mar its elegance: Einstein himself thought in 1917 that a cosmological constant Λ had to be brought in, and it was only the astronomical evidence for an expanding universe, presented by Hubble in 1928, that provided the possibility that $\Lambda = 0$. A weak Brans–Dicke scalar field contribution is still allowed; so is Rosen's bimetric theory, and so on.

The situation in the small is entirely different. There is not a bit of evidence for the theory in this region. Indeed, as long as it is not given a finite quantum formulation, there is just no theory. *We are thus faced with the task of developing a finite quantum field theory of the gravitational interaction. As a boundary condition we require Einstein's equations to appear as the "limit in the large" of the quantized theory's equations.*

Our experience in developing relativistic quantum field theory (RQFT) is based upon two and possibly three successes: quantum electrodynamics (QED) and its recently experimentally vindicated unification with the weak (Fermi) interactions within the new quantum asthenodynamics (QAD); in addition, I should perhaps include the as yet not proven theory of quantum chromodynamics (QCD). The last two, of course, form the subject of most papers in this colloquium. Let me, however, just draw the reader's attention to what happened in going from Maxwell's classical theory to QED and then to QAD: beyond the mere modification of Maxwell's equations as implied by QED (various counterterms necessary to the renormalization procedure), the weak currents (and especially the newly discovered neutral ones) may be regarded as a new and previously unsuspected short-range component of electrodynamics!

It may therefore be concluded that *in going from Einstein's theory to the RQFT of gravity (QGD) we may also encounter new short-range effects.* Some of these may involve known but as yet unrelated interactions; some may be entirely new. In the preceding chapter Dr. Freedman has described supergravity, a candidate QGD, which indeed involves such new components: the supersymmetry current, coupled to the "gravitino" (the spin 3/2 field), and in the case of "extended supergravity" some components resembling QED and QCD, though as yet nothing like the (now related) weak interactions. Supergravity is also the only known extension of Einstein's theory that preserves a feature distinguishing this theory from all three quantized ones (QED, QAD, QCD — the *gauge* RQFTs): linearity of the Lagrangian as expressed in the gauge curvatures.

In the alternative approach presented here, the Einstein–Hilbert (linear) classical Lagrangian is considered as an *effective macroscopic limit of a quadratic quantum Lagrangian. The quantum Lagrangian is thus taken to be quadratic in the curvature, as in all three gauge theories.* There is an entire class of theories that have adopted this hypothesis, and some of them have even been claimed as renormalizable and possibly finite quantized theories,[2] although the proofs are as yet very incomplete. I shall return to this issue.

The quadratic theories chosen for consideration here involve several quadratic terms, with only one term surviving macroscopically and reproducing all the experimentally confirmed results of Einstein's theory. The other terms represent a *confined* interaction, and there is a possibility that *they might represent strong interactions and contribute to quark-confinement* (see the next section of this volume, on quantum chromodynamics).

The three successfully renormalized theories are gauge theories. In my paper at the Einstein centennial celebration at the Institute for Advanced Study in Princeton,[3] I discussed the reasons for a reformulation of gravity as a gauge theory of the Poincaré group. In a linear formulation, this is the Einstein–Cartan theory, or supergravity if we enlarge the Poincaré group to its graded extension. Spin is gauged "normally" in the Poincaré case (that is, it may form the fiber in a torsion-free principal bundle) but not exactly so in supergravity[4] where the base manifold preserves constant torsion. The gauging of translations and supertranslations is done as an anholonomized general coordinate transformation (AGCT) generated by the parallel-transport operators, with curvatures and torsions modifying the original Poincaré or graded-Poincaré commutators.

The gauging of spin is necessitated by the inclusion of half-integer spin fields in the source terms of Einstein's theory. It is interesting that spin on the other hand plays a dynamical role in hadron "flavor" dynamics: the 1964–65 emergence of a highly successful static (and then a collinear) $SU(6)$ symmetry can be summed up as spin-unitary spin [$SU(3)$] independence. At the time, the current-algebra interpretation of charge-densities was generalized to space-components of the asthenodynamic currents. A relativistic definition of quark-spin was thus based on the observation of Gamow–Teller transitions. Within Einstein–Cartan gravity, these can already be replaced by gravitational matrix elements coupled to torsion and reducing to a spin-spin term in the energy-momentum tensor. Hadronic approximate spin-independence may then emerge out of asymptotic freedom, that is, very-short-range vanishing of the effective QCD coupling. Quarks then behave as free fields, with their relativistic spin being further preserved in the presence of gravity. This is a minimalist view, as against a more daring hypothesis *according to which the spin-interaction in a quadratic gravitational Lagrangian is to be identified with a part of the strong interaction itself, that part that is displayed in spin-independence.*

As we shall see here, extending the Poincaré to the affine group generalizes these arguments to two additional regularities displayed by the strong interactions: scaling and Regge trajectories.

Another possible hint at the gravitational nature of strong interactions is provided by the success of their characterization as extended geometric structures: *relativistic strings, dual models* (strings with spin-excitations), and *bags* or lumps. The dynamics of such structures involve geometrical groups such as $SL(2,R)$ or $SA(2,R)$ for strings and $SL(4,R)$ or $SA(4,R)$ for bags. If these groups be gauged, it would seem difficult to distinguish between the relevant gauge fields and those of pure gravity. Indeed, the MIT bag model has been hindered by the impossible task of writing a special relativistic Lagrangian term for the volume-preserving pressure field other than through highly nonlinear structures.[5] This is an affine situation and thus involves an extension of general relativity rather than of special relativity.

A theory of "strong gravity" had indeed been hypothesized in the sixties by Salam and collaborators on an ad hoc basis, assuming that the f⁰(1250) spin-2 field is a "strong graviton."[6] In the present treatment, all the gauge fields are derived "from first principles;" that is, they are fixed by the gauge group and the structure of the Lagrangian.

It is also worth noting that the actual model I use applies the result of "teleparallelism" theories,[7] which were first studied by Einstein himself as candidate unified field theories. They failed in that role, but as I show may yet provide a viable alternative for gravity itself.

2. The Poincaré and Affine Groups and Gauges

The abstract affine group $SA(4,R)$ on a four-dimensional (flat) manifold (or on the tangent manifold to an affine L_4) is given by the following:

$$[J_{ab}, J_{cd}] = i(\eta_{ad} J_{bc} - \eta_{ac} J_{bd} + \eta_{bc} J_{ad} - \eta_{bd} J_{ac}) , \tag{7.1}$$

$$[J_{ab}, T_{cd}] = i(-\eta_{ad} T_{bc} - \eta_{ac} T_{bd} + \eta_{bc} T_{ad} + \eta_{bd} T_{ac}) , \tag{7.2}$$

$$[T_{ab}, T_{cd}] = i(\eta_{ad} J_{bc} + \eta_{ac} J_{bd} + \eta_{bc} J_{ad} + \eta_{bd} J_{ac}) , \tag{7.3}$$

$$[J_{ab}, P_c] = i(\eta_{bc} P_a - \eta_{ac} P_b) , \tag{7.4}$$

$$[T_{ab}, P_c] = -i(\eta_{ac} P_b + \eta_{bc} P_a - \tfrac{1}{2}\eta_{ab} P_c) , \tag{7.5}$$

$$[P_a, P_b] = 0 , \tag{7.6}$$

$$J_{ba} = -J_{ab} , \tag{7.7}$$

$$T_{ba} = T_{ab} , \tag{7.8}$$

$$\mathrm{Tr}(J_{ba}) = \mathrm{Tr}(T_{ba}) = 0 , \tag{7.9}$$

where Tr stands for the trace in a 4×4, x/p representation. J_{ab} are the Lorentz transformations (angular momenta), T_{ab} the shears, P_a the translations. Equations (7.1), (7.4), (7.6), (7.7) define the Poincaré subgroup $ASO(1,3)$, Eqs. (7.1) and (7.7) its $SO(1,3)$ (Lorentz) subgroup. Equations (7.1), (7.2), (7.3), (7.7), (7.8), (7.9) define $SL(4,R)$ the group of special (unimodular) linear transformations on an L_4. Quantum mechanics involves a complex Hilbert space and its ray representations (that is, it involves moduli only, not the complex amplitudes themselves). This is equivalent to using the double-covering for groups classifying the Hilbert space states. Since the aim in this study is to produce a quantized theory rather than to study the classical geometry of space-time, we have to replace $SO(1,3)$ by $\overline{SO}(1,3) = SL(2,C)$ and take the double-covering $\overline{ASO}(1,3) = ISL(2,C)$ of the Poincaré group as well. This is trivial, since it does not involve additional structure. However, when going to $\overline{SL}(4,R)$ and $\overline{SA}(4,R)$ we make use of the recently identified and constructed[8] double-covering of the linear groups (the metalinear groups) involving the $\overline{T}_{ab}$.

Under $SL(2,C)$ as given by (j_1, j_2) eigenvalues, the various generators split into:

$$J_{ab} \sim (1, 0) + (0, 1) ,$$

$$P_a \sim (\tfrac{1}{2}, \tfrac{1}{2}) , \qquad (7.10)$$

$$T_{ab} \sim (1, 1) .$$

Gauging noninternal groups such as these is best done by referring first to the group manifold itself.[9] With the coordinates x as parameters for P_a, the generalized Euler angles of $SL(2,C)$ Ξ^{ab}, and the parameters of shear ξ^{ab}, we have a nineteen-dimensional manifold (ten-dimensional for Poincaré). If the Lagrangian is made locally invariant under $\overline{SL}(4,R)$ (or $SL(2,C)$ in the Poincaré case), the ξ^{ab} and Ξ^{ab} factor out and the fundamental forms, connections, and curvatures involve dx only; that is, they are on the L_4. However, the P_a have to be replaced by parallel-transport operators as follows (D_μ^H is taken over the homogenous subgroup H; D_μ^G would involve momentum space):

$$D_a = v_a^\mu D_\mu^H = e_a^\mu (\partial_\mu + \Gamma_\mu^{cd} j_{dc} + \Lambda_\mu^{cd} t_{dc}) , \qquad (7.11)$$

where v_a^μ is an inverse-tetrad, Γ_μ^{cd} the Lorentz connection (gauge potential), and Λ_μ^{cd} the shear connection (gauge potential). j_{dc} and t_{dc} are representations of J_{dc} and T_{dc}, and the inversion in (dc) is a matter of sign conventions. The gauge potential of the D_a translations is e^a_μ, the tetrad field. Note that each one of J_{ab}, P_a, and T_{ab} gives rise to a separate spin 2 (and even spin 3 in Λ_μ^{cd}) gauge field. The D_a replace the P_a in (7.4) and (7.5), but (7.6) is replaced by (the minus is a sign convention)

$$[D_a, D_b] = -F_{ab}{}^c D_c + F_{ab}{}^{cd} J_{dc} + F_{ab}{}^{cd} T_{cd} , \qquad (7.12)$$

where the anholonomized curvatures (field strengths) are

$$F_{ab}{}^A = v_a^\mu v_b^\nu F_{\mu\nu}{}^A , \qquad (7.13)$$

and A stands for all values of the generators.

$$F_{\mu\nu}{}^a := \partial_\mu e_\nu^a - \partial_\nu e_\mu^a + \Gamma_{\mu b}{}^a e_\nu^b - \Gamma_{\nu b}{}^a e_\mu^b$$
$$+ \Lambda_{\mu b}{}^a e_\nu^b - \Lambda_{\nu b}{}^a e_\mu^b \qquad (7.14)$$

is the "torsion," or translation-gauge field strength.

$$F_{\mu\nu a}{}^b = \partial_\mu \Gamma_{\nu a}{}^b - \partial_\nu \Gamma_{\mu a}{}^b + \Gamma_{\mu c}{}^b \Gamma_{\nu a}{}^c - \Gamma_{\nu c}{}^b \Gamma_{\mu a}{}^c$$
$$- \Lambda_{\mu c}{}^b \Lambda_{\nu a}{}^c + \Lambda_{\nu c}{}^b \Lambda_{\mu a}{}^c \qquad (7.15)$$

is the "curvature," and the shear strength is given by

$$\widetilde{F}_{\mu\nu a}{}^b := \partial_\mu \Lambda_{\nu a}{}^b - \partial_\nu \Lambda_{\mu a}{}^b - \Gamma_{\mu c}{}^b \Lambda_{\nu a}{}^c + \Gamma_{\nu c}{}^b \Lambda_{\mu a}{}^c$$
$$- \Gamma_{\mu a}{}^c \Lambda_{\nu c}{}^b + \Gamma_{\nu a}{}^c \Lambda_{\mu c}{}^b . \qquad (7.16)$$

For any matter field $\psi(x)$ spanning a Hilbert space carrying a representation of $\overline{SA}(4,R)$ or of $ISL(2,C)$, the infinitesimal variations are given by the following equation ($\varepsilon^a = e_\mu^a \varepsilon^\mu$ is an anholonomized or gauge-potential dependent general coordinate transformation, related to "shifts" and "lapses"):

$$\psi'(x) = (1 - \varepsilon^a(x) D_a + \varepsilon^{ab}(x) j_{ab} + \widetilde{\varepsilon}^{ab}(x) t_{ab}) \psi(x) . \qquad (7.17)$$

The matter-Lagrangian $\mathcal{L}_\psi(\psi, \partial_\mu \psi)$ is gauged minimally by replacing the constant frame ∂_μ^a and field gradients $\partial_\mu \psi$,

$$\partial_\mu^a \to e_\mu^a , \quad \partial_\mu \psi \to D_\mu \psi . \qquad (7.18)$$

Note that the gauge potentials transform according to:

$$\delta e_\mu^a = - D_\mu^G \varepsilon^a - \varepsilon^c F_{c\mu}{}^a ,$$
$$\delta \Gamma_{\mu b}{}^a = - D_\mu^G \varepsilon^a{}_b - \varepsilon^c F_{c\mu b}{}^a , \qquad (7.19)$$
$$\delta \Lambda_{\mu b}{}^a = - D_\mu^G \widetilde{\varepsilon}^a{}_b - \varepsilon^c \widetilde{F}_{c\mu b}{}^a ,$$

where the affine covariant derivative is given by the following equations (D_μ is taken over $\overline{SL}(2,C)$ only):

$$D_\mu^G \varepsilon^a = D_\mu \varepsilon^a + e_\mu^c \varepsilon_c^a + e_\mu^c \tilde{\varepsilon}_c^a - \Lambda_{\mu\ c}^{\ a} \varepsilon^c ,$$

$$D_\mu^G \varepsilon^{ab} = D_\mu \varepsilon^{ab} + \Lambda_{\mu\ c}^{\ a} \tilde{\varepsilon}^{cb} - \Lambda_{\mu\ c}^{\ b} \tilde{\varepsilon}^{ca} , \qquad (7.20)$$

$$D_\mu^G \tilde{\varepsilon}^{ab} = D_\mu \tilde{\varepsilon}^{ab} + \Lambda_{\mu\ c}^{\ a} \varepsilon^{cb} + \Lambda_{\mu\ c}^{\ b} \varepsilon^{ca} .$$

For covariant tensors, η_A:

$$D_\mu^G \eta_A = \partial_\mu \eta_A + f_{BA}^E \rho^B \eta_E ,$$

the Bianchi identities are as follows (where $[\mu\, \nu\rho]$ denotes total antisymmetrization):

$$\partial_{[\mu} F_{\nu\rho]}^{\ a} + \tfrac{1}{2} \Gamma_{[\mu}^{\ ac} F_{\nu\rho]}^{\ c} - \tfrac{1}{2} \Lambda_{[\mu}^{ac} F_{\nu\rho]}^{\ c} -$$
$$\tfrac{1}{2} e_{[\mu}^c (F_{\nu\rho]}^{\ ac} - \tilde{F}_{\nu\rho]}^{\ ac}) = 0 ,$$

$$\partial_{[\mu} F_{\nu\rho]}^{\ ab} - \Gamma_{[\mu}^{\ ac} F_{\nu\rho]}^{\ bc} + \Lambda_{[\mu}^{ac} F_{\nu\rho]}^{\ bc} = 0 , \qquad (7.21)$$

$$\partial_{[\mu} \tilde{F}_{\nu\rho]}^{\ ab} - \Gamma_{[\mu}^{\ ac} F_{\nu\rho]}^{\ bc} + \Lambda_{[\mu}^{bc} F_{\nu\rho]}^{\ ac} = 0 .$$

3. The Equations of Motion

Let $V(e, e_\mu^{\ a}, \Gamma_\mu^{\ ab}, \Lambda_\mu^{\ ab})$ be the Lagrangian of the gravitational fields themselves. We define their canonical momenta[10] ($e = \det e_\mu^{\ a}$):

$$\pi_a^{\mu\nu} = \frac{\partial V}{\partial(\partial_\nu e_\mu^{\ a})} = \frac{\partial V}{\partial(F_{\nu\mu}^{\ a})} ,$$

$$\pi_{ab}^{\mu\nu} = \frac{\partial V}{\partial(\partial_\nu \Gamma_\mu^{\ ab})} = \frac{\partial V}{\partial(F_{\nu\mu}^{\ ab})} , \qquad (7.22)$$

$$\tilde{\pi}_{ab}^{\mu\nu} = \frac{\partial V}{\partial(\partial_\nu \Lambda_\mu^{\ ab})} = \frac{\partial V}{\partial(\tilde{F}_{\nu\mu}^{\ ab})} ,$$

and get the following generalized variational equations,

$$D_\mu \pi_a{}^{\nu\mu} - E_a{}^\nu = e\, \theta_a{}^\nu ,$$
$$D_\mu \pi_{ab}{}^{\nu\mu} - E_{ab}{}^\nu = e\, \sigma_{ab}{}^\nu , \qquad (7.23)$$
$$D_\mu \widetilde{\pi}_{ab}{}^{\nu\mu} - \widetilde{E}_{ab}{}^\nu = e\, \varsigma_{ab}{}^\nu ,$$

where for a gauge potential $\omega_\mu{}^A$

$$E_A{}^\nu = \frac{\partial V}{\partial \omega_\nu{}^A} + (\Gamma_\mu \pi^{\nu\mu})_A , \qquad (7.24)$$

and the right-hand side represents the conserved currents — that is $\theta_a{}^\nu$ the energy momentum tensor, $\sigma_{ab}{}^\nu$ the spin current, and $\varsigma_{ab}{}^\nu$ the shear current (intrinsic)

$$e\, \theta^\mu{}_a = v^\mu{}_a \mathcal{L}_\psi - \frac{\partial \mathcal{L}_\psi}{\partial(\partial_\mu \psi)} D_a \psi ,$$
$$e\, \sigma_{ab}{}^\mu = - \frac{\partial \mathcal{L}_\psi}{\partial(\partial_\mu \psi)} J_{ab} \psi , \qquad (7.25)$$
$$e\, \tau_{ab}{}^\mu = - \frac{\partial \mathcal{L}_\psi}{\partial(\partial_\mu \psi)} t_{ab} \psi ,$$

and the conservation laws are now given by the modified algebra

$$D_\mu(e\theta_a{}^\mu) = F_{a\mu}{}^{bc} e\sigma_{bc}{}^\mu + \widetilde{F}_{a\mu}{}^{bc} e\tau_{bc}{}^\mu + F_{a\mu}{}^b e\theta_b{}^\mu , \qquad (7.26)$$

$$D_\mu(e\sigma_{ab}{}^\mu) - e\theta_{[ab]} = 0 ,$$
$$D_\mu(e\tau_{ab}{}^\mu) - e\theta_{(ab)} = 0 . \qquad (7.27)$$

The last two subsist in the global case and serve to define total angular momentum and shear, beyond their intrinsic parts. Note that the first term of $E_A{}^\nu$ of (7.24) does not exist in an internal gauge theory, since the gauge Lagrangian contains only field strengths. It is the modified $P_a \to D_a$ due to the noninternal action that is responsible for the gravitational field energy-momentum tensor density $E_a{}^\mu$ (or the Hamiltonian $E_0{}^0$ analogous to $g_0{}^0 V - \partial_0 \phi \cdot \pi$),

$$E_a{}^\mu := v_a{}^\mu V - F_{av}{}^c \pi_c{}^{\nu\mu} - F_{av}{}^{cd} \pi_{cd}{}^{\nu\mu} - \widetilde{F}_{av}{}^{cd} \widetilde{\pi}_{cd}{}^{\nu\mu} .$$

Its spin density is, in the Poincaré case,

$$E_{ab}{}^\mu := \pi_{[ba]}{}^\mu \, , \qquad (7.28)$$

and its shear

$$\widetilde{E}_{ab}{}^\mu := \widetilde{\pi}_{(ab)}{}^\mu \, . \qquad (7.29)$$

Notice that in Einstein–Cartan (linear V) theory, $V \neq V(F_{\mu\nu}{}^a, F_{\mu\nu}{}^{ab})$ and we get the equations

$$F_{a\mu}{}^{cd} \pi_{cd}{}^{\mu\nu} - v_a{}^\nu V = e\theta_a{}^\nu \, ,$$
$$D_\mu \pi_{ab}{}^{\nu\mu} = e\tau_{ab}{}^\nu \, . \qquad (7.30)$$

Choosing

$$\pi_{ab}{}^{\nu\mu} = \tfrac{1}{2} \frac{e}{l^2} (v_a{}^\nu v_b{}^\mu - v_b{}^\nu v_a{}^\mu) \, , \qquad (7.31)$$

$$V = \tfrac{1}{2} F_{\mu\nu}{}^{ab} \pi_{ab}{}^{\nu\mu} = \frac{e}{2l^2} v_{[a}{}^\nu v_{b]}{}^\mu F_{\mu\nu}{}^{ab} = \frac{1}{2l^2} R \, , \qquad (7.32)$$

(7.30) becomes

$$F_{ca}{}^{\mu c} - \tfrac{1}{2} v_a{}^\mu F_{cd}{}^{dc} = l^2 \theta_a{}^\mu \, ,$$
$$\tfrac{1}{2} F_{ab}{}^\mu + v_{[a}{}^\mu F_{b]c}{}^c = l^2 \sigma_{ab}{}^\mu \, . \qquad (7.33)$$

The first equation of (7.33) is Einstein's (l^{-2} is Newton's constant); the second is an algebraic relation between a modified torsion tensor and the spin. One can solve that equation for the $\Gamma_\mu{}^{ab}$ of (7.14) and thus go over to "second-order formalism." If $\sigma_{ab}{}^\mu = 0$, this just reduces to the Christoffel formula. In either case, we see that Einstein–Cartan theory involves only one propagating gravitational field. This is just because its V contains only one kinetic term (the $\partial_\mu \Gamma_\nu{}^{ab} - \partial_\nu \Gamma_\mu{}^{ab}$), the other potential thus representing a nonphysical field. That we chose to express $\Gamma_\mu{}^{ab}$ in terms of $e_\mu{}^a$ (rather than the opposite) is due to the "boundary condition" represented by the principle of equivalence $\Gamma_\mu{}^{ab} \to 0$, $e_\mu{}^a \to \partial_\mu{}^a$.

We now turn to quadratic Lagrangians with both $e_\mu{}^a$ and $\Gamma_\mu{}^{ab}$ propagating. We thus do not deal with quadratic Lagrangians involving as usual[11] the square of the curvature by itself, or together with the Einstein–Hilbert linear term. These still involve only $\partial_\mu \Gamma^{ab}$ and not $\partial_\mu e_\nu{}^a$. We deal first with a Poincaré gauge, and use the Lagrangian,[12]

$$V = \frac{e}{4l^2}\left(-F^{\mu\nu}{}_a F_{\mu\nu}{}^a + 2 F^{\mu c}{}_c F_{\mu d}{}^d\right)$$

$$+ \frac{e}{4\alpha}\left(-F_{\mu\nu a}{}^b F^{\mu\nu a}{}_b\right), \tag{7.34}$$

where α is a dimensionless coupling. The second term is similar (even in dimensions) to an ordinary Yang–Mills Lagrangian and should therefore correspond to an asymptotically free gauge field $\Gamma_\mu{}^{ab}$ and, it is hoped, to confinement. The first term contains Newton's constant and should produce long-range gravity of the Newton–Einstein type. The canonical momenta are then

$$\pi_a{}^{\mu\nu} = \frac{e}{l^2}\left(F^{\mu\nu}{}_a + 2 v_a{}^{[\mu} F^{\nu]c}{}_c\right),$$

$$\pi^b{}_a{}^{\mu\nu} = \frac{e}{\alpha} F_{ab}{}^{\mu\nu}, \tag{7.35}$$

and the equations of motion are

$$D_\mu\{e(F^{\nu\mu}{}_a + 2 v_a{}^\nu F^{\mu c}{}_c)\} - l^2 E_a{}^\nu = e\, l^2\, \theta_a{}^\nu,$$

$$D_\mu(e F^{\nu\mu}{}_{ab}) + \frac{\alpha}{l^2} e(F^\nu{}_{ab} + v_{[b}{}^\mu F_{a]c}{}^c) = e\, \alpha\, \sigma_{ab}{}^\nu, \tag{7.36}$$

Note that

$$E_c{}^c = 0, \quad E_{ab} = \tfrac{1}{2} F_{\mu\nu[a} \pi_{b]}{}^{\nu\mu} = \frac{e}{l^2} F^c{}_{[ba]} F_{cd}{}^d. \tag{7.37}$$

The dynamics are thus given by the coupled nonlinear partial differential equations of second order (7.36), for $e_\mu{}^a$ and $\Gamma_{\mu a}{}^b$. If we stop at second derivatives of these potentials, the equations decouple and take the form

$$\partial\partial e \sim l^2 \theta,$$

$$\partial\partial \Gamma \sim \alpha \sigma. \tag{7.38}$$

In the weak-field limit

$$e_\mu^a = \delta_\mu^a + \tfrac{1}{2} h_\mu{}^a, \quad \Gamma_\mu{}^{ab} \ll 1, \tag{7.39}$$

and defining

$$\gamma_{\mu\nu} := h_{\mu\nu} - \tfrac{1}{4}\eta_{\mu\nu} h_\rho{}^\rho \quad \text{(symmetric, traceless)}$$

$$a_{\mu\nu} := h_{[\mu\nu]} \quad \text{(antisymmetric)} \tag{7.40}$$

$$\Gamma_\mu := \Gamma_{\nu\mu}{}^\nu \, ,$$

using the de Donder and Lorentz gauges

$$\partial_\mu \gamma^{\mu\nu} = 0 \, , \quad \partial_\mu \Gamma^{\mu\nu\rho} = 0 \, , \tag{7.41}$$

yields the equations

$$\Box \Box \gamma = -2\alpha(\theta - 2\partial_\mu \sigma^\mu) - 2 l^2 \Box \theta \, ,$$

$$\Box \partial_\mu \Gamma^\mu = \frac{\alpha}{2}(\theta - 2\partial_\mu \sigma^\mu) \, . \tag{7.42}$$

For a spinning point particle of mass m one finds

$$\gamma(r) = -\frac{l^2}{2\pi}\frac{m}{r} + C_1 - \frac{\alpha}{4\pi} mr + C_2 r^2 \, ,$$

$$a(r) = 0 \tag{7.43}$$

in a region $r_{min} \leqslant r \leqslant r_{max}$. Also,

$$\partial_\mu \Gamma^\mu = \frac{\alpha}{8\pi}\frac{m}{r} - \frac{3}{2} C_2 \, . \tag{7.44}$$

C_1 and C_2 will be fixed by the boundary conditions. Note that both Γ^μ and the curvature

$$F_{\mu\nu\rho}{}^\mu = 2\, \partial_{[\mu} \Gamma_{\nu]\rho}{}^\mu$$

depend on α. In the limit $\alpha \to 0$ (no Lorentz gauge) the curvature vanishes and we are left with the translational gauge. In (7.43) the term $\sim r$ appears to indicate confinement. If $\alpha \to 0$ outside of that region, we are left with a Newtonian limit just as in Einstein's theory. It has been shown[13] that this theory possesses the Schwarzschild solution. More recently, it was shown[14] that in the parametrized post-Newtonian approximation, the first departure from Einstein's theory is in fifth order in v/c. Thus all known experimental tests (including the calculations relating to binary pulsars and gravitational radiation) do not distinguish between our theory and Einstein's.

Note that although (7.34) is quadratic in the field strengths (that is, zero-forms) $F_{\mu\nu}{}^a$, the action integrand of the translational part is the exterior product of a 3-form by a 1-form.[15]

$$A = -\frac{1}{2l^2} \int (F^a \wedge e^b) \wedge {}^*(F_b \wedge e_a) - \frac{1}{2\alpha} \int F^{ab} \wedge {}^*F_{ab} \; . \tag{7.45}$$

The translational part, which emerges when $F^{ab} \to 0$, represents a special value of the Weitzenböck invariants. Once $F^{ab} = 0$, one can select an anholonomic frame in which $\Gamma_\mu{}^{ab} \to 0$ as well. One is then left with a simple curl,

$$F_{\mu\nu}{}^a \to 2 \, \partial_{[\mu} e_{\nu]}{}^a =: \Omega_{\mu\nu}{}^a \; , \tag{7.46}$$

also known as the "object of anholonomity." There are three independent quadratic invariants[16] one can build out of (7.46):

$$\begin{aligned} I_1 &:= e \, \Omega_{\mu\nu\rho} \, \Omega^{\mu\nu\rho} \; , \\ I_2 &:= e \, \Omega_{\rho\nu\mu} \, \Omega^{\mu\nu\rho} \; , \\ I_3 &:= e \, \Omega_{\mu\nu}{}^\nu \, \Omega^{\mu\rho}{}_\rho \; . \end{aligned} \tag{7.47}$$

Hehl et al. have noted that, for $F^{ab} \to 0$,

$$V \to \frac{1}{l^2} \, (-I_1 + 2I_3) \; . \tag{7.48}$$

They showed that any translational gauge Lagrangian

$$\begin{aligned} V &:= \tfrac{1}{2}(d_1 I_1 + d_2 I_2 + d_3 I_3) \; , \quad \text{for} \quad d_1 = -1 - \tfrac{1}{2} d_2 \\ & \qquad\qquad\qquad\qquad\qquad\qquad\qquad d_3 = 2 \end{aligned} \tag{7.49}$$

has a Newtonian limit, just as (7.34) or, more exactly, (7.48).

4. The Affine Gauge

The equations of motion (7.23) and conservation laws (7.25) include the case of a $\overline{GA}(4,R)$ noninternal gauge. In the metric-affine theory[17] there is still only one propagating gravitational field, as in the Einstein–Cartan case. No explicit Lagrangian has as yet been discussed for propagating $e_\mu{}^a$, $\Gamma_\mu{}^{ab}$, $\Lambda_\mu{}^{ab}$. One possibility would be to take $v = e$, the measure, which is $\overline{GL}(4,R)$ invariant as a *constrained* Lagrangian. This would have the advantage of connecting directly with the Nambu string, of which it would be a four-dimensional extension: the string Lagrangian is the $SL(2,R)$ measure, representing the surface spanned by an evolving string. Here we would have either an evolving lump of matter, or an evolving universe, depending upon the size of the confined region. In the first case, this solution would be embedded in a Freidman universe, in which Γ_μ^{ab} and Λ_μ^{ab} do not propagate. $\overline{GA}(4,R)$ is then represented nonlinearly, with the metric field as Goldstone boson.[18]

It was in the course of searching for a representation of matter with intrinsic shear that I was able to correct an error that had plagued studies of gravity for some fifty years. When the electron spin was discovered and after Dirac discovered the electron's (special) relativistic wave equation, Fock and Ivanenko and Hermann Weyl immediately showed how Lorentz spinors could be represented in general relativity in the "tangent" frame. Cartan proved that one could not have an equation like those of Dirac's, to describe half-integer spins on curved space (but assuming a finite number of components), and most textbooks disregard the assumption and refer to Cartan in stating that one is forced to place spin in an anholonomic frame (the tetrads). The possibility that world-covariant spinors with an infinite number of components might still exist was thus disregarded. This was also due to an error in another theorem of Cartan's, stating that "$SU(2)$, $SL(2,C)$ and $SL(2,R)$ admit no linear many-valued representations." This is true of the compact $SU(2)$, and of $SL(2,C)$, whose maximal compact subgroup is $SU(2)$ (which thus fixes its topology). The error about $SL(2,R)$ was due to the fact that $SL(2,R) = \overline{SO}(1,2)$ just as $SU(2) = \overline{SO}(3)$ and $SL(2,C) = \overline{SO}(1,3)$. This only implies that double-valued representations of $\overline{SO}(1,2)$ become single-valued in $SL(2,R)$. However, $SO(1,2)$ like its maximal compact subgroup $O(2)$ is infinitely many-valued, which is also true of $SL(2,R)$. There thus exists

$$\overline{SL}(2,R) = \overline{\overline{SO}}(1,2) ,$$

which is indeed the linear simple subgroup of the double-covering of the two-dimensional covariance group (for diffeomorphisms).

I have since constructed the unitary irreducible spinor representations[19] of $\overline{SL}(3,R)$ and recently D. Sijacki and I have done so also[20] for $\overline{SL}(4,R)$. These "bandors" fit the phenomenological description of the nucleon or any hadron, with excitations at $\Delta J = 2$ intervals in the simplest cases. One can now write a covariant derivative with such an infinite matrix, and I hypothesize that this is the correct physical treatment for a phenomenological hadron. It could thus represent an alternative to writing the interaction with the quarks and gluons and assuming an as yet unknown description of the hadron as a bound state.

The $\overline{GL}(4,R)$ representations give the skeleton structure for a "manifield" theory, which is as yet unwritten. Presumably, as intrinsic shear cannot be conserved in special realtivity, the only allowed equations of motion will be the $\overline{GL}(4,R)$ gauge equations themselves. This might thus go somewhat beyond the case of the $J = 3/2$ Rarita–Schwinger field, which as we saw requires the presence of the gravitational field (and supersymmetric coupling) for a physical description of its electromagnetic interactions only.

I refer the reader to the references in note 8 for the details of the classification and construction of $\overline{SL}(3,R)$ and $\overline{SL}(4,R)$ bandors. The $\overline{SL}(3,R)$ bandors appear as the representations of the stability subgroup of $\overline{SA}(4,R)$, provided we impose the condition

$$W_i = \tfrac{1}{2}(T_{0i} + J_{0i}) , \qquad i = 1, 2, 3 ,$$

$$W_i \mid \text{physical states} > = 0 . \tag{7.50}$$

This condition is identical to the "gauge condition" used to remove ghosts from the $\overline{SA}(2,R)$ string theory. Generally speaking, the bandor representations appear as the four-dimensional extensions of the spectrum of states defined by the string. It remains to be shown that such representations indeed correspond to the confined solutions of our V and fit ranges of the order of one Fermi.

To complete the system of excitations, my colleagues and I have also investigated[21] the (infinite) graded Lie algebras generated by adjoining to $\overline{SL}(3,R)$ or $\overline{SL}(4,R)$ a set of operators behaving as a bandor under these groups. It turns out that one can close the algebra over a set (G,S,E) where G is $\overline{SL}(3,R)$ or $\overline{SL}(4,R)$, S is a system of operators behaving as a bandor with $J = \tfrac{1}{2}$ as its lowest state and E is such a system with $J = 1$ (alternatively $J = 0$) as its lowest state. The commutators are of the type:

$$\{S_m^j, S_{m'}^{j'}\} \subset E_{m+m'}^{j+j'} , \tag{7.51}$$

$$[S_m^j, E_{m'}^{j'}] = 0 , \tag{7.52}$$

$$[E_m^j, E_{m'}^{j'}] = 0 . \tag{7.53}$$

Equation (7.51) thus extends ordinary supersymmetry, which can indeed be recovered by a specific contraction[22] of this algebra. Such an extension of the algebraic constraints imposed by the symmetry might be useful in further restricting the types of allowed counterterms in the renormalization procedure.

I have sketched a number of ideas and partial results that could provide the elements for a new affine theory of gravity. Affine theories go back to Eddington and the early years of general relativity. However, physics now possesses entirely new tools, and physicists aim at a quantum theory with what would appear to represent a different setting and new freedoms, though with new constraints too.

Acknowledgment

Research supported in part by the U.S.-Israel Binational Science Foundation.

NOTES

1. Note that, despite unanimity in the physics community, both general and special relativity are still very much under attack in every country by a pseudoscientific flat-earth-type sector, like almost no other field in science (with the possible exception of Darwin's theory of evolution). This is due in part, in my opinion, to difficulties in admitting the breakdown in the concept of simultaneity, for the "commonsense" educated man. The other reason is the theory's evocation of the entire *universe* as explained by *one man,* a highly provocative image for some minds.

2. Some examples of theories with quadratic terms: R. Utiyama and B. DeWitt, J. Math. Phys. **3,** 608 (1962); C. N. Yang, Phys. Rev. Lett. **33,** 445 (1974); S. Deser, P. van Nieuwenhuizen, and H. S. Tsao, Phys. Rev. **D10,** 3337 (1974); and the study of renormalization in K. S. Stelle, Phys. Rev. **D16,** 953 (1977); E. E. Fairchild, preprint.

3. In the *Albert Einstein Centennial Celebration,* edited by Harry Woolf (Addison-Wesley Advanced Book Program, Reading, Mass., 1980).

4. V. P. Akulov, D. V. Volkov, and V. A. Soroka, JETP Lett. **22,** 396 (1975); B. Zumino, *Proceedings of the Conference on Gauge Theories and Modern Field Theory,* edited by R. Arnowitt and P. Nath (MIT Press, Cambridge, Mass., 1976), p. 255; J. Wess and B. Zumino, Phys. Lett. **66B,** 361 (1977); J. Thierry-Mieg and Y. Ne'eman, Ann.Phys. **112,** 555 (1979); Y. Ne'eman, *Proceedings of the 19th International Conference on High Energy Physics,* Tokyo, 1978, edited by S. Homma, M. Kawaguchi, and H. Miyazawa (Physical Society of Japan, 1979), p. 552.

5. K. Johnson, Phys. Lett. **78B,** 259 (1978).

6. C. J. Isham, A. Salam, and J. Strathdee, Phys. Rev. **D3,** 867 (1971).

7. A. Einstein, Sitz. Preuss. Akad. Wiss. (Berlin), Phys.-Math. Kl., pp. 217, 224 (1928); R. Weitzenbock, Sitz. Preuss. Akad. Wiss. (Berlin), Phys.-Math. Kl., p. 466 (1928); R. Zaycoff, Z. f. Phys. **53,** 719 (1929), **54,** 590, 738 (1929), and **56,** 717 (1929); W. Scherrer, Z. f. Phys. **141,** 374 (1955); K. Hayashi and T. Shirafuji, Proceedings of the GR8 Conference, Waterloo, Canada, 1977, Abstracts Vol., p. 180; C. Moller, Kgl. Dan. Vid. Selok. Mat. Fys. Medd. **39** (13) (1978).

8. Y. Ne'eman, Proc. NAS **74,** 4157 (1977), and Ann. Inst. Henri Poincaré **A28,** 369 (1978); Y. Ne'eman and D. Sijacki, Proc. NAS **76,** 561 (1979), and Ann. Phys. (N.Y.) **120,** 292 (1979), and Erratum (1980).

9. Y. Ne'eman and T. Regge, Riv. Nuovo Cimento, **1** (5) (1978); J. Thierry-Mieg and Y. Ne'eman, op. cit. in n. 4. See also Ne'eman in op. cit. in n. 3.

10. P. von der Heyde, Phys. Lett. **58A,** 141 (1976); F. W. Hehl et al., Phys. Lett. **78B,** 102 (1978).

11. See references cited in n. 2.

12. See references cited in n. 10.

13. Ibid.

14. Schweizer and Straumann, preprint; J. Nitsch and F. W. Hehl, preprint.

15. H. Rumpf, Z. Naturforsch. **33a,** 1224 (1978).

16. F. W. Hehl, J. Nitsch, and P. von der Heyde, GRG Einstein Centenary Volume, Plenum, New York, to be published.

17. F. W. Hehl, G. D. Kerlick, and P. von der Heyde, Phys. Lett. **63B,** 446 (1976); F. W. Hehl, E. A. Lord, and Y. Ne'eman, Phys. Lett. **71B,** 432 (1977); E. A. Lord, Phys. Lett. **65A,** 1 (1978); F. W. Hehl, E. A. Lord, and Y. Ne'eman, Phys. Rev. **D17,** 428 (1978).

18. See Ne'eman and Sijacki, op. cit. in n. 8.

19. Ne'eman, op. cit. in n. 8.

20. Ne'eman and Sijacki, op. cit. in n. 8.

21. Y. Ne'eman and T. N. Sherry, Phys. Lett. **76B,** 413 (1978); Y. Ne'eman and D. Sijacki, J. Math. Phys. (forthcoming); D. Sijacki, forthcoming.

22. Ibid.

Discussion

Following the Paper by Y. NE'EMAN

I. Segal (MIT): The deviations in the cosmic background radiation as described by Ward and Richards are consistent with a substantial angular momentum. Will that arise in a natural way for many of the variations that you have indicated?

Ne'eman: I do not know, but it seems implausible. It is possible that the small, short-range effects that I was describing might have nothing to do with strong interactions, but would still be there physically and involve very short-range effects. However, I do not see how this would affect the background radiation, which arises when the universe is too large for these ranges.

S. Lindenbaum (Brookhaven and City College of New York): I wonder: Can you get out of this the coupling constants, or the effective strength of the various interactions as a function of energy? Suppose your strong gravity was strong indeed and you had something electromagnetic and perhaps eventually weak. Can you foresee showing how these things behave as a function of energy?

Ne'eman: I think in supergravity that ought to be the case if we succeed. In the $GL(4,R)$ case there is a source term. Electromagnetism and weak interactions would still come in on the right-hand side of the equations. Now, on the other hand, most probably there could be some constraints relating the two coupling constants of gravity and strong interaction. For example, it is clear that the expansion I gave for the potentials works only between some lower and upper range. Hopefully, we may get some numbers and know which range it is. But to constrain the weak-electromagnetic couplings requires a theory that unifies all forces. At some stage we hoped supergravity might do it. It might still provide the answer especially if one can go beyond 0(8) and have a large enough internal group, as Dr. Freedman was saying. In that case there definitely could be relations between all the couplings, and ideally, they should all be Clebsch–Gordan coefficients. However, the gravitational coupling is dimensional and can thus not be a Clebsch–Gordan coefficient. This may correspond to some spontaneous breakdown effects that would introduce the dimensions of the Planck constant.

H. J. Lipkin (Weizmann Institute): You spoke of this linear potential as perhaps being confining. It is not clear how color fits into this, because you want a force that will confine quarks within a color singlet hadron but leave no residual force between two hadrons. The normal gravitational force does not saturate and just acts between all particles. Is this considered in this framework?

Ne'eman: I have no idea how color comes in. I forgot to mention that as an open problem. There was at some stage a very faint hope it would come in naturally. Working with $\overline{GA}(4,R)$ the four-dimensional affine group, the "little" (or stability) group if we pick a momentum in the fourth direction is $\overline{SL}(3,R) \times T_3 = \overline{SA}(3,R)'$. The T_3 does not exist in the Poincaré group, where only $\overline{0}(3,R)$ is left due to the antisymmetricity of the Lorentz group matrices. In $SL(4,R)$ we have both symmetric and antisymmetric matrices and thus have three additional generators in the fourth row, which commute with time translations. Now, to use $\overline{SL}(3,R)$, as the interesting part of the little group, let the T_3 generators' action on physical states vanish. It turns out that this is the contion known as "the gauge condition" in strings, for $\overline{SL}(2,R) \times T_2$. Coming back to color, I entertained faint hopes that because this is $\overline{SL}(3,R)$ which has a "3" in it, one might, via some triality feature, somehow connect with the triality of color, and have color act naturally. I cannot say that it cannot be done, but I have not been able to do it. The finite representations of $SL(3,R)$ are the same as those of $SU(3)$, but if you take the infinite ones, needed for $\overline{SL}(3,R)$, they look as if triality has been "averaged" there. It is a very funny situation in which there is some kind of "squeezing" of triality. It goes like this: if you start a representation with $J = 0$ plus $J = 2$, then, in $SU(3)$ that is the six-dimensional representation (2.0) which has triality $\tau = -1$. Then, if you add the next stage $J = 4$ you get the fifteen-dimensional (4.0), which has $\tau = 1$. Adding $J = 6$ we get (6.0) with $\tau = 0$. We keep varying the triality. The $\overline{SL}(3,R)$ ladder representation with lowest spin $J = 0$ decomposes into $0 \oplus 2 \oplus 4 \oplus 6 \oplus \ldots$ so that triality keeps changing along the representation, if you cut it somewhere and go over to $SU(3)$. What then is its own triality? Notice that one could still just add color and work with $SL(12,R)$ instead of $SL(4,R)$, but this looks "expensive."

PART IV: QUANTUM CHROMODYNAMICS

In 1961, $SU(3)$ symmetry (now denoted as "flavor" $SU(3)$) was identified as an approximate symmetry of the strong Hamiltonian, with the baryons grouped in octets.[1] This could be explained by the existence of a "fundamental" triplet of quark fields[2] $(u,d,s,)$ with $m(u) = m(d) = m(s)$ in the symmetry limit. For constituent quarks, $m(u) \sim m(d) \sim 300$ MeV, $m(s) \sim 400$ MeV. For the quark fields (or current quarks) $m(u) \sim 4$ MeV, $m(d) \sim 7$ MeV, $m(s) \sim 140$ MeV, and the symmetry limit corresponds to massless fields for u, d.

$SU(3)$ and the quark model led to $SU(6)$ and a rich set of results[3] such as the ratio of magnetic moments $\mu(p)/\mu(n) = -3/2$. For high-energy amplitudes, the quark model again yielded numerous predictions, such as

$$\frac{\sigma(\pi N)}{\sigma(NN)} \xrightarrow[\nu \to \infty]{} \frac{2}{3} .$$

In 1969, deep inelastic electron nucleon scattering displayed scaling. The high-energy results could also be related to such scaling.[4]

To satisfy the spin-statistics correlation, the quark model had to invoke a further $SU(3)$ symmetry[5] (exact, in QCD; approximate in the Pati–Salam model) of "color." All known baryons are assigned to antisymmetric singlets of that group. The mesons are assigned to a contracted singlet.

What was needed was thus a dynamical theory that could do the following:

1. Explain spin-independence ($SU(6)$) and other features of an "independent particle" quark model;
2. Exhibit scaling and related effects;
3. Exhibit saturation at qqq and $q\bar{q}$;
4. Ordain quark confinement, since no free quark has been observed to date in a verified experiment.

In 1965 Nambu had suggested that $SU(3)_{color}$, if gauged ($=$ QCD), would provide an explanation for characteristic (3) in the preceding list. However, when the Yang–Mills interaction was renormalized by 't Hooft[6] and as a result he and others discovered asymptotic freedom,[7] (1) and (2) were explained as well. Note however that *there exists to date no detailed QCD derivation of (1)*, especially of the transition between current quarks and constituents. However, the consensus among theoreticians assumed that this was implied, and that the only unproved feature was (4). Since 1974, a vast effort has gone into proving quark confinement. The chapters in this part relate to this topic.

Nambu reviews the issue, using a transition to a "string" formalism and a classical source method that relates to the 't Hooft-Polyakov monopole. 't Hooft introduces new resummation methods and a topological analysis meant to extend the reach of perturbation theory. His analysis tends to produce confinement as a "color magnetic superconductor." He conjectures the existence of new structures around 1,000 GeV.

Dashen describes a plasmalike approach developed at Princeton. This suggests that aside from the short-range asymptotically free region and the long-range confinement there exists an intermediate region dominated by instantons, effective color magnetic dipoles. Lipkin clarifies the assumptions made when trying to distinguish between QCD and the Pati-Salam (or Han-Nambu quarks) models. He also derives a variety of new results relating to constituent quarks.

NOTES

1. Y. Ne'eman, Nucl. Phys. **26**, 222 (1961). See also M. Gell-Mann and Y. Ne'eman, *The Eightfold Way* (W. A. Benjamin, New York, 1964).
2. H. Goldberg and Y. Ne'eman, Nuovo Cim. **27**, 1 (1963) and IAEC Report IA-725 (1962); M. Gell-Mann, Phys. Lett. **8**, 14 (1964); G. Zweig, CERN Reports Th 401, 412 (1964).
3. F. Gürsey and L. Radicati, Phys. Rev. Lett. **13**, 175 (1964). See also F. Dyson, *Symmetry Groups* (W. A. Benjamin, New York, 1965), and J. J. J. Kokkedee, *The Quark Model* (W. A. Benjamin, New York, 1967).
4. L. Gomberoff et al., Phys. Lett. **52B**, 74 (1974).
5. O. W. Greenberg, Phys. Rev. Lett. **13**, 598 (1964); M. Han and Y. Nambu, Phys. Rev. **139**, 1006 (1965).
6. G. 't Hooft, Nucl. Phys. **B33**, 173 (1971), and **B35**, 167 (1971).
7. D. Gross and F. Wilczek, Phys. Rev. Lett. **30**, 1343 (1973); H. D. Politzer, Phys. Rev. Lett. **30**, 1346 (1973).

IV. Quantum Chromodynamics

Session Chairman's Remarks

M. GOLDHABER, Brookhaven National Laboratory

Gerald Holton has reminded us that Einstein thought of theories as free inventions of the mind, loosely based on facts. But according to the test of different theories, these inventions go through filters that you might call preconceived notions. And Einstein has sometimes been accused of having preconceived notions. Perhaps the completely open mind cannot bounce an idea around. Modern theorists have gone further than Einstein in his free inventions based on facts and have built inventions freely, loosely connected with previous theories. But, as Einstein has often reminded us, these theories must finally come down to earth and stand comparison with the facts. I am sure that all the theorists who have contributed to this part of this volume are well aware of this.

8. QUARK CONFINEMENT: THE CASES FOR AND AGAINST

Yoichiro Nambu

The title of this paper was suggested to me by Professor Ne'eman, presumably because I have worked in the past on both sides of the fence regarding the observability of quarks. My thinking on this matter has taken a zigzag course, partly influenced by my theoretical and maybe philosophical prejudices, but also following the general progress of particle physics, both experimental and theoretical.

The quark model was introduced in 1964 by Gell-Mann and by Zweig.[1] Originally they were conceived as carriers of internal quantum numbers, now called flavor, that gave order to the spectroscopy of hadrons. But gradually and steadily the quarks have gained more and more reality. Although actual quarks have not been directly and convincingly observed yet, it is generally accepted now that quarks are on an equal level with the leptons to form a set of fundamental material particles.

In the development of the quark model from the initial phase to the present, the following four evolutionary stages may be pointed out:

1. Incorporation of parastatistics[2] or color quantum number;[3]
2. The successes of the quark-parton model;[4]
3. The discovery of new flavors;[5]
4. The successes and promises of the gauge theories of flavor[6] and color dynamics.[7]

But I do not think I have to elaborate on them.

The only curious element in the development concerns the direct observability of quarks. Stable isolated quarks have not been seen so far. Besides, evidence is very strong that quarks tend to be confined within hadrons, unlike all other known bound systems. This is certainly consistent with the unique antiscreening properties of quantum chromodynamics (QCD). These same properties also lead to the asymptotic freedom aspect of QCD in the short-distance regime, and together they give a strong support for QCD as the basis of strong interactions.

A majority of theorists then elevate the unobservability to a matter of principle and try to deduce permanent confinement of quarks from QCD. This is a new situation that we have not

---

Yuval Ne'eman (ed.), To Fulfill A Vision: Jerusalem Einstein Centennial Symposium on Gauge Theories and Unification of Physical Forces
ISBN 0-201-05289-X
Copyright © 1981 by Addison-Wesley Publishing Company, Inc., Advanced Book Program.
All rights reserved. No part of this publication may be reproduced, stored in a retrieval system, or transmitted, in any form or by any means, electronic, mechanical photocopying, recording, or otherwise, without the prior permission of the publisher.

known before. It also raises a philosophical question: In what sense are quarks real when they cannot be observed? Actually I do not think the matter is as serious as it sounds. Confinement may be compared to the two poles of a bar magnet, which always come as a pair and cannot be isolated from each other.

First, the latest news about quark search.

Fairbank's group[8] has recently reported further evidence for fractional charge of $\sim \pm e/3$ on niobium spheres, following his earlier results published a few years ago. Out of a total of nine samples reported, he has seen the fractional charge on three of them. But the charge seems to have a tendency to come and go, indicating that, if the effect is real, there must be large numbers of quarks loosely attached to the spheres.

I have to hasten to add, however, that the data are not conclusive. An important issue like this must be settled one way or the other by several independent experiments. I understand that a recent report by Morpurgo[9] is also of inconclusive nature.

Next, about integrally charged quarks.

This possibility was first raised partly with a view to making quarks "invisible".[10] Such quarks may or may not be stable, and in either case would look like ordinary hadrons with large masses. Evidence is negative for such objects, whether stable or unstable. They have not been seen directly, nor have the inclusive cross-sections in accelerator experiments shown any dramatic rise due to excitation of colored states, which are naturally expected in this kind of model.

A recent example of such experimental tests is muon pair production[11] due to the Drell-Yan $\pi^{\pm} + N$ reactions. The ratio of cross-sections π^+/π^- should equal the ratio of square averages of u quark to d quark. This ratio is 4 for fractional charge and 2 for integral charge. The experimental data for high dimuon mass clearly tend to 4. One could, however, always say that color threshold has not been reached yet.

In the framework of gauge theories with spontaneous breakdown, integral charge is not an unnatural scenario if it arises from a mixing of color and flavor gauge symmetries. This is a possibility most vigorously advocated by Pati and Salam.[12] In this case, quark charges are effectively integral only at low energies, and thus one can escape many of the traps the integer charge model has encountered. Still, there exist colored states; the scheme should be evaluated in all its aspects as a possible model for unification of color and flavor.

From the foregoing information I am inclined to conclude that there is no hard evidence to suggest that isolated quarks exist, although, on the theoretical side, it is possible, or may even be natural, to construct a gauge theory in which both color and flavor are spontaneously broken and colored states are observable.

However, the predictions of the standard quantum chromodynamics in the asymptotic, high-energy regime are showing impressive agreement with experiment, with respect for example to structure functions for inclusive reactions, and to jet phenomena at very high energies.

What we do not know about for sure is the low-energy regime, and in particular, whether QCD can result in absolute quark confinement, and if it does, what is the precise mechanism for it. The possibility of confinement is, of course, very real[13] because of the antiscreening behavior of unbroken non-Abelian gauge fields, and most of the subsequent theories of confinement appeal to this fact at one point or another, if only indirectly.

Two most attractive and plausible theories of confinement within the framework of QCD are the following:

1. Wilson's lattice gauge theory[14] in which confinement is automatic but sacrifices the continuity of space-time in an essential way. The lattice size is of the order of the hadronic size, not of the order of the Planck length.
2. Theories based on topologically nontrivial media (instanton gas,[15] magnetic superconductor[17]).

It is not too difficult to demonstrate the possibility of confinement in terms of a *Gedanken* experiment. For this purpose, drop a magnetic monopole in a superconductor.[17] The magnetic field will form an Abrikosov flux tube extending to infinity, and thus the system will have infinite energy. But a pair of monopoles can be joined by a finite tube. This will simulate the bag model of hadrons for low angular momentum states and the string model for high angular momentum states, both of which are quite successful phenomenological models in their respective domains.

Such a possibility does already exist in the Weinberg–Salam theory of unified weak and electromagnetic interactions. In this theory, the Higgs field, being an $SU(2)$ doublet, cannot form 't Hooft–Polyakov monopoles, but a pair of them joined by a string made of Z° may be formed.[18] In other words, the monopoles exist, but are confined in the Weinberg–Salam theory and have mass scales of the order of TeV. If such systems can somehow be produced, they will simulate many of the properties of ordinary hadrons although the dynamics involves flavor rather than color. The interaction is nevertheless strong because magnetic charge is inversely related to electric charge.

Unfortunately, this analogy may not be directly relevant to chromodynamics because the latter does not invoke a breaking of color, and furthermore, quarks carry electric-type charges rather than magnetic. So, one would have to carry out a duality operation that interchanges electric and magnetic charges. This is a program advocated by Mandelstam and is also inherent in the lattice gauge theory. Since the two charges are inversely related to each other, it is also a natural procedure for handling the strong coupling limits of each other. General arguments clarifying the meaning of electric-magnetic duality have been put forward by 't Hooft.[19] I will confine myself to my own thoughts and proposals on the confinement problem.

One approach is to establish a direct connection between the string model and the natural stringlike construct in QCD, i.e., the nonlocal path-ordered phase factor

$$U(x/\sigma/y) = P \exp[i \int A_\mu dx^\mu] ,$$
$$A_\mu = gA_\mu^a \lambda^a/2 \tag{8.1}$$

taken along a path σ between points x and y.[20] This phase factor plays several important roles: one can form with it gauge-invariant operators like the Wilson loop and the meson operator:

$$M_{\alpha\beta} = \bar{q}_\alpha(x) U q_\beta(y) , \tag{8.2}$$

where q_α is the quark field, α representing the Dirac and flavor indices. It also generates the fields $F_{\mu\nu}$ through infinitesimal deformations of σ:

$$\delta U/\delta\sigma_{\mu\nu}(z) = P(iF_{\mu\nu}(z), U) , \tag{8.3}$$

where $d\sigma_{\mu t} = dz_\mu \wedge dz_t$, dz_t being tangent to σ at the point z on σ.

One can repeat this operation at two distinct points z and z' independently. Then letting $z' \to z$, one arrives at a second-order form,

$$\sum_\mu \delta^2 U/\delta\sigma_{\mu t}\delta\sigma^{\mu t}(z) = -\sum_\mu P(F_{\mu t}F^{\mu t}(z), U) . \tag{8.4}$$

But such a limit is smooth only if the field equation

$$D_\mu F^{\mu t} = j^t = 0 \tag{8.5}$$

is satisfied.

An interesting point here is that Eq. (8.4) bears a resemblance to the set of equations, originally due to Virasoro, for the wave funtion $\psi[\sigma]$ of a quantized string.[21] In the present framework, ψ may be considered to be the matrix element of U, or of the operator (8.2)

$$\psi_{\alpha\beta}[\sigma] = \langle 0|M_{\alpha\beta}|\mu\rangle , \tag{8.6}$$

representing a meson μ. These Virasoro equations can be cast in the form:

$$\sum_\mu \delta^2 \psi/\delta\sigma_{\mu t}\delta\sigma^{\mu t}(z) = (1/2\pi\alpha')\psi , \tag{8.7}$$

where α' is the slope of the Regge trajectory. Thus, one can see that

$$\sum_\mu F_{\mu t}F^{\mu t}(z) = -(1/2\pi\alpha')^2 1 , \quad z \varepsilon \sigma \tag{8.8}$$

is a sufficient condition for the equivalence of the two equations. In addition, ψ must also satisfy the usual Dirac equation with respect to the endpoints:

$$(\gamma \cdot \partial_x + m)\psi = \psi(\gamma \cdot \partial_y - m) = 0 . \tag{8.9}$$

Roughly speaking, Eq. (8.8) means that there is an electric-type color flux of fixed magnitude along the string. Equation (8.5) amounts to a condition that quarks should not lie on the string, because otherwise the string would break. These relations make our intuitive picture more precise, but they leave open the question whether and how conditions like (8.8) follow from dynamics.

I will, therefore, turn to this question.

This time I would like to see how the color fields respond to external sources. Although the Wilson loop is a possible representation of an external source, it is difficult to manipulate. So I want to seek alternatives.

In the Abelian case, a classical point source gives a piece in the action

$$\int A_\mu(x)\frac{dx^\mu}{d\tau} d\tau = \int j_\mu(x)A^\mu(x)d^4x ,$$

$$j_\mu(x) = \int \frac{dy_\mu}{d\tau} \delta^4(y-x)d\tau, \quad \partial_\mu j^\mu = 0 \tag{8.10}$$

that is gauge-invariant.

Let us assume, in non-Abelian cases, that the source is described by a classical trajectory j and a normalized scalar field $\phi^a(x)$, $\phi^a\phi^a = 1$ to reflect its internal quantum number. For simplicity, we work with $SU(2)$, so that ϕ is an isovector $\vec{\phi}$. By an external source I mean that the trajectory j_μ is fixed, but not $\vec{\phi}$. The latter is to be integrated over, together with the gauge field, in the sense of function integration.

The naive interaction $J_\mu^a A^{\mu a} = j_\mu \phi^a A^{\mu a}$, is not gauge-invariant. It turns out, however, that a gauge-invariant action can be formed if one introduces a (semi-infinite) Dirac sheet having j_μ as its boundary:

$$I_{int} = \tfrac{1}{2} \int [2\partial_\mu(A_\nu \cdot \vec{\phi}) - \partial_\mu\vec{\phi} \times \partial_\nu\vec{\phi} \cdot \vec{\phi}]do^{\mu\nu}$$

$$= \tfrac{1}{2} \int (\mathbf{F}_{\mu\nu} - D_\mu\vec{\phi} \times D_\nu\vec{\phi}) \cdot \vec{\phi}\, do^{\mu\nu} \,, \tag{8.11}$$

where $do^{\mu\nu}$ is a 2-form representing an element of the sheet. From this follow three Euler equations:

1. $\delta I/\delta A_\nu = D_\mu F^{\mu\nu} - J^\nu = 0 \,,$ \hfill (8.12)

2. $\delta I/\delta\phi = D_\mu j^\mu = j^\mu D_\mu \phi = 0 \,,$ \hfill (8.13)

3. $\delta I/\delta\Omega_{\mu\nu\lambda} = \sum_{cycl} D_\mu\vec{\phi} \times D_\nu\vec{\phi} \cdot D_\lambda\vec{\phi}$

   $\qquad\qquad = \sum_{cycl} \partial_\mu\vec{\phi} \times \partial_\nu\vec{\phi} \cdot \partial_\lambda\vec{\phi}$

   $\qquad\qquad = \varepsilon_{\mu\nu\lambda\rho} k^\rho \,,$

   $\partial_\rho k^\rho = 0, \quad \int k_\rho d^3s^\rho = \pm 4\pi \,.$ \hfill (8.14)

The third equation refers to the variation of a sheet element:

$$\delta\Omega_{\mu\nu\lambda} = do_{\mu\nu} \wedge \delta x_\lambda.$$

Interestingly, the integrand of Eq. (8.11) is precisely the gauge-invariant combination given by 't Hooft[22] which is associated with a monopole in the presence of a "Higgs field" ϕ.

Equation (8.14) characterizes the monopole current k_ρ as the world line of a topological singularity of ϕ, if there is one. Whenever $\delta\Omega$ encloses a monopole, I jumps by $\pm 4\pi$, so the Dirac sheet remains unphysical so long as the monopole does not lie on the sheet. This is the so-called Dirac's veto condition for monopoles.

One might argue that what we have here is nothing but an $SU(2)$ theory broken down to $U(1)$ by a Higgs mechanism. This is not true, however, because the latter implies $D_\mu\phi \sim 0$ away from monopoles, which we do not assume.

A slightly different formulation of the external source problem is as follows. This formulation is also more suitable for extension to $SU(n)$. Let u be a normalized auxiliary field in the fundamental representation of $SU(n)$, so that

$$u^+u = 1 ,$$

$$u^+\lambda^a u \equiv \phi^a, \phi^a\phi^a = 1 . \tag{8.15}$$

Then

$$\frac{i}{2}u^+D_\mu u = \tfrac{1}{2}A_\mu^a\phi^a + A_\mu^0 ,$$

$$A_\mu^0 = \frac{i}{2}u^+\partial_\mu u, \quad D_\mu = \partial_\mu - \frac{i}{2}A_\mu^a\lambda^a \equiv \partial_\mu - i\mathcal{A}_\mu \tag{8.16}$$

is gauge-invariant, so one can construct the coupling to a source:

$$I' = \frac{i}{2}\int u^+D_\mu u j^\mu d^4x ,$$

$$I'' = \int [\partial_\mu(u^+\mathcal{A}_\nu u) + i\partial_\mu u^+ \partial_\nu u]d\sigma^{\mu\nu} \tag{8.17}$$

$$= \tfrac{1}{2}\int (u^+\mathcal{F}_{\mu\nu}u + 2iD_\mu u^+ D_\nu u)d\sigma^{\mu\nu} . \tag{8.18}$$

These two forms differ by

$$\tfrac{1}{2}\int (u^+[\partial_\mu, \partial_\nu]u + [\partial_\mu, \partial_\nu]u^+ u)d\sigma^{\mu\nu} \neq 0 . \tag{8.19}$$

The point is that when ϕ has a monopole singularity along a world line k, Eq. (8.14), the corresponding u must have a singular sheet bordered by the former, and on the sheet Eq. (8.19) is not zero. Now one can show that

$$I' = I'' \bmod (2\pi), \text{ and}$$

$$I = 2 I'' \text{ (valid for } SU(2)) . \tag{8.20}$$

The factor 2 is due to a difference in the normalization of the auxiliary field.

It has thus turned out that there are two alternative forms. A Dirac sheet is not necessary if I' is adopted. The difference between the two actions I' and I'' is as follows: In the case of I'', the electric source j is accompanied by a Dirac sheet, and the latter feels the monopoles represented by singularities of ϕ. On the other hand, if I' is adopted, the source j does not need a sheet, but this change is due to the fact that a term is added that depends on the Dirac sheets accompanying the monopoles. Thus one sees an interesting duality.[22] Because of the magnetic charge quantization, however, the sheet appears to be unobservable in any event.

What relevance does this have to the question of confinement? Here my point is that if the action has an inherent dependence on a sheet, the possibility of confinement would be more easily realizable. With this in mind, let us reexamine the case of action I'' carefully. Although the quantum action factor $\exp[i I''_{\text{int}}]$ is sheet-independent, the function space gets constrained if the Dirac veto condition is to be observed. But, such a constraint has measure zero, and thus would not produce a finite effect. This last point, however, is not entirely obvious; it could be wrong if the entropy of the function space plays a nontrivial role. Setting aside this reservation, I would like to propose the following scheme that allows the quantization condition to be violated.

Instead of a single field u, take several of them such that only their sum is normalized,

$$\Sigma u_n^+ u_n = 1 \ . \tag{8.21}$$

Let us adopt the action I'', which now becomes a sum $\Sigma I''_n$. (Note that the relation (8.20) is valid whether or not u is normalized everywhere.) As each I''_n will, in general, have its own sheet, and the singularities of different u_n's occur independently, the change of I'' as a sheet crosses one of them will be

$$\Delta I'' = \pm 2\pi u_n^+ u_n \text{ for some } n \ . \tag{8.22}$$

Since this is not properly quantized, the quantum action now explicitly depends on the sheet configuration.

How can one justify the preceding *Ansatz*? One might say that the source current $\Sigma u_n^+ D_\mu u_n$ stands for the sum $\Sigma q_\alpha^+ D_\mu q_\alpha$ over the quark spinor indices, or the sum $\Sigma u_f^+ D_\mu u_f$ over different flavors. Each u_n then picks up a different one-parameter subgroup of $SU(n)$ along the world line of the source, supplemented by a suitable sheet.

My last task will be to argue that the proposed form of action can actually lead to confinement.

Let us thus borrow from the usual Euclidean field theory and consider the partition function

$$Z = \int \exp[-\frac{1}{g^2} I_0 + i I_{\text{int}}] \mathscr{D}(A) \mathscr{D}(\phi \text{ or } u)$$
$$\approx \exp[-\frac{1}{g^2} F] \ . \tag{8.23}$$

There still is an i in front of I_{int}. Assuming a thermodynamic limit, F is given by the minimum

$$F = (I - g^2 S)_{\min} \ , \tag{8.24}$$

where S is the entropy of the function space.[23] Suppose that in the absence of I_{int}, the vacuum is populated with topological singularities of the fields, and in particular, the sheet singularities of u_n, or line singularities of the corresponding ϕ_n. As these singularities are randomly distributed, there should be no dependence on the sheet to the first order when I_{int} is turned on; that is

$$\langle I_{\text{int}} \rangle_0 = 0 \ . \tag{8.25}$$

The presence of I_{int}, however, will polarize the medium, and one expects a nonzero effect in the second order. This will increase, rather than decrease, F because u_{int} is imaginary, and its magnitude will be proportional to the surface area of the sheet if the polarization is a short-range effect. Alternatively, one might have argued that the Dirac veto condition excludes the monopole singularities of ϕ from the neighborhood of the sheet, so that entropy decreases accordingly, or the free action F increases.

The foregoing line of reasoning can be demonstrated in explicit mathematical form. Make the plausible *Ansatz* that the entropy density s of the auxiliary field is effectively represented by

$$s = aV_\mu V^\mu + b\Omega_{\mu\nu}\Omega^{\mu\nu} , \qquad (8.26)$$

where $\Omega_{\mu\nu}$ is a smoothed out version of $D_\mu u^+ D_\nu u - D_\nu u^+ D_\mu u$, and $V_\mu = \frac{1}{2}\, \varepsilon_{\mu\nu\lambda\rho}\partial^\nu\Omega^{\lambda\rho}$. In other words, I am suggesting that the entropy of the function space may be effectively represented as an additional piece in the Lagrangian involving local fields. I have made a particular choice here for this additional piece, which happens to be a modified form of the Kalb-Ramond theory of relativistic hydrodynamics.[24] Such a model is known to lead to the string model when $\Omega_{\mu\nu}$ is coupled to a Dirac sheet.

In conclusion I would like to say that within the framework of gauge theory, quark confinement is still an open question. My last exercise was an attempt to convince myself that test quarks can be confined if monopoles dominate the vacuum. The resulting model is a simple realization of Mandelstam's ideas. But if this is not the case, confinement might be only partial, and the Princeton model[25] might be quite relevant, where the dielectric constant can be very small, but not necessarily zero.

I will add two additional remarks which are by-products of the ideas that have been developed in this paper:

1. The medium populated with nonquantized monopoles has a simple Abelian analog, which I will call Aharonov-Bohm medium.[26] It is a medium in which magnetic flux lines having improper magnitudes are distributed either randomly or in a regular fashion. Such a medium is of interest in its own right and may even have some practical implications.
2. In Eq. (8.14) the world lines of monopoles are characterized as the lines of singularity in the field $\vec{\phi}$. The variation of the isospin direction from point to point is largely irrelevant, except for its global properties characterized by these singularities. This offers an alternative to the usual way of describing mass points. For example, the electric-type external current j_μ defined by Eq. (8.10) could also be thought of as a line of singularity of another auxiliary field $\vec{\chi}$. In this way the symmetry, or duality, between electric and magnetic current would become more explicit.

The foregoing observation gives rise to further interesting questions. Einstein, in his attempt at a unified description of gravity and matter, initiated the idea that the mass points are singularities of the metric field.[27] The 't Hooft-Polyakov description of monopoles has much in common with the Einstein description, with the additional property that the internal quantum numbers are attributed to topology. Is it possible, in view of the duality just mentioned, that the conventional Noether current is equivalent to a topological current in a dual description? This question is related to the following more basic one. In quantum theory, the mass points are replaced by quantized fields, and nonclassical statistics is instituted. This replacement is, of course, of an entirely different nature from the one under discussion. What, then, would transition to quantum theory mean in the corresponding dual picture?

Acknowledgment

This work was supported in part by the NSF: Contract No. PHY78-01224.

NOTES

1. M. Gell-Mann, Phys. Lett. **8**, 214 (1964); G. Zweig, CERN Preprints TH 401, 402 (1964) (unpublished).
2. O. W. Greenberg, Phys. Rev. Lett. **13**, 598 (1964).
3. M.-Y. Han and Y. Nambu, Phys. Rev. **139B**, 1006 (1965); A. Tavkhelidze, in *Seminar on High Energy Physics and Elementary Particles* (International Atomic Energy Agency, Vienna, 1965), p. 763; P. G. O. Freund, Phys. Lett. **15**, 352 (1965); Y. Miyamoto, Prog. Theor. Phys. Suppl., extra number **187** (1965).
4. R. P. Feynman, Phys. Rev. Lett. **23**, 1415 (1969).
5. J. J. Aubert et al., Phys. Rev. Lett. **33**, 1404 (1974); J. J. Augustin et al., Phys. Rev. Lett. **33**, 1406 (1974); S. W. Herb et al., Phys. Rev. Lett. **39**, 252 (1977).
6. S. Weinberg, Phys. Rev. Lett. **29**, 1264 (1967); A. Salam, in *Elementary Particle Physics*, edited by N. Swartholm (Almquist and Wiksells, Stockholm, 1968), p. 367.
7. D. Gross and F. Wilczek, Phys. Rev. Lett. **30**, 1343 (173); H. D. Politzer, Phys. Rev. Lett. **30**, 1346 (1973).
8. G. S. La Rue et al., Phys. Rev. Lett. **42**, 142 (1979).
9. G. Morpurgo, report at the 4th Europe Phys. Soc. General Conference, forthcoming.
10. See references cited in n. 3.
11. C. E. Hogan et al., Phys. Rev. Lett. **42**, 948 (1979).
12. J. C. Pati and A. Salam, Phys. Rev. **D8**, 1240 (1973), and **D10**, 275 (1974).
13. S. Weinberg, Phys. Rev. Lett. **31**, 494 (1973).
14. K. Wilson, Phys. Rev. **D10**, 2445 (1974).
15. A. M. Polyakov, Phys. Lett. **59B**, 82 (1975), and Nucl. Phys. **B120**, 429 (1977); C. Callan et al., Phys. Rev. Lett. **66B**, 375 (1977).
16. S. Mandelstam, Phys. Rep. **23C**, 245 (1976); Polyakov, op. cit. in n. 15.
17. Y. Nambu, Phys. Rev. **D10**, 4262 (1974); G. Parisi, Phys. Rev. **D11**, 970 (1975).
18. Y. Nambu, Nucl. Phys. **B130**, 505 (1977).
19. G. 't Hooft, Utrecht preprint, 1979.
20. Y. Nambu, Phys. Lett. **80B**, 372 (1979); J. L. Gervais and A. Neveu, Phys. Lett. **80B**, 255 (1979).
21. See, for example, C. Rebbi, Phys. Rep. **12C**, 1 (1974).
22. G. 't Hooft, Nucl. Phys. **B79**, 276 (1974); J. Arafune et al., J. Math. Phys. **16**, 433 (1975).
23. See, for example, Y. Nambu, Physica **96A**, 89 (1979).
24. M. Kalb and P. Ramond, Phys. Rev. **D9**, 2273 (1974); Y. Nambu, in *Quark Confinement and Field Theory*, edited by D. R. Stump and D. H. Weingarten (John Wiley, New York, 1977), p. 1.
25. See references cited in n. 15.
26. Y. Aharonov and D. Bohm, Phys. Rev. **115**, 485 (1959).
27. A. Einstein, L. Infeld and B. Hoffmann, Ann. Math. **39**, 1, 66 (1938); A. Einstein, "Autobiographical Notes" in *Albert Einstein, Philosopher–Scientist*, edited by P. A. Schilpp (Library of Living Philosophers, Evanston, Ill., 1949); Centennial Edition (Open Court, La Salle, Ill., 1979), p. 85.

Discussion

Following the Paper by Y. NAMBU

Chairman, M. GOLDHABER, Brookhaven National Laboratory

H. J. Lipkin (Weizmann Institute): A minor point, on the question of using $\pi^\pm$ on Carbon to check whether the quarks are of Gell-Mann–Zweig or Han–Nambu type. I think it is the same for the two models, because the process involves here electromagnetic currents acting once between initial and final states, which are both $SU(3)$ colored singlets. So it is only the colored singlet part of the current that counts, and that is the same for the two theories.

Nambu: You are right. I forgot to mention that you can always say that the color threshold is very high, so that we are not exciting the colored state yet. In that case both give us the same answer.

Goldhaber: It would be very good to think of very explicit tests that compare predictions for fractionally charged quarks and integer-charged quarks, and perform these tests.

9. BEYOND PERTURBATION EXPANSION

G. 't Hooft

The first remarkable successes of relativistic quantum field theory were the accurate predictions concerning the interactions of electrons with an electromagnetic field. It turned out to be useful to represent the various terms of the formulas as "Feynman diagrams," of which Fig. 9.1 presents examples.

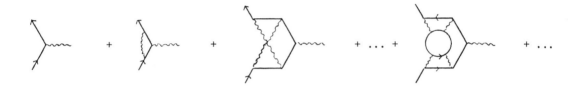

Fig. 9.1 Diagrammatic representation of the electron gyromagnetic ratio.

The diagrams of Fig. 9.1 represent various terms in the gyromagnetic ratio g_e of the electron:

$$\frac{g_e}{2} = 1 + \frac{\alpha}{2\pi} - (0.328\ldots)\left(\frac{\alpha}{\pi}\right)^2 + 1.5\left(\frac{\alpha}{\pi}\right)^3 + \ldots . \tag{9.1}$$

Here $\alpha = e^2/4\pi\hbar c \cong 1/137$, is the fine structure constant. Accidently, α is quite small and this is why the series (9.1) converges rapidly to a number that is now known to some nine decimal places. The more complicated diagrams give high powers of α/π and can therefore be ignored. Strictly speaking, the infinite series (9.1) does diverge, but only after 137 or so terms. In practice this formal difficulty is of no importance.

We have the same situation in the presently fashionable models for weak and electromagnetic interactions. They usually have two important expansion parameters (a third, the Higgs coupling constant, is still unknown but usually also assumed to be small):

Yuval Ne'eman (ed.), To Fulfill A Vision: Jerusalem Einstein Centennial Symposium on Gauge Theories and Unification of Physical Forces
ISBN 0-201-05289-X

Copyright © 1981 by Addison-Wesley Publishing Company, Inc., Advanced Book Program.
All rights reserved. No part of this publication may be reproduced, stored in a retrieval system, or transmitted, in any form or by any means, electronic, mechanical photocopying, recording, or otherwise, without the prior permission of the publisher.

$$g = e/2 \sin \theta_w ,$$

and (9.2)

$$g' = e/2 \cos \theta_w ,$$

where θ_w is the so-called Weinberg angle and is such that both g and g' are still quite small. Therefore these theories too have the welcome property that complicated diagrams may be ignored.

In contrast with these we have the presently widely favored theory for the strong interactions. This theory, dubbed "quantum chromodynamics," describes the forces that keep "quarks" together into tightly bound systems — the protons, neutrons, pions, and so forth. In any attempt to perform perturbation expansions here the strong coupling constant appears, which is not small. Thus, in the series that replace (9.1) expansion parameters occur that are not small. One then has to deal with the problem that all these expansion series have a strictly vanishing radius of convergence.

It is claimed that these difficulties in doing computations with the theory are of a purely technical, mathematical origin. The physical equations are well defined. What are these equations? They are strikingly simple — and beautiful! They can all be obtained from one single expression for the Lagrangian density[1]:

$$\mathcal{L} = -\frac{1}{4g^2} G_{\mu\nu} G_{\mu\nu} - \sum_i \bar{\psi}_i (\gamma D + m_i) \psi_i . \tag{9.3}$$

Here $G_{\mu\nu}$ are a non-Abelian matrix generalization of the Maxwell fields $F_{\mu\nu}$ and ψ_i the quark fields; m_i are the bare quark masses.

As for its unique structure this system compares very well with quantum electrodynamics, the set of Maxwell equations, and indeed also with general relativity: the pure gauge fields, represented by the first part of the Lagrangian, satisfy already nonlinear equations, just like the Einstein equations for the metric in the vacuum. Other "matter fields" (the quarks) act as sources for the gauge fields. The theory differs from the general theory of relativity in the way it has been obtained: not one stroke of genius but a painfully long struggle was needed. It has not one superb inventor, but a long list of authors, each of whom gave some small but important contribution that led to its discovery.[2] Just imagine a distant civilization on some planet in another galaxy where special and general relativity were only obtained after much struggle. One brilliant investigator there might invent quantum chromodynamics in one piece and would then become as famous as our Albert Einstein, and his theory as legendary as his.

Apart from the quark masses the only free parameter is the mass scale. This is because the gauge coupling constant g must be chosen differently at different energy scales μ:

$$g = g(\mu) , \tag{9.4}$$

where the function $g(\mu)$ satisfies a known differential equation.[3] Essentially it reads as follows:

$$\frac{\mu d}{d\mu} g^2(\mu) = \beta_1 g^4(\mu) , \qquad (9.5)$$

(with small, unimportant higher-order modifications)

$$\beta_1 = \frac{1}{8\pi^2} (\tfrac{2}{3} N_f - 11) . \qquad (9.6)$$

Here N_f is the number of quark flavors relevant at the energies of order μ (it runs from 2 to 5, probably more).

β_1 is negative, so if we plot $g(\mu)$ versus μ we get a logarithmically decreasing function. This implies that at extremely high energies we have a relatively small expansion parameter at our disposal. But if we wish to explore the bound state spectrum then we must realize that the bound state masses are determined by those values of μ where $g(\mu) = O(1)$, so perturbation theory cannot be reliable.

Can we improve perturbation theory? One very remarkable attempt was initiated by Lipatov[4] and is called "Borel resummation." Let us consider any Green function in a field theory, expanded with respect to a coupling constant g. Often only even powers occur:

$$\Gamma(g^2) = a_0 + a_1 g^2 + a_2 g^4 + \ldots . \qquad (9.7)$$

Let us consider another function:

$$F(z) = a_0 + a_1 z/1! + a_2 z^2/2! + \ldots . \qquad (9.8)$$

There are reasons to believe that $F(z)$ has a finite radius of convergence, so that it may perhaps be well defined as an analytic function in some domain of complex values for z. Formally we have

$$\Gamma(g^2) = \int_0^\infty F(z) e^{-z/g^2} dz/g^2 , \qquad (9.9)$$

although the function

$$\Gamma'(g^2) = \int_0^\Lambda F(z) e^{-z/g^2} dz/g^2 , \qquad (9.10)$$

has the same asymptotic expansion as Γ. We will have to *define* $F(z)$ as the function that generates $\Gamma(g^2)$ through Eq. (9.9). As such it might have discontinuities on the positive real z-axis.

Is there a way to determine $F(z)$ completely for all positive real z? Perturbation expansion determines F for all z with $|z| < z_0$, where z_0 can be computed by the methods of Lipatov and others.[5] Unfortunately it turns out that there are two kinds of difficulties both of which lead to singularities (branch points) precisely on the positive real axis (Fig. 9.2).

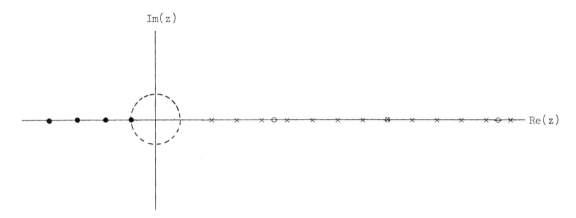

Fig. 9.2 The complex z-plane. The case of quantum chromodynamics with $N_f = 3$. Dashed line: radius of convergence of expansion for $F(z)$; o = branch points due to instantons; × = branch points due to infrared effects; • = branch points due to renormalization effects.

The first difficulty is due to the instantons[6] (field configurations with finite action that solve the classical field equations in Euclidean space). They give rise to branch points or discontinuities on the positive real axis. Tunneling phenomena[7] occur with typical amplitudes going like $e^{-8\pi^2/g^2}$, so the branch points are at $z = 8\pi^2 n$, with n integer. However, since those tunneling phenomena due to instantons are computable,[8] the correct behavior of $F(z)$ near these branch points should be computable as well, such that the contributions of these branch points to the integral (9.9) can be uniquely defined. So, although they are tough, I think that the problems due to instantons can be handled.[9] The nicest feature of the treatment of instantons via the Borel resummation formulation is that instanton effects are unambiguously separated from perturbative effects: the instantons determine the nature of the branch point singularities of F away from the origin, and perturbation expansion determines the behavior of F close to the origin. We need all this information to determine F completely.

Much less transparent are the difficulties due to infrared divergences in the theory. It can be proven that for quantum chromodynamics in an infinite volume $F(z)$ has other branch points besides the ones produced by the instantons.[10] They are indicated by crosses in Fig. 9.2. A physical explanation of these has been given by Parisi.[11] Composite operators such as $O_1 = G_{\mu\nu} G_{\mu\nu}$, $O_2 = G_{\mu\nu} D^2 G_{\mu\nu}$, and so on, have canonical dimension of (mass)^4, (mass)^6, and so on. Within perturbation expansion they cannot have a nonvanishing vacuum expectation value, because nothing can be constructed out of the dimensionless coupling constant with such dimensions. (We may here ignore the quark masses.) But in view of Eq. (9.5) the quantity $n/e^{\beta_1 g^2}$ (where $\beta_1 < 0$), has a dimension of (mass)^n. Therefore the (unknown) vacuum expectation values of $O_{1,2,\ldots}$ give new features (branch points) at $z = 4/-\beta_1, 6/-\beta_1, \ldots$.

In fact, the vacuum expectation values of O_i obey infinite series of self-consistency equations that cannot be obtained from perturbation theory alone. This infinite series of unknown numbers

$<O_1>$, $<O_2>$, ..., obstructed until now all attempts to compute $F(z)$ reliably. Since vacuum expectation values have everything to do with phase transitions, our problem here is, I think, related to the quark-confinement problem.

For completeness I mention a third class of branch points due to ultraviolet renormalization effects, at $z = 2n/\beta_1$, $n > 0$. For $\beta_1 < 0$ they occur at the negative real axis and therefore they are of little importance when we wish to integrate from $z = 0$ to $z = \infty$; see Eq. (9.9).

Difficulties with the Borel resummation techniques led me to consider one further intermediate step: gauge theory in a finite box, with by preference periodic boundary conditions. It is here in particular that peculiar topological features become apparent, besides those topological properties that are connected to the ordinary instanton.[12] The group of color-gauge rotations that can be performed on the 3 × 3 matrices of the color-gauge fields resembles in some aspect the ordinary rotations in three-dimensional real space. The feature I want to bring to your attention can be demonstrated with a glass of orange juice.

Is it possible to hold the glass in one hand and rotate it over 720° with respect to oneself, without spilling any of its contents? The answer is yes, but it is remarkable that after 360° one has a twist in one's arm, and after another 360°, in the same direction, the twist disappears. Clearly, by watching the twist in the arm one can at any time tell whether the glass rotated by an odd or an even multiple of 360°.

In color space of quantum chromodynamics one can rotate similarly, but the twists disappear only after rotations by 1,080° (3 × 360°). The physical way in which the twists manifest themselves here is the possibility to construct locally stable magnetic vortex lines. However, three of these magnetic vortex units can combine to give something without any twist. This implies that it is possible to construct in the Hilbert space of color-gluons objects with three basic units of pure magnetic charge. In gauge theories with spontaneous symmetry breakdown such magnetic monopoles can be constructed explicitly. They have a computable mass and magnetic charge. In quantum chromodynamics one may assume that these magnetic charges move collectively having a formally negative mass-squared (magnetic superconductor). To be precise, there are three possible phases that a color-gauge theory could be in: electric superconductivity ("Higgs phase," only short-range strong forces), magnetic superconductivity ("confinement phase"), and a phase with neither electric nor magnetic superconductivity ("Coulomb," or "Georgi-Glashow" phase, with long-range nonconfining forces and massless gluons).

Quantum chromodynamics, as I believe it describes the forces between quarks, cannot be in the "Coulomb" phase, because strongly interacting, massless, photonlike gluons are not seen experimentally. Neither can it be far out in the Higgs phase; then strong forces between quarks would be too short of range and too many free quarks would be produced. Neither can it be close to the dividing line between two phases, because then long-range correlations would imply light physical gluonic quanta that also are not seen. Therefore, experimental evidence suggests strongly that quantum chromodynamics is in the magnetically superconducting phase.

In the magnetic superconductor a phenomenon analogous to the Meissner effect takes place. Electric fields can only penetrate the system by forming narrow quantized vortex tubes. These vortex tubes act as strings. Every quark must have an endpoint of such a string attached to it, because it carries one unit of electric flux. The strings produce a permanent binding force between the quarks. The binding must be absolute.[13]

As yet such topological arguments have not provided very accurate calculational procedures that can substitute for the usual perturbative approaches. But clearly they have given us insight in the qualitative behavior of the system. Quantum chromodynamics tends to become a respectable field theory. It gives an answer to the question whether quarks exist: they definitely exist but cannot be isolated. If ever fractionally charged particles will be found experimentally then those should not be identified with the quarks that make up hadrons; they are something else such as fractionally charged leptons or integer-charged quarks, to be incorporated in a future grand unification scheme.

The knowledge obtained with this theory may have a wider range of applicability than just the ordinary strong interactions. There are reasons to believe that nature might repeat itself. Particles that at present are considered elementary (such as the Higgs particle and perhaps also the quarks and leptons) may turn out to be composite again. Binding forces much like quantum chromodynamics can keep those new constituents ("quinks") together. The unit of mass determining the new Regge slopes of these bound states is presumably a few thousands of GeV.

Would we be violating a peace treaty if a new quantum chromodynamical settlement were to be built in the region of 1,000 GeV, usually considered to be a desert? The familiar "grand unified theories" postulate the absence of any structure between 100 GeV and 10^{16} or so GeV. This requires an "unnatural" cancellation of the renormalization corrections to the Higgs mass parameter. If we wish to avoid this unnaturalness, then new color-gauge theories in the TeV region are unavoidable. However, models of this kind are not easy to construct. At superhigh energies our particles seem to be governed by paradoxes. Presumably many more Einsteins are needed in the future to disentangle the undoubtedly interesting physical laws that lie ahead of us.

NOTES

1. C. N. Yang and R. Mills, Phys. Rev. **96**, 191 (1954); R. Shaw, Ph.D. thesis, Cambridge University; M. Y. Han and Y. Nambu, Phys. Rev. **139**, B1006 (1965); H. J. Lipkin, in *Nuclear Physics,* Les Houches (Gordon and Breach, New York, 1968), pp. 634-38; M. Gell-Mann, H. Fritzsch, and H. Leutwyler, Phys. Lett. **47B**, 365 (1973); S. Weinberg, Phys. Rev. Lett. **31**, 494 (1973); J. Kogut and L. Susskind, Phys. Rev. **D9**, 3501 (1974); K. G. Wilson, Phys. Rev. **D10**, 2445 (1974).
2. See the references in these notes for example.
3. H. D. Politzer, Phys. Rev. Lett. **30**, 1346 (1973); D. J. Gross and F. Wilczek, Phys. Rev. Lett. **30**, 1343 (1973).
4. L. N. Lipatov, Pisma JETP **24**, 179 (1976), and Leningrad preprints; E. Brezin, J. C. LeGuillou, and J. Zinn-Justin, Phys. Rev. **D15**, 1544, 1558 (1977); G. Parisi, Phys. Lett. **66B**, 167 (1977).
5. Ibid.
6. A. A. Belavin et al., Phys. Lett. **59B**, 85 (1975).
7. G. 't Hooft, Phys. Rev. Lett. **37**, 8 (1976), and Phys. Rev. **D14**, 3432 (1976).
8. Ibid.
9. W. Y. Crutchfield, State University of New York, Stony Brook, preprint ITP-SB-78-59.
10. G. 't Hooft, Erice lecture notes 1977.
11. G. Parisi, Nucl. Phys. **B150**, 163 (1979).
12. G. 't Hooft, Nucl. Phys. **B153**, 141 (1979).
13. G. 't Hooft, Nucl. Phys. **B138**, 1 (1978).

Discussion

Following the Paper by G. 'T. HOOFT

Chairman, M. GOLDHABER, Brookhaven National Laboratory

S. L. Glashow (Harvard U.): I hate to make such a fiddling comment after such a spectacular lecture, but I would suggest that at the head of your list of the many fathers of quantum chromodynamics you might possibly include Professors Yang and Mills, who invented the idea of gauge theories in the first place.

Goldhaber: It is nice to see so many parts of the composite Einstein here. Are there any other questions?

J. Segal (MIT): Since for very simple cases, like the anharmonic oscillator, one has an essential singularity when the coupling constant is zero, would you not anticipate that that would be the case here, and would not that cause some problems?

't Hooft: It is certainly true. In fact, it is generally believed that there is an essential singularity in the origin of the coupling constant plane, and that is one of the ways to express mathematically the fact that the perturbation expansion in terms of the coupling constant does not converge. It is a fundamentally bad expansion, although mathematically it is defined to be an asymptotic expansion only. This is the reason that the procedure of Borel summation was introduced. In particular in the case of the anharmonic oscillator which you mention, the Borel function $f(z)$ has a finite radius of convergence. And indeed, the anharmonic oscillator is one of the fine examples where you can apply this Borel procedure. So, it is just an example, precisely, where such procedures would have been totally successful. In the anharmonic oscillator, you would have phenomena similar to instanton effects but not the complications that I mentioned briefly and called "renormalons."

H. Harari (Weizmann Institute): Concerning your last remark on the quinks: The electron and the neutrino were introduced as elementary objects in the third stage of your description and they survived that stage and continued to be elementary in the fourth stage. Do you expect them to be still elementary in the fifth stage, or are they made out of quinks?

't Hooft: I wish I could answer that question. Of course, I have absolutely no idea, but chances are that they will be composite just as well, because, you know, they take part in the weak interactions. The energy scale I mentioned, 1,000 GeV, would be the energy scale when the Higgs particle becomes strongly interacting. Since the Higgs particle can couple directly to the fermions, I have a reason to expect that fermions would not escape this time at the fifth step. Of course we still have to await future developments to find out.

Goldhaber: Doesn't 1,000 GeV seem a little small? At the Interesecting Storage Rings they have already center of mass energies of almost 1,000 GeV.

't Hooft: I will just mention that 1,000 is $\sim 80 \times \sqrt{137}$. This is how I got the figure. It is certainly only an order of magnitude.

Y. Ne'eman (Tel Aviv U.): Another argument for believing that the electron and the neutrino will not survive the next step: nature appears to work like a Xerox machine, producing electron and its neutrino, muon and its neutrino, tau and its neutrino. That shows compositeness, if anything.

't Hooft: Yes. One of the basic problems, and also one of the motivations behind this idea of quinks is precisely the Xerox copy machine, the suggestion that perhaps there will be an infinity of Xerox copies. That clearly reminds one of Regge trajectories, of resonances, and so on. I thank you for this remark. I also want to mention that gauge theories very likely play an extremely crucial role in all this, no matter how far it goes down.

Ne'eman: The need to keep going further down this ladder (quarks, quinks, and so forth) is referred to Lenin in some countries. He said that "the electron is inexhaustible."

10. TOWARD A THEORY OF HADRON STRUCTURE

Roger Dashen

Over the 1970s it has become increasingly clear that hadrons (the nuclear particles, neutrons, protons, and so on) are composed of more fundamental objects called quarks. In the latter half of the decade we have, in fact, seen the development of a specific theory of how the quarks interact and bind to form the observed hadrons. This theory, called quantum chromodynamics (QCD), is a gauge theory and has many features in common with ordinary electrodynamics.[1] However, it is a non-Abelian gauge theory and in several ways is profoundly different from electrodynamics. Because I do not want to get into too many details, I will emphasize the analogies rather than the differences.

The quarks carry color charges and produce color electrical fields. As in electrodynamics, the vacuum acts like a (Lorentz-invariant) material medium with a dielectric constant ε and a magnetic susceptibility $\mu = 1/\varepsilon$. The unusual thing about QCD is that, due to non-Abelian effects, the dielectric constant of the vacuum is less than 1. This means that when color charges are placed in the vacuum they are antiscreened (that is, they get bigger) and interact more strongly. Antiscreening is the origin of both the great hopes for QCD as a theory of hadrons and the difficulties we face in coping with the mathematics of the theory. The lore is that antiscreening is so strong that quarks can never really escape from one another and are permanently bound into colorless hadrons. The problem is to see, in at least a qualitative way, how it is that this happens.

The approach that my collaborators, C. Callan and D. Gross, and I have taken is to start at small distances where the theory is manageable and try to work our way out to larger scales. At the shortest distances, the dielectric properties of the vacuum are governed entirely by the perturbative zero-point oscillations of the gauge field. These zero-point oscillations antiscreen and, as one increases the scale, the effective coupling g grows. When g reaches a value between 1 and 2, a new effect appears which *rapidly* drives g up to about 10. I will discuss the detailed nature of this effect in a moment, but first I would like to point out what it means. A rapid rise in g to some large but finite value is certainly not by itself a proof of confinement. However, it may well be the bridge between weak-coupling perturbative QCD and the strong-coupling (lattice) version of the theory. If so, confinement will follow directly from the strong-coupling expansion. I would like to suggest therefore that there are three regions in QCD: 1) the weak-coupling regime at short distances, 2) a

---

Yuval Ne'eman (ed.), To Fulfill A Vision: Jerusalem Einstein Centennial Symposium on Gauge Theories and Unification of Physical Forces
ISBN 0-201-05289-X
Copyright © 1981 by Addison-Wesley Publishing Company, Inc., Advanced Book Program.
All rights reserved. No part of this publication may be reproduced, stored in a retrieval system, or transmitted, in any form or by any means, electronic, mechanical photocopying, recording, or otherwise, without the prior permission of the publisher.

region of intermediate coupling where the effects to be described herein occur, and 3) a strong-coupling regime where the color fields are completely random and true confinement occurs.

To explain why the rapid rise in g occurs, I have to backtrack a bit. Three years ago a peculiar tunneling phenomenon was discovered in gauge theories.[2] Perturbation theory turns out to contain a countable set of discrete "vacua," and consequently the perturbative vacuum is highly degenerate. This degeneracy is lifted by nonperturbative tunnelings. The tunneling transitions can be treated by a WKB approximation, and one finds that they are events localized in both space and time. It is helpful to think of these tunneling events as fictitious particles in four space dimensions; 't Hooft coined the name "instanton." Looked at in this way, the instanton or tunneling field is a four-dimensional color magnetic dipole. But dipoles are just the kind of objects that produce a coupling renormalization in the ordinary electrodynamics of continuous media. In particular, magnetic dipoles lead to an $\varepsilon = 1/\mu$ which is less than 1 and consequently the tunneling fluctuations will reinforce the antiscreening effect of zero-point oscillations. The effect turns on rapidly and a detailed calculation yields the rapid rise in g mentioned previously.

The analogy with magnetic dipoles allows one to use all the machinery of classical statistical mechanics. In particular, methods for computing the susceptibility of a paramagnetic gas can be used to compute the instanton-induced coupling renormalization in QCD. The analogy also suggests something else. A dense enough dipole gas undergoes a van der Waals liquid-gas phase transition, and one might guess that the instantons in the QCD vacuum are, in fact, in a liquid phase. It turns out that they are, but that *locally* a strong enough color electric field can cause a local transition to a (dilute) gaseous phase.[3] Thus, inside a hadron where there are strong local color fields, one can have the dilute phase in equilibrium with the liquid phase outside. It turns out that this produces a picture of hadrons that is for practical purposes identical to the MIT bag model.[4] On a qualitative level, the bag model has now been derived from QCD.[5] Further work is required before more quantitative predictions can be made.

Finally, there is a whole class of instanton effects that I have not discussed. They have to do with light quarks and spontaneous chiral symmetry breaking and are reasonably well understood but need to be integrated into the picture given here. It is hoped that this will be accomplished over the next few years.

NOTES

1. Only the most recent references are given. For a review of QCD see, H. Pagels and W. Marciano, Phys. Reports **36C**, 137 (1978).
2. A Belavin et al., Phys. Lett. **59B**, 85 (1975); G. 't Hooft, Phys. Rev. Lett. **37**, 8 (1976); C. Callan, R. Dashen, and D. Gross, Phys. Lett. **63B**, 334 (1976); R. Jackiw and C. Rebbi, Phys. Rev. Lett. **37**, 172 (1976).
3. C. Callan, R. Dashen, and D. Gross, Phys. Rev. **D19**, 1826 (1979).
4. A. Chodos et al., Phys. Rev. **D9**, 3471 (1974).
5. Callan, Dashen, and Gross, op. cit. in n. 3.

Discussion

Following the Paper by R. DASHEN

K. Bleuler (Bonn U.): I think that was a very impressive justification of the famous bag model. May I ask you several questions about the bag? Is it now natural that we have mostly three or two quarks only? Or could you also construct very heavy bags? My other question is: What about the Zweig rule? Does the Zweig rule, which is very important for the interaction between bosons and nucleons, does that follow naturally from your viewpoint?

Dashen: I wish I could answer the question. I do not really know. Most of the things I am talking about were only discovered last summer and we have made some attempt to try to understand them better. If you would like, the MIT bag model could either be improved or its limitations seen, without any further progress. I have nothing to add to the question whether there might be 4-quark bags or things like that; I do not know.

H. Pagels (Rockefeller U.): In this very dense liquid phase, there is essentially some kind of constant density that you subtract out. Then on top of that there are going to be enormous fluctuations of some kind or another. Are there any observable consequences of these fluctuations? And cannot these fluctuations be gigantic and destroy the type of approximations that you made to get this boundary? I would expect that there are many fluctuations in this dense liquid phase. And if you cannot control those, how can you justify the bag picture that you gave here?

Dashen: Let me go back to something I did not say because there was not enough time. There are huge fluctuations in the vacuum. The vector potential is essentially at random when the coupling constant is so large. However, if you imagine a very large infrared fluctuation coming in and sweeping by a bag, the bag is now a relatively small perturbation on it. You might imagine you could "linearize" in this situation. Locally, you could linearize the effect of the bag on the large fluctuation. If you do that you will see that what you have is a Meissner effect, that the magnetic permeability inside is equal to one and is very large outside. If you look at a superconductor, where you have magnetic permeability, one outside and zero inside, the ratio is the same. Infinity over one is the same thing as one over zero. When you calculate this Meissner effect the vacuum looks like a perfect ferromagnet and the bag is like a little chunk of superconductor. What happens is that fluctuations are just refracted around it. That is not a perfect argument, because to really handle things, one needs to get away from this quasi-Abelian approach. But it looks reasonable not to worry because, although the fluctuations outside are large and completely unknown at this point, they just do not come inside, and therefore do not affect most of the properties of the hadrons. Let me admit one problem. I don't know if the surface energy of this is positive or negative. Surface energy is normally positive, but in type-2 superconductors it is negative. If the surface energy were negative you would tend to get very kinky bags. So, it is to be hoped that the answer is that there is a fairly large positive surface tension that keeps the sides of the bag tight.

11. HOW CAN WE MEASURE THE MASS, CHARGE, AND MAGNETIC MOMENT OF CONFINED QUARKS?

Harry J. Lipkin

1. Introduction

Once upon a time physicists believed that nucleons and pions were elementary like electrons and photons, and that Yukawa's theory of nuclear forces was the analog of quantum electrodynamics (QED) for strong interactions. Then the Δ, ρ and other resonances were discovered and showed that neither nucleons nor pions were elementary, but that both had a composite structure. Today both pions and nucleons seem to be made of the same basic building blocks: spin 1/2 quarks bound by colored gluons. But perhaps history will repeat itself. Maybe a talk at the next Einstein conference will begin "Once upon a time physicists believed that quarks and gluons were elementary and that quantum chromodynamics (QCD) was the analog of quantum electrodynamics for strong interactions. Then . . . ??"

But this talk discusses only today's physics, with the new QXD model for everything, where X = A, B, C, D, E, F, G, etc. So far only models for X = C, E, F, and G have been confirmed, but the others are expected to be discovered eventually. However, the original non-Abelian gauge model for hadron dynamics has faded away — that is, the gauge theory of strong interactions mediated by the octet of vector mesons ρ, ω, and K^* coupled to conserved vector currents. The $SU(3)$ group of Gell-Mann and Ne'eman is now called flavor and dismissed as an irrelevant complication in the QCD description of strong interactions.

The question whether quarks are permanently confined or have a large mass and will eventually be liberated is still open. But in any case they are at present unavailable for direct measurement of their properties. All values of quark parameters come from experiments on quarks bound in hadrons. This paper discusses the nature of the information obtained by such measurements on bound quarks. Almost everything we know about the electron could be determined from measurements on bound electrons in atoms without ever seeing a free electron. But measurements on bound quarks are not so simply interpreted.

Both electrons and quarks are particles with very long range (probably infinite) Coulomb or color fields containing energy and momentum densities over all space. Representing these objects as point particles with well-defined values of mass, charge, and magnetic moment seems at first a drastic oversimplification. Yet the point particle gives a good description of the electron. An alternative picture considers the point particle as a simplified description of a complicated system. The measured parameters are "effective" charges, masses, and magnetic moments, as in the case of electrons in a lattice or nucleons in the nuclear shell model. This picture is more appropriate for quarks, because of two important differences between quarks and electrons:

1. $\alpha_s \gg \alpha$;
2. QCD is non-Abelian, in contrast to QED.

The large value of α_s means that effects small enough to be neglected in QED are not small in QCD. The non-Abelian nature of QCD gives new effects arising from asymptotic freedom and infrared slavery (confinement) that makes an unambiguous definition of quark properties much more difficult. Consider, for example, the contribution of the long-range part of the field of an electron or quark to its mass. The energy density in the Coulomb field of an electron continues out to infinity. A moving electron also has a magnetic field, Poynting vector, and momentum density at large distances. Yet the motion of an electron in an external field is well described by the motion of a point mass with all the charge and inertia localized at a point.

The contribution of the field to the mass of the electron cannot be calculated because of ultraviolet divergences. But the energy in the field of an electron-positron pair separated by a finite distance differs from the energy of two isolated electrons by a finite amount that can be calculated. The change in the inertial mass of an electron-positron pair as a function of the distance between them is just the Coulomb interaction between the two particles. But it is also given by the change in the energy of the electric field, and comes primarily from the field at large distances. The mass of an electron-positron pair is less than the mass of two electrons because the fields of the electron and positron cancel at large distances and the energy in the field is decreased. The question of what is the mass of an electron bound in positronium is very subtle and not easily answered. It leads into the question of the meaning of mass. One possible answer is the quantity that enters into the magnetic moment of a particle. For an electron in an atom or positronium, this is the mass of a free electron. However, this result depends upon the Lorentz character of the potential.[1] With a Lorentz scalar potential the mass parameter appearing in the magnetic moment is not the mass of the free particle, but an effective mass that includes binding effects.

The color field of an isolated quark is generally believed to be very different from the Coulomb field of an electron, due to the non-Abelian nature of QCD. Unlike the photon, which has no electric charge and cannot emit other photons, the colored gluon of QCD carries a color charge and can emit colored gluons. The very soft gluons in the color field at large distances from the quark can emit more very soft gluons, and the theory has very strong infrared divergences. Thus the color field of an isolated quark is believed either to remain finite at large distances or to go to zero much more slowly than a Coulomb field. The energy in this field at large distances is very large and the mass of an isolated quark would be very large. In theories with quark confinement this mass is infinite and isolated quarks are unobservable.

Quarks bound in color singlet hadrons do not have fields at large distances, because the fields of the constituents cancel one another as in positronium. The motion of quarks in a hadron should therefore not exhibit the infinite inertia due to the color field at large distances, which is no longer present. One might expect that low-energy hadron spectroscopy could be described in terms of the motion of "constituent quarks" in which each quark carries with it a share of the color field, and the effective mass of a constituent quark is roughly equal to its share of the total mass. This has been found in fact to be true, and rather surprising because it works much too well.[2]

In deep inelastic processes, where a large momentum is transferred suddenly to a quark, the quark recoils too fast for the color field to follow it. The effective mass of the "current quark" that recoils is very small, and the color field that remains behind must be considered as an independent gluon constituent in the hadron.

Thus a quark has an effective mass that depends upon its environment and upon the type of measurement performed. Most mass measurements involve energy and momentum transfer to the quark. The "effective mass" then depends upon how much of the color field around the quark recoils with it. This effective mass is zero in deep inelastic scattering when practically no color field recoils with the quark and infinite for an isolated quark when the entire color field must recoil. In low-energy spectroscopy where the nonrelativistic quark model describes the phenomenology very well, the quark seems to have an effective mass equal to its share of the hadron mass.

The infrared effects arising from the large value of α_s and the non-Abelian nature of QCD also cause difficulties in measuring the color and electric charge of a quark. In QED there is the Weizsacker–Williams picture of a fast electron accompanied by a cloud of virtual photons, which can be emitted as bremsstrahlung when the electron is scattered. The number of photons emitted in a scattering process is infinite, but the energy and momentum carried is small and finite. Finite values of cross-sections are obtained for realistic processes in which inelastic scattering with energy losses below a small finite amount is included with the elastic scattering. In QCD these bremsstrahlung effects can be expected to be much larger, with a large number of soft infrared gluons emitted during a scattering process. These may still carry only small finite amounts of energy and momentum as in QCD, so that neglecting the infrared radiative corrections is also a good approximation for the kinematical variables. But gluons carry color charge that is quantized and cannot be small. Thus the color of a quark is changing constantly and cannot be easily measured because of the constant emission and absorption of infrared gluons.[3]

If quarks are permanently confined, color is unobservable because all states are color singlets, and no preferred direction in color space can be defined. The color of a given quark then changes at an infinite frequency. If free quarks and other states that are not color singlets exist at a high mass, then the color of a quark changes at a frequency related to the color threshold and can be observed only in experiments that have a built-in energy scale that is higher than the color threshold — that is, a time scale that is short compared to the period of the color oscillations. These oscillations can be compared to neutrino oscillations. In models like the Han–Nambu model where the electric charge of the quark depends upon its color, the electric charge of a quark also undergoes oscillations. Experiments with a characteristic time scale long compared with these oscillations will then measure an effective quark charge that is its average over color. Only experiments that have a built-in time scale short compared to the oscillations can observe the instantaneous charge of the quark and detect a color dependence.

We now consider in detail two examples of measurement of properties of bound quarks. Section 2 considers the effective mass of constituent quarks as measured by hadron magnetic moments and masses and shows that a simple picture gives a remarkable agreement with experiment. Section 3 considers the measurement of the charge of the quark and shows that measurement of a color dependence of the charge is very difficult below the threshold for excitation of states that are not color singlets.

2. Masses and Magnetic Moments

Let us consider the masses and magnetic moments of bound quarks as revealed by two recent experiments at Fermilab. Despite the emphasis on theory at this Einstein conference, we should not forget the importance of experiment. The present discussion takes a "Galilean approach," which learns about nature from experimental observations rather than from reading words written by great establishment theorists like Aristotle. The quark model has grown out of such experimental observations against the opposition of the Aristotelian establishment, who always found weighty reasons why quarks could not exist and dismissed the agreement with experiment.

The quark model is very ad hoc. It lacks a fundamental theoretical basis, yet it provides a very good description of much experimental data. Today theorists believe that the fundamental basis will eventually come from QCD, and there are many indications that this is indeed true. But there are many slips between cup and lip, and a measure of skepticism is always in place. The approach and methods of QCD may be used to guide our intuition in discussing the quark model, but we should not attempt any rigorous derivation and should always be looking at experimental data to see what nature is trying to sell us.

A "Newtonian approach" is used here in discussing constituent quarks. Newton described the motion of the earth around the sun with great precision without ever having heard of asymptotic freedom. The gravitational field of the earth is indeed asymptotically free, given by Coulomb's law with an effective or "running" coupling constant proportional to the mass of the earth at large distances, but decreasing at short distances less than the radius of the earth to zero at the center. This asymptotic freedom was irrelevant to Newton, who described the earth as a point mass with a pure Coulomb field. This drastic and unwarranted assumption for short-distance phenomena ignores, for example, the fact that we exist. But it is certainly adequate and can be rigorously justified for precise calculations of the earth's orbit.

In the same spirit we may assume the nucleon to be made of three very simple constituent quarks. In reality they must be much more complicated and include virtual gluons, quark-antiquark pairs, and so on. But a wide variety of phenomena are successfully described by these simple constituent quarks. The fundamental theory must have an explanation justifying this simple picture, but so far there is no good fundamental theory and no satisfactory explanation. All we know is that the model works.

One interesting experimental result from Fermilab is the discovery[4] of the new fifth heavy quark in the bound states now called the Υ and Υ', with the surprising equality of the Υ-Υ' mass splitting to the mass splitting in the charmonium system, namely the ψ-ψ' splitting. Nature seems be telling us that hadron mass splittings are simpler than expected: they do not depend very strongly on the flavor or mass of the quarks of which they are composed. This principle was incorporated formally in the logarithmic potential model of Quigg and Rosner,[5] which has this mass scaling property. The general philosophy of this model can be used without taking it too seriously in detail, and we can assume that flavor dependence of many mass splitting effects can be neglected. This might be called "rolling off the log."

A second interesting result from Fermilab is a new measurement[6] of the magnetic moment of the Λ to a precision of 1 percent that agrees with quark model predictions[7] of this moment to 1 percent. This agreement is completely unexpected, since the quark model is not expected to be that good.

The simple-minded calculations of baryon magnetic moments in the quark model[8] treat the baryons as three quarks in a relative s-wave, with an $SU(6)$ wave function and with the total spin and magnetic moment given by simple vector addition of the quark spins and Dirac[9] magnetic moments. The magnetic moment of a quark of flavor f is

$$\mu_f = q_f(M_p/m_f) \text{ nm} , \qquad (11.1)$$

where q_f and m_f are the charge and mass of the quark of flavor f, M_p is the proton mass, and nm stands for nuclear magnetons.

The Λ, p, and n magnetic moments ae then

$$\mu_\Lambda = \mu_s = -(1/3)(M_p/m_s) , \qquad (11.2)$$

$$\mu_p = (8/9)(M_p/m_u) + (1/9)(M_p/m_d) = (M_p/m_u) , \qquad (11.3)$$

$$\mu_n = -(4/9)(M_p/m_d) - (2/9)(M_p/m_u) = -(2/3)(M_p/m_u) , \qquad (11.4)$$

where we neglect the difference between m_u and m_d. Combining Eqs. (11.3) and (11.4) gives the well-known successful prediction for the neutron magnetic moment,

$$\mu_n = (-2/3)\mu_p = -1.86 \text{ nm} , \qquad (11.5)$$

in excellent agreement with the experiment value of -1.91.

The Λ magnetic moment cannot be related directly to the nucleon magnetic moments without some assumption regarding the difference between m_s and m_u. The very drastic assumption that this quark mass difference is exactly equal to the hadron mass difference,

$$m_s - m_u = M_\Lambda - M_p , \qquad (11.6)$$

leads to a very interesting prediction,

$$\mu_\Lambda = (-1/3)[(1/\mu_p) + (M_\Lambda - M_p)/M_p]^{-1} = -0.61 \text{ nm} . \qquad (11.7)$$

Another prediction is obtained by assuming that the ratio of quark magnetic moments μ_u/μ_s is obtainable from hadron spin splittings like the ratio $(M_\Lambda - M_N)/(M_{\Sigma^*} - M_\Sigma)$. This ratio, which is unity in the $SU(3)$ limit, is directly related to the ratio of quark magnetic moments under the assumption that the spin splittings come from a "color magnetic" interaction[10] proportional to the color magnetic moments of the quarks, which are in turn proportional to electromagnetic moments. The result obtained is

$$\mu_\Lambda = -(\mu_p/3)(M_{\Sigma^{*+}} - M_{\Sigma^+})/(M_{\Delta^+} - M_p) = -0.61 \text{ nm} . \qquad (11.8)$$

Both predictions (11.7) and (11.8) are in remarkable agreement with the new experimental value $\mu_\Lambda = -0.6138 \pm 0.0047$ nm. That they are also in remarkable agreement with each other suggests a new relation between hadron masses and the proton magnetic moment. Eliminating μ_Λ between (11.7) and (11.8) gives

$$[(M_{\Delta^+} - M_p)/(M_{\Sigma^{*+}} - M_{\Sigma^+})] - 1 = \mu_p(M_\Lambda - M_N)/M_N \ . \tag{11.9}$$

This peculiar relation is in excellent agreement with experiment. The left-hand side is 0.523; the right-hand side is 0.528. This unorthodox combination of hadron mass differences and the proton moment has a simple physical interpretation. The $SU(3)$-breaking quark mass parameter $(m_s - m_u)/m_u$ is computed in two ways. The left-hand side uses the quark mass *ratio* (m_s/m_u) obtained from hadron spin splittings. The right-hand side uses the quark mass *difference* $(m_s - m_u)$ obtained from hadron strangeness splittings but needs the proton moment to provide a quark mass scale relating the mass difference to a mass ratio. Thus Eq. (11.9) says that the quark mass ratio and the quark mass difference determined in two different ways from hadron masses are consistent at the 1 percent level with the quark mass m_u determined from the proton mass and magnetic moment.

The success of these relations suggests a review of the underlying physics and its implications for hadron models. We first consider the flavor dependence of the mass splittings, which plays a crucial role in both predictions (11.7) and (11.8). The nuclear shell model approach to baryon masses of Federman, Rubinstein, and Talmi[11] in 1966 considered the low-lying baryon octet and decuplet to all have the same "shell-model" wave functions with their mass degeneracy split only by the strange quark mass difference $m_s - m_u$ and by a residual two-body interaction. This led to a relation between hadron masses in good agreement with experiment:

$$M_\Delta - M_N = (1/2)(2M_{\Sigma^*} + M_\Sigma - 3M_\Lambda) \ . \tag{11.10}$$

The left-hand side is 307 MeV; the right is 294 MeV, thus providing support for the assumption that baryon splittings are described by two-body forces. De Rujula, Georgi, and Glashow[12] assumed a specific form for the spin dependence of the interaction; namely a color-magnetic hyperfine interaction having the form

$$V_{hf} = \mu_i^c \mu_j^c \vec{\sigma}_i \cdot \vec{\sigma}_j \, v(r_{ij}) \ , \tag{11.11}$$

where μ_i^c denotes the color magnetic moment of quark i. The particular form (11.11) immediately leads to the prediction (11.8) and relates the triplet and singlet effective matrix elements by the eigenvalues of the operator $\vec{\sigma}_i \cdot \vec{\sigma}_j$.

This assumption is also crucial for the prediction (11.7). The interaction (11.11) does not contribute to the difference $M_\Lambda - M_N$. This is the reason to choose the $\Lambda - N$ mass difference for the right-hand side of (11.6) rather than $\Sigma - N$ or $\Sigma^* - \Delta$. The decuplet mass splitting has previously been used for the quark mass difference because the equal mass spacing has been interpreted as indicating that decuplet mass splittings are simpler than octet splittings. However, Eq. (11.11) shows that the decuplet splitting involves a complicated interplay of both the quark mass differences (11.6) and the spin splittings, while use of the $\Lambda - N$ mass difference eliminates effects of spin splittings.[13]

The assumption that meson spin splittings are due to a similar interaction but with a different strength led to a relation between baryon and meson masses[14] also in surprising agreement with experiment,

$$M_\Lambda - M_N = m_s - m_u = (3/4)(M_{K^*} - M_\rho) + (1/4)(M_K - M_\pi) \ . \tag{11.12}$$

The left-hand side is 177 MeV and the right is 180 MeV. Thus there seems to be experimental justification of the use of the hadron mass difference (11.12) as a quark mass difference in deriving the prediction (11.7).

We now have three independent relations between hadron masses and moments that are experimentally confirmed at the 1 percent level, namely relations (11.7), (11.8) and (11.12). What have we put in to get these results?

1. Hadron mass differences are identified with quark mass differences in a very simple way, neglecting flavor dependence of binding effects and kinetic energies.
2. The quark masses in (11.12) are identified with the masses appearing in the electromagnetic and color-magnetic moments of the quarks.
3. The hyperfine interaction is assumed to have the form (11.11).

However, it is still a big step further to use the mass difference of Eq. (11.6) as the mass parameter in the magnetic moments and to obtain results valid to a few percent. This assumption is generally not valid in conventional models. In the De Rujula–Georgi–Glashow model Eq. (11.6) does not hold because the hadron masses include additional terms like kinetic energies that are inversely proportional to quark masses and do not cancel in the difference (11.6). The model in the reference cited in note 2 avoids these terms by the use of scaling properties of the Quigg-Rosner logarithmic potential model. In this model kinetic energies and mass splittings in the hadron spectrum are independent of the quark mass and cancel out of mass differences like (11.6). This can be seen explicitly by the use of the virial theorem. The success of the relations (11.7)–(11.9) at this level indicate that the "quasinuclear colored quark model" of note 2 and the three basic assumptions just listed should be taken more seriously than indicated by their crude derivations. The underlying physics is that the *same quark mass parameter* appears in the simplest possible way in the electromagnetic moments, the color-magnetic moments, and the hadron mass splittings. That electromagnetic and color moments should depend upon the same mass parameter is not surprising. But the value of the magnetic moment is not expected to be determined to 1 percent by the mass parameter that enters hadron mass splittings and includes binding energies as well as quark masses.

3. Measurement of Colored Quark Charges below Color Threshold

One might think that the electric charge of the quark is observable directly from measurements of the electromagnetic couplings of hadrons. However, the coupling of the electromagnetic current to a color singlet hadron depends only on the color-averaged quark charge $\langle Q \rangle_c$ — that is, the average over color of the charges of quarks having identical values for all other quantum numbers. A color singlet state is completely symmetric in color space and can give no information about a particular preferred direction in this space, for example the charge of a red quark rather than that of a blue quark. The color-averaged quark charge does not contain information on whether quarks of all colors have the same charge or whether the charge depends upon color. A measurement of the color-averaged *square* of the quark charge $<Q^2>$ should give this information. If $<Q^2>_c$ *differs from* $<Q>_c^2$, there must be a color dependence of the quark charge. However, it is not simple to devise experiments on hadrons that measure $<Q^2>_c$ for a given quark in a hadron, as will be shown.

One of the difficulties in interpreting results of simple parton-type models in cases where color symmetry is important arises from peculiar quantum-mechanical coherence effects of the type that disturbed Einstein and are appropriate for discussion at this conference. As an example consider a model in which strong interactions exactly conserve color symmetry, but electromagnetic interactions break the symmetry by giving different electric charges to the red, white, and blue quarks. The weak interaction can also break the color symmetry but might be simply expressed in a different basis from the red, white, and blue quarks that are eigenstates of electric charge. For example, there might be purple and lavender quarks, defined as two orthogonal linear combinations of red and blue quarks, rotated by a Colorbibbo angle (private communication).

Suppose that a color singlet meson is given a high momentum transfer by a strong interaction that sends the quark in the meson to the moon, while the antiquark remains on earth. Since strong interactions conserve color, the system is still in a color singlet state. If an astronaut on the moon measures the electric charge of the quark and finds that it is red, then the antiquark on the earth must also be red, and similarly for blue and white. But if the astronaut does a weak interaction experiment and finds that the quark was purple or lavender, then the antiquark on the earth must also be purple or lavender. Thus whether the density matrix describing the antiquark on earth is diagonal in the red-blue or purple-lavender basis depends upon whether the astronaut on the moon chooses to measure an electromagnetic or a weak property of the quark on the moon. This is a manifestation of the famous paradox of Einstein, Podolsky and Rosen.

Such coherence properties arise in parton models where non-Abelian symmetries are present. One generally draws diagrams in which one parton absorbs a photon and behaves as if it were free during the interaction. However, it is only as free as the quark on the moon in the previous example. If it comes from a color-singlet hadron state, a measurement of its electric charge by a photon absorption process affects the properties of the rest of the system, even though there is no interaction. This effect appears in the relative phases of the contributions of the three diagrams where the photon is absorbed by a red, a white, and a blue quark, respectively. If this phase information is ignored and the contributions of the diagrams are added incoherently, the results obtained can have serious errors.

An example of the importance of relative phases in cases where internal symmetries are present has been pointed out in the deep inelastic production of exclusive final states on a pion target by the isovector component of the photon.[14] Since the initial state has odd G parity, only final states with odd numbers of pions can be produced. However, a parton model in which individual quark partons absorb the photon and the amplitudes are added incoherently loses the G parity information and gives equal production of states with even and odd numbers of pions. The G parity information is contained in the relative phase of the contributions from pairs of diagrams that go into each other under the G transformation — that is, those in which the current is absorbed by a quark and by its G-conjugate antiquark.

Distinguishing between fractionally charged and integrally charged colored quark models has been shown to be difficult below color threshold.[15] However, recent proposals for measuring the quark or gluon charge in deep inelastic scattering processes have not taken this difficulty properly into account and give the misleading impression that their results are valid below the threshold for producing states that are not color singlets.[16] We consider here the case of second-order electromagnetic transitions and show the following:

1. The contribution to the transition matrix of any component of the electromagnetic current that is not a color singlet is inversely proportional to the threshold energy for the excitation of states that are not color singlets. Thus any model that obtains a finite contribution independent of the color threshold from the nonsinglet part of the current should be viewed with suspicion.
2. Models that give a second-order transition matrix element proportional to the average of the square of the quark charge rather than to the square of the average are shown to neglect a set of terms of the same order of magnitude as the terms that are kept.

The essential physics underlying the argument is that the color excitation threshold defines a characteristic length (or time) that is very short and that is ignored in the "short-distance" expansions generally used in describing deep inelastic proceses. This point is generally overlooked because the non-Abelian nature of color is not fully understood and violates the intuition from Abelian QED commonly used in the conventional models. The physical implications are discussed in detail in a recent paper.[17]

The transition probability for a second-order electromagnetic transition between two states $|A\rangle$ and $|B\rangle$ is proportional to the square of the transition matrix element,

$$M_{AB} = \sum_X \frac{\langle B|J|X\rangle \langle X|J|A\rangle}{E_X - E_A} , \quad (11.13)$$

where J is some component of the electromagnetic current operator and the summation is over a complete set of intermediate states $|X\rangle$. This result is exact to second order in electromagnetism and to all orders in strong interactions if the states $|A\rangle$, $|B\rangle$, and $|X\rangle$ are the exact eigenstates of the strong interaction Hamiltonian and E_A, E_B, and E_X are the exact energy eigenvalues.

We can now prove theorems valid to all orders in strong interactions because they depend only on the following general properties of the exact solutions and not specific details:

1. Color is an exact symmetry of strong interactions. Thus the states $|A\rangle$, $|B\rangle$, and $|X\rangle$ in Eq. (11.13) are all color eigenstates.
2. A color threshold exists so that the lowest state that is not a color singlet has an excitation energy E_{col} above the initial state $|A\rangle$ in Eq. (11.13). Thus for any state $|Y\rangle$ that is not a color singlet,

$$E_Y - E_A \geq E_{col} . \quad (11.14)$$

The current J can be written as the sum of a color singlet component J_0 and a remainder ΔJ that has no color singlet component, as follows:

$$J = J_0 + \Delta J. \qquad (11.15)$$

In models like the Han–Nambu model where the quark charge depends upon color, the additional component in the quark current transforms under color like members of a color octet. However, let us keep our treatment general and include the possibility of contributions from higher color multiplets as well.

The transition matrix element (11.13) separates into two terms with contributions from J_0 and ΔJ, respectively,

$$M_{AB} = M^0_{AB} + \Delta M_{AB}, \qquad (11.16)$$

where

$$M^0_{AB} = \sum_{X_0} \frac{\langle B| J_0 |X_0\rangle \langle X_0| J_0 |A\rangle}{E_{X_0} - E_A}, \qquad (11.17)$$

$$\Delta M_{AB} = \sum_Y \frac{\langle B| \Delta J |Y\rangle \langle Y| \Delta J |A\rangle}{E_Y - E_A}, \qquad (11.18)$$

and the states $|X_0\rangle$ are all color singlets, while the states $|Y\rangle$ are all color nonsinglets.

If the Schwarz inequality and the inequality (11.14) are applied to the summation (11.18), we obtain

$$\Delta M_{AB} = M_{AB} - M^0_{AB} \leq \frac{\sqrt{\langle B||\Delta J|^2|B\rangle \langle A||\Delta J|^2|A\rangle}}{E_{\text{col}}}. \qquad (11.19)$$

Thus the contribution to the transition matrix element (11.13) from the color-nonsinglet component of the current is bounded by an expression inversely porportional to E_{col} that goes to zero as E_{col} becomes infinite. Any model that obtains finite contributions from ΔJ independent of E_{col} should be checked carefully for unjustified assumptions causing a violation of the theorem (11.19). The contribution (11.19) may still be appreciable in comparison with the leading term (11.17), but only if the main contributions to (11.17) come from intermediate states above the color threshold. The usual deep inelastic processes do not seem to satisfy this condition. Some possibilities of this nature have been discussed.[18]

Let us examine the common approximation made in the standard models and see how it fails. Let us assume that the current is entirely due to quarks at this point. The extension to gluon and other contributions will be discussed. In a model with red, white, and blue quarks, the current can be written as the sum of the red, white, and blue quark currents,

$$J = Q_R J_R + Q_W J_W + Q_B J_B = \sum_{i=1}^{n} Q_i J_i , \tag{11.20}$$

where Q_R, Q_W and Q_B are the electric charges of the red, white, and blue quarks, respectively, and J_R, J_W, and J_B are the vector currents of these quarks normalized to give an electromagnetic current with the quark charges appearing as external multiplicative factors. The summation on the right-hand side is written in a general form to apply not only to quark currents but also to more general cases like currents from gluons, which have more states. For the case of quark currents the sum is over the three colors and $n = 3$. We disregard flavor, which is an irrelevant complication for the present argument.

With this expression for the current, the transition matrix element (11.13) can be written

$$M_{AB} = \sum_{i,j=1}^{n} \sum_{X} \frac{Q_i Q_j \langle B| J_i |X\rangle \langle X| J_j |A\rangle}{E_X - E_A} . \tag{11.21}$$

The conventional models, like the parton model, all neglect the terms in Eq. (11.21) with $i \neq j$ that do not have the same color in both current operators. This leads to the parton model approximate relation for the matrix element (11.13),

$$P_{AB} = \sum_{i=1}^{n} Q_i^2 \sum_{X} \frac{\langle B| J_i |X\rangle \langle X| J_i |A\rangle}{E_X - E_A} . \tag{11.22}$$

The terms with $i \neq j$ neglected in Eq. (11.21) are

$$\Delta P_{AB} = \sum_{i=1}^{n} Q_i \sum_{\substack{j=1 \\ j \neq i}}^{n} Q_j \sum_{X} \frac{\langle B| J_i |X\rangle \langle X| J_j |A\rangle}{E_X - E_A} , \tag{11.23}$$

where

$$M_{AB} = P_{AB} + \Delta P_{AB} . \tag{11.24}$$

The essential physical assumption underlying the neglect of the terms (11.23) is that the two electromagnetic interactions occur on the same parton, which behaves like a free particle. Thus the state of this parton does not change, including its color, and the same component of the current (11.20) must act at both vertices. We shall now see that this assumption is generally not valid, and that the terms (11.23) neglected in the naive models are of the same order as those kept (11.22).

The expressions (11.22) and (11.23) have contributions from intermediate states that are both color singlets and nonsinglets. These two types of contributions can be separated as in Eq. (11.16)–(11.18):

$$P_{AB} = P^0_{AB} + \sum_\lambda P^\lambda_{AB} , \qquad (11.25)$$

$$\Delta P_{AB} = P^0_{AB} + \sum_\lambda \Delta P^\lambda_{AB} , \qquad (11.26)$$

where

$$P^0_{AB} = \sum_{i=1}^n Q_i^2 \sum_{X_0} \frac{\langle B|J_i|X_0\rangle \langle X_0|J_i|A\rangle}{E_{X_0} - E_A} , \qquad (11.27)$$

$$P^\lambda_{AB} = \sum_{i=1}^n Q_i^2 \sum_{Y_\lambda} \frac{\langle B|J_i|Y_\lambda\rangle \langle Y_\lambda|J_i|A\rangle}{E_{Y_\lambda} - E_A} , \qquad (11.28)$$

$$\Delta P^0_{AB} = \sum_{i=1}^n Q_i \sum_{\substack{j=1 \\ j \ne i}}^n Q_j \sum_{X_0} \frac{\langle B|J_i|X_0\rangle \langle X_0|J_j|A\rangle}{E_{X_0} - E_A} , \qquad (11.29)$$

$$\Delta P^\lambda_{AB} = \sum_{i=1}^n Q_i \sum_{\substack{j=1 \\ j \ne i}}^n Q_j \sum_{Y_\lambda} \frac{\langle B|J_i|Y_\lambda\rangle \langle Y_\lambda|J_j|A\rangle}{E_{Y_\lambda} - E_A} , \qquad (11.30)$$

where the intermediate states Y that are not color singlets have been separated into sets of states Y_λ that all transform in the same way under color $SU(3)$. With the quark currents (11.20) only the octet representation of $SU(3)$ occurs and there are two independent states. These can be chosen as the third and eighth components of the octet in the standard notation with λ matrices, as discussed in the paper cited in note 3. For our purposes the general expressions (11.24)–(11.26) and (11.27)–(11.30) are sufficient.

The matrix elements of the different components of the current J_i are related by the Wigner-Eckart theorem,

$$\langle X_0|J_i|A\rangle = \langle X_0|J_j|A\rangle , \qquad (11.31)$$

$$\langle Y_\lambda|J_i|A\rangle = C^\lambda_{ij} \langle Y_\lambda|J_j|A\rangle , \qquad (11.32)$$

where C^λ_{ij} is a coefficient of order unity depending upon the particular color $SU(3)$ quantum numbers of the state Y_λ and the colors i and j.

Substituting the relations (11.31), (11.32) into the relations (11.27)–(11.30) gives

$$\Delta P^0_{AB} = P^0_{AB} \left[\sum_{i \neq j} Q_i Q_j / \sum_i Q_i^2 \right] , \qquad (11.33)$$

$$\Delta P^\lambda_{AB} = P^\lambda_{AB} \left[\sum_{i \neq j} Q_i Q_j C^\lambda_{ij} / \sum_i Q_i^2 \right] , \qquad (11.34)$$

$$M^0_{AB} = P^0_{AB} + \Delta P^0_{AB} = \left(\sum_i Q_i \right)^2 \sum_{X_0} \frac{\langle B| J_i |X_0\rangle \langle X_0| J_i |A\rangle}{E_{X_0} - E_A} . \qquad (11.35)$$

This result shows that the individual terms (11.27) and (11.28) retained in the naive models are of the same order of magnitude as the neglected terms (11.29) and (11.30). Thus the neglect of the "cross-terms" (11.26) can be justified only if a cancellation occurs between the various terms. However, the leading term (11.29) in the sum comes from color-singlet intermediate states, while the remaining terms all come from nonsinglet intermediate states. These terms are suppressed by energy denominators according to Eq. (11.19) and cannot cancel the leading term, which does not have a suppression.

The sum of the leading terms (11.35) differs in an essential way from the parton model expression (11.22). It is proportional to the square of the average quark charge, which is the same in all models, rather than to the average of the square. The average squared charge differs in the different models and could be used to distinguish between them. However, all claims that naive models measure squared charge are based on the neglect of the cross-terms (11.26).

The treatment in Eqs. (11.20)–(11.35) is easily extended to the case where the electromagnetic vertices involve gluons instead of quarks. Gluon current components J_i can be defined for the eight colored gluon states. The summation in Eq. (11.20) is now applied to the gluon current with $n = 8$. The remaining treatment in Eqs. (11.21)–(11.35) then follows in exactly the same way with the gluon interpretation of the current and $n = 8$ everywhere. Note, however, that only the color-octet component of the current couples to gluons, so that Eqs. (11.27), (11.29), (11.31), (11.33), and (11.35) all vanish in the gluon case and all terms are suppressed by the color threshold. The conclusions are the same as for the case of quark currents. The cross-terms neglected in the naive models are of the same order as the terms kept.

The results (11.19) and (11.33)–(11.35) can also be expressed in a time-dependent formulation that exhibits the loopholes in treatments based on the operator-product expansion. The matrix element of the time-dependent operator product that appears in these treatments can be written as follows:

$$M_{AB}(t) = \langle B| J(t) J(0) |A\rangle = M^0_{AB}(t) + \Delta M_{AB}(t) , \qquad (11.36)$$

where

$$M^0_{AB}(t) = \sum_{X_0} \langle B| J(0) |X_0\rangle \langle X_0| J(0) |A\rangle e^{-i(E_{X_0} - E_A)t/\hbar} , \qquad (11.37)$$

$$\Delta M_{AB}(t) = \sum_Y \langle B| \Delta J(0) |Y\rangle \langle Y| \Delta J(0) |A\rangle e^{-i(E_Y - E_A)t/\hbar} , \qquad (11.38)$$

and the intermediate states $|X_0\rangle$ and $|Y\rangle$ are color singlets and nonsinglets, respectively, as in Eqs. (11.16)–(11.18).

It immediately follows that $\Delta M_{AB}(t)$ has no Fourier components below $E_{col}/\hbar$,

$$\int \Delta M_{AB}(t)e^{i\psi t}dt = 0 \quad \text{if } |\omega| < E_{col}/\hbar \,. \tag{11.39}$$

Any approximation of $\Delta M_{AB}(t)$ by the operator-product expansion that introduces such Fourier components must be in error. The contribution of gluon currents to Fourier components of $M_{AB}(t)$ below $E_{col}/\hbar$ is seen to vanish, since these appear only in $\Delta M_{AB}(t)$.

Similarly we can define

$$P_{AB}(t) = \sum_{i=1}^{n} Q_i \langle B| J_i(t)J_i(0) |A\rangle \,, \tag{11.40}$$

$$\Delta P_{AB}(t) = \sum_{\substack{i=1\\j=1\\i\neq j}}^{n} Q_iQ_j \langle B| J_i(t)J_j(0) |A\rangle \,. \tag{11.41}$$

For Fourier components of $\Delta P_{AB}(t)$ and of $P_{AB}(t)$ below $E_{col}/\hbar$ only color-singlet intermediate states contribute. The Wigner–Eckart theorem then shows that these Fourier components of $P_{AB}(t)$ and $\Delta P_{AB}(t)$ satisfy Eq. (11.33) and are of the same order of magnitude,

$$\int \Delta P_{AB}(t)e^{i\omega t}dt = \int P_{AB}(t)e^{i\omega t}dt \left[\sum_{i\neq j} Q_iQ_j / \sum_i Q_i^2 \right] \tag{11.42}$$

if $|\omega| < E_{col}/\hbar$.

Thus any approximation that neglects the cross-terms $\Delta P_{AB}(t)$ cannot be valid for Fourier components below $E_{col}/\hbar$.

Further insight into the role of the cross-terms (11.41) is obtained by noting that the expressions (11.37)–(11.41) can be applied to inclusive hadron production in electron-positron annihilation. For this case the states $|A\rangle$ and $|B\rangle$ are both taken as the vacuum and the cross-section is proportional to the single Fourier component corresponding to the total energy of the system. The naive parton model result (11.40) seems to give the well-known result that the cross-section is proportional to the number of colors in the case where all quark charges are equal. However, below color threshold only color-singlet intermediate states contribute, as shown by Eq. (11.39). This introduces an additional factor of $1/n$ in the matrix elements (11.40) due to the projection into the color-singlet sector. This exactly cancels the other factor n and gives a result independent of the number of colors.

The well-known result is regained by including the contribution from the cross-terms (11.41). These terms have the same factor $1/n$ in the matrix element from the projection into the color-singlet intermediate states, but the factor multiplying the matrix element is $n(n-1)$. Thus this contribution is exactly $n-1$ times the contribution from the direct terms (11.40) and gives a total contribution proportional to n.

In this particular case the naive parton model expression gives the correct result when it is evaluated incorrectly, namely by using the incorrect assumption that there is no n dependence in the matrix elements. This n dependence is exactly canceled by the cross-terms, which are omitted.

Our principal results (11.19), (11.33)–(11.35), (11.39), and (11.42) show that the color threshold E_{col} plays an important role in second-order electromagnetic transitions in suppressing the contributions from any component of the electromagnetic current that is not a color singlet. The naive models that obtain results of the form (11.22) proportional to the sum of the squares of the quark charges neglect cross-terms of the form (11.23) without justification. In the simple parton picture the cross-terms are neglected because the two electromagnetic interactions are assumed to take place on a single "active" parton, while the rest of the system is considered as inert spectators. The active parton is assumed to remember its color between the two electromagnetic interactions. This assumption breaks down because color cannot be localized and color oscillations analogous to neutrino oscillations occur, as discussed in my paper cited in note 2. These color oscillations define a time scale determined by E_{col} and short compared to the lifetime of the intermediate state in experiments at energies far below E_{col}. The physics of this argument is discussed in detail in the paper in note 2. Here I simply point out the rigorous results (11.19), (11.33)–(11.35), (11.39), and (11.42). Any model that claims to measure quark charges below color threshold should be checked for violations of these theorems.

Acknowledgment

This work was supported in part by the Israel Commission for Basic Research and the United States–Israel Binational Science Foundation.

NOTES

1. H. J. Lipkin and A. Tavhelidze, Phys. Lett. **17**, 331 (1965).
2. H. J. Lipkin, Phys. Lett. **74B**, 399 (1978), and Phys. Rev. Lett. **41**, 1629 (1978).
3. H. J. Lipkin, Nucl. Phys. **B155**, 104 (1979).
4. S. W. Herb, et al., Phys. Rev. Lett. **39**, 252 (1977); W. R. Innes et al., Phys. Rev. Lett. **39**, 1240 (1977).
5. C. Quigg and J. L. Rosner, Phys. Lett. **71B**, 153 (1977).
6. L. Schachinger et al., Phys. Rev. Lett. **41**, 1348 (1978).
7. Schachinger et al., ibid.; A. De Rujula, H. Georgi, and S. L. Glashow, Phys. Rev. **D12**, 147 (1975).
8. H. J. Lipkin, "Why Are Hyperons Interesting and Different from Nonstrange Baryons?" in *Particles and Fields*, edited by H. J. Lubatti and P. M. Mockett (University of Washington Press, Seattle, 1975), p. 352.
9. O. W. Greenberg, Phys. Rev. Lett. **13**, 598 (1964).
10. De Rujula, Georgi, and Glashow, op. cit. in n. 7.
11. P. Federman, H. R. Rubinstein, and I. Talmi, Phys. Lett. **22**, 208 (1966).
12. De Rujula, Georgi, and Glashow, op. cit. in n. 7.
13. Lipkin, op. cit. in n. 2.
14. H. J. Lipkin, Phys. Rev. Lett. **28**, 63 (1971), and in *Particle Physics*, Irvine Conference, 1971, edited by M. Bander, G. L. Shaw, and D. T. Wong (American Institute of Physics, New York, 1972), p. 30.
15. Ibid.; H. J. Lipkin, in *Experimental Meson Spectroscopy*, Proceedings of the 5th International Conference on Experimental Meson Spectroscopy, Boston, 1977, edited by E. von Goeler and R. Weinstein (Northeastern University Press, Boston, 1977) p. 38; H. J. Lipkin and H. R. Rubinstein, Phys. Lett. **76B**, 324 (1978).

16. E. Witten, Nucl. Phys. **B120**, 189 (1977); H. K. Lee and J. K. Kim, Phys. Rev. Lett. **40**, 485 (1978).
17. Lipkin, op. cit. in n. 3.
18. Lipkin, 1977 op. cit. in n. 15; Lipkin and Rubinstein, op. cit. in n. 15; Lipkin, op. cit. in n. 3.

Discussion

Following the Paper by H. J. LIPKIN

Chairman, M. GOLDHABER, Brookhaven National Laboratory

S. Weinberg (Harvard U.): I am not sure whether this is a question for Dr. Lipkin or for some of the previous speakers, but a number of you today have discussed the possibility that color is not trapped, and Roger Dashen specifically said that maybe the gluon is a massless particle that could be produced, but for complicated dynamical reasons it has a somewhat weak coupling to ordinary hadrons so that it is produced very slowly, and we do not see it even though it is massless. An argument occurred to me while the papers were being presented, an experimental argument against this, and I wonder if any of the speakers would like to comment. If the gluon exists and is massless, and is rather weakly coupled, it is still hard to imagine that it would not be coupled strongly enough for the sun to be opaque to it. A gluon striking the sun will probably be absorbed rather than get through. It would have to be very weakly coupled for that not to be true. Now if it is massless, and the sun is opaque to it, then the sun as a blackbody would be radiating gluons just as much as it is radiating photons. Hence, for its given surface temperature of 5,800 degrees it would be losing energy at a rate nine times as great as we normally think it is. I do not know whether astrophysical theorists are flexible enough to accommodate that, but I would doubt it.

Goldhaber: They would have the neutrino problem to a higher degree.

R. Gatto (U. Geneve): It is interesting that this wonderful approach for not very massive quarks, which one would expect to be relativistic, does not seem to work so well for the charmed quarks. In the first part of your talk, you had these beautiful numerical relations between magnetic moments and masses. Strangely, you would expect them to be better in the charm sector, but from experiment at this time, they are not. For instance, the expectation that the hyperfine splitting is giving the mass difference between the psuedoscalars and the vectors.

Lipkin: This I regard as an open problem because the charm–anticharm pseudoscalar has not yet been found. There are rumors from Stanford Linear Accelerator Center that the 2.8 does not exist. I would say that there is one paradox that still has to be explained and that is the whole question of mixing in the pseudoscalar mesons, and why the mass of the η and the η' are what they are. In particular, if you say that there is some kind of annihilation interaction in which $u\bar{u} \to s\bar{s}$ I would like to find a model that can predict unambiguously just the sign of this mass shift. The most naive models in which you say that it mixes with some other state give the wrong sign, and I think that as long as we do not understand the mass of η and η' the same annihilation interaction must be present in the η_c' and it will have a diagonal term $c\bar{c} \to c\bar{c}$, and therefore it will give you a mass shift. If you believe it has the same sign as η, η', and the same relative magnitude compared to the hyperfine splitting, then this is a shift that will push the mass up and be roughly of the same order as the hyperfine splitting. Therefore, you cannot really calculate where the η_c should be with any reliability.

C. Cooper (Technion, Haifa): Does this mean that in the spectroscopic regime you can actually get a really unambiguous value for quark masses?

Lipkin: Oh yes. In fact, I got to Fermilab last summer and the first thing I heard was that there was the new measurement of the lambda magnetic moment and the experimentalists who do not have a theoretician's scruples say: "Well, once you have the magnetic moment you can calculate the quark mass." So they got 330 MeV for the u and d and 510 for the s. I saw the difference was 180 MeV, which was exactly the same as I had calculated for the lambda nucleon and the mesons, and so I realized that this all fits together. I could have predicted that number in advance if I would have really believed what I was doing with a "crazy" approximation.

J. Lindenbaum (Brookhaven, and City College of New York): Speakers this morning have referred to the fact that the quark may be composed of other so-called elementary particles. I fully agree with that. The question I would like to ask is how one tells when this happens? Well, for one thing, when you reach the magical number of order 100, it seems to have happened in the past that you were dealing with a composite of a more elementary particle. That was true of the elements, the nuclei, the elementary particles and so on. But, being more specific, when you find what you think is a new quark, like the upsilon, you might ask: Indeed, is that a new quark or is it some kind of composite itself? To my limited knowledge, one way of telling is unless you can show there is a new quantum number involved, the chances are that you are playing with a composite. For example, the lambda, the sigma, the K, and so on, they really did not involve a new quark because they all had just the same new quantum number. So, I would like to hear from the two speakers involved if they have some ideas as to how experimentally one could tell whether one has new quarks or not? Furthermore, even if one has so-called new quarks, if one finds too many of them one must consider the implication that a new, more fundamental building block would explain the quark family.

Lipkin: I think we went through this with charm. The ψ was found and people had all kinds of models for it, but what really clinched it as being a $c\bar{c}$ state was when the charmed mesons were found. And I think that in the case of the upsilon if it really is the new $b\bar{b}$ quark, then there will have to be new mesons found with new quantum numbers. If experiments do not find them, then we will be in the same state as we were when people were looking for charm and had not found it yet and theorists were inventing new models for the ψ. The theorists will be a little more wary with the upsilon, because that will look a little bit like history repeating itself, and they will wait until experimentalists have really tried hard to find the new quantum number. Meanwhile, the spectrum of the upsilon excitations looks very much like just another quark.

PART V: FLAVOR AND ASTHENODYNAMICS

The term *flavor* was introduced to distinguish between two sets of quantum numbers carried by quarks. Seen in that light, flavor (as distinguished from *color*) G_F denotes the set of quantum numbers characterizing the hadron spectrum: $SU(3)$ for the world of (u,d,s), $SU(4)$ once charm was added (the $(u,d,s,c,)$ system). With the discovery of the Υ at 9.5 GeV and the conjecture that it represents a "$b\bar{b}$" state, with b a fifth quark, we go on to $SU(5)$ and most probably $SU(6)$ for the conjectured (u,d,s,c,b,t) set. We list the quarks according to their masses, whether ultraviolet masses of current quarks or the (mostly larger) constituent quark masses (of the less understood infrared limit).

The concept can be extended to the leptons, where at least the charged ones display a mass spectrum (e,μ,τ). The situation for the neutrinos $(\nu_e, \nu_\mu,$ and the hypothetical $\nu_\tau)$ is less clear.

Gauge asthenodynamics (the electroweak $SU(2) \times U(1) = G_w$) involves an entirely different set of quantum numbers. It distinguishes between right- and left-handed states, its $SU(2)$ connects u_L with $d_L' = (\cos\theta\, d_L + \sin\theta\, s_L)$, or c_L with $s_L' = (\sin\theta\, d_L - \cos\theta\, s_L)$ in the 4-quark picture. It is possible to generalize flavor itself into a chiral group $G_{FL} \times G_{FR}$ and have G_w as a subgroup. However, many theories do not aim at such a structure. "Grand unification" theories, as discussed in this and the following part of this volume, deal with gauge groups containing G_w G_s, where G_s is the gauge group of QCD, generally $SU(3)_{\text{color}}$. Some early examples tried to reproduce parts of G_F, but with the experimental discovery of sequential G_w sets ("families," or "generations"), the chances of a grand unification group yielding all six quarks and six leptons have dwindled. Thus, the groups discussed in Part VI, to the extent that these theories are still "on," should probably be aiming at reproducing one family (for example, $SU(5)$). The additional multiplicity is then relegated to a "seriality" quantum number, which is discussed in Harari's and Gatto's chapters. These authors explore the possible mass matrices yielding the six masses and four angles, making up ten out of Glashow's seventeen unexplained parameters.

Glashow's chapter is a very frank assessment of our gauge theories. His doubts with respect to the great "particle desert" between 100 GeV and 10^{14} GeV are shared by many. It is also rather surprising that the enthusiasm evoked by the success of G_w and perhaps G_s should have led to a certain underemphasis of the complexity and importance of the completely unexplained G_F. Finally,

---

Yuval Ne'eman (ed.), To Fulfill A Vision: Jerusalem Einstein Centennial Symposium on Gauge Theories and Unification of Physical Forces
ISBN 0-201-05289-X

Copyright © 1981 by Addison-Wesley Publishing Company, Inc., Advanced Book Program.
All rights reserved. No part of this publication may be reproduced, stored in a retrieval system, or transmitted, in any form or by any means, electronic, mechanical photocopying, recording, or otherwise, without the prior permission of the publisher.

G_F relates to the only systematics and symmetries that we observe directly, and one cannot evade the issue by claiming that it is all subsidiary and unimportant. Is there really direct proof that all that we see results from an arbitrary set of trivially coupled Higgs fields? We dare conjecture that flavor physics may be all the more interesting for being that complex. The discussions of the mass matrix pattern in an ad hoc approach as exemplified by Gatto and Harari serve to underscore the mystery of flavor.

There are two recent ideas touching upon the topics of this part. The first is that of a more "primitive" particle model, from which both quarks and leptons would arise as composites. This possibility is mentioned by Harari, who has recently suggested an extremely simple model.[1] His "rishons" are T and V, with charges $+1/3$ and 0, respectively. They have spin $1/2$. Their 3-rishon combinations are $(TTT) = \bar{e}^+$, $3 \times (TTV) = u$, $3 \times (TVV) = \bar{d}$, $(VVV) = v_e$. Other families represent excitations. It is not clear whether there shouldn't have been other states with spin $3/2$ even prior to such excitations. The question of chiralities is also unclear.

We have suggested a less radical model,[2] but Harari's has greater simplicity. Both G_s and G_w gauges would then be phenomenological. It is also remarkable that the universe is symmetric in rishon-number! Such a model is of course highly speculative, but it illustrates the fact that our supposed "fundamental" gauges might be phenomenological, or should be reformulated sometime in terms of new basic constituents.

The other recent idea is a suggested solution[3] to what Harari has called "simple unification." This is the introduction of a new type of gauge symmetry, the supergroup $SU(2/1)$. This can only be a symmetry of the *quantum* Lagrangian, including Faddeev–Popov ghosts, since its graded Lie algebra changes the statistics without changing the spin: it thus relates fields to ghosts. The representations are doubled by a fermionic involution ε, $\varepsilon^2 = 1$, that transforms a representation of $n_{\text{Fermi}} + m_{\text{Bose}}$ components into $n_{\text{Bose}} + m_{\text{Fermi}}$. The quantum numbers are exactly those arbitrarily assumed in the Weinberg–Salam model: $\mathbf{4}$:(u_L, d_L, d_R, u_R) with correct assignments of $SU(2)$ and $U(1)$ for all states, and a "prediction" that for integer charges the representation $\mathbf{4}$ goes into $\mathbf{3}$: (v_L, e_L^-, e_R^-) losing its fourth state. The Faddeev–Popov ghosts C^a and the Higgs fields ϕ^i make an octet $\mathbf{8}$:(C^a, ϕ^i) precisely. The theory invokes a new set of $J = 1$ ghosts ξ_μ^i fitting together with the vector mesons in the adjoint $\mathbf{8}$:(W_μ^a, ξ_μ^i). An analysis of our loop diagrams shows that it is these new ghosts ξ_μ^i that *preserve the Weinberg angle*. Its original value is $\sin^2\phi_w = 0.25$, which yields 0.23 when renormalized by the ~ 100 GeV masses of the spontaneous symmetry breakdown. The Higgs mass is predicted to be around 220–250 GeV. Two of Glashow's seventeen parameters are thus fixed.

NOTES

1. H. Harari, Phys. Lett. **86B**, 83 (1979); M. A. Shupe, Phys. Lett. **86B**, 87 (1979).
2. Y. Ne'eman, Phys. Lett. **82B**, 69 (1979).
3. Y. Ne'eman, Phys. Lett. **81B**, 190 (1979).

V. Flavor and Asthenodynamics

Session Chairman's Remarks

R. BACHER, California Institute of Technology

The following part of this volume deals with weak and electromagnetic interactions and with flavor. Philosophers, historians, and psychologists at this conference have quoted Einstein saying that "the most incomprehensible thing about our world is that it is comprehendable." This comment, I noticed, was not made by any of the particle physicists — not that we find our world uncomprehendable, but the new things to be explained have expanded so rapidly that it certainly has confronted us with many problems. One does not need to go back very far, and in an earlier chapter Lipkin went back part of the way to a period when it seemed quite different from the way it seems today. I remember with some clarity the end of 1945, when most physicists were returning from having done various other things during the war. A good many distinguished nuclear physicists expressed the firm opinion that if we knew the proton-proton scattering and the neutron-proton scattering really well, up to 10 million electron volts, we would understand the atomic nucleus. In retrospect, it seems almost incredibly naive, and illustrates what has often been done in the past, and that is to underestimate very seriously the richness of nature. Whether Dr. 't Hooft has changed that direction with his extrapolations (in Chapter 9) as to how imaginative one can be about the future remains to be seen, but perhaps this is a change in trend. Of course, after the war it did not take long before unstable particles started coming along, the nucleon showed some structure, and all sorts of excited states. We spent two decades in accumulating an enormous number of unstable particles that many of the attendants at this conference helped to unravel and put in some sort of order. But particle physicists in general, in spite of the complexity that has been shown in this field for a long time, have persisted really very hard in trying to get a degree of order. And a surprising degree of order has been achieved. One of the problems, of course, has been that particularly in recent years, the experimental discoveries have come out at such a pace that, rather like the White Queen in Alice in Wonderland, the theorists have had to run as fast as they could just to stay where they are. Nevertheless, the advance has been and is today most impressive.

12. OLD AND NEW DIRECTIONS IN ELEMENTARY PARTICLE PHYSICS

Sheldon L. Glashow

1. Atomic Synthesis

A time came when people began to believe in atoms. Simply and inexorably the laws of chemical combination emerged from the atomic hypothesis. A successful theory of heat, foreseen too soon by Daniel Bernoulli, was truly born. Yet, until this century there were those who would doubt atomic reality. Brownian motion — false evidence, at first, for the force of vitalism — was ultimately to be explained by an Einstein who did believe in atoms. Doubters were silenced. But, it soon became evident that the atom itself could not be truly fundamental. Let us review some early indicators of atomic structure.

There were so many different elements, each presumed to consist of a different type of atom — too many, certainly, for them to be the most fundamental constituents of matter. None of the four Greek elements had survived scrutiny, yet the notion of a very small number of truly basic constituents remained (and remains) compelling.

Prout, in 1813, saw the tendency of atomic weights to be nearly integral multiples of that of hydrogen. He suggested that all atoms were built up of hydrogen atoms. We now know that he was essentially right, but his hypothesis has had its ups and downs. (It was, by the way, an alleged small departure from Prout's law that led Lord Rayleigh to the studies that culminated in the astounding, but overdue, discovery of argon and of an unanticipated eighth column of Mendeleev's table.) The underlying simplicity that Prout sensed was there, but it took more than a century to root out.

The periodic table was correct. It successfully predicted the existence and the properties of scandium, gallium, and germanium. The properties of atoms varied systematically. Does not order imply structure?

The establishment of a theory of electromagnetism was quickly followed by the discovery of the electron. The discoverer himself realized that the electron must be a constituent of the atom, and that the atom must be held together electrically. But where did the bulk of the atom's mass and its positive electric charge reside?

Yuval Ne'eman (ed.), To Fulfill A Vision: Jerusalem Einstein Centennial Symposium on Gauge Theories and Unification of Physical Forces
ISBN 0-201-05289-X
Copyright © 1981 by Addison-Wesley Publishing Company, Inc., Advanced Book Program.
All rights reserved. No part of this publication may be reproduced, stored in a retrieval system, or transmitted, in any form or by any means, electronic, mechanical photocopying, recording, or otherwise, without the prior permission of the publisher.

Rutherford did the first truly important scattering experiments. Thereby he established the most important tool in the workshop of microphysics. And he found the atomic nucleus. All the ingredients of the atom were then known. With the discovery of quantum mechanics, a theory of atomic structure was established for all time. Why is the sky blue and copper red? How do rabbits multiply and cells divide? All this, and more, follows (with some hard work) from the rules.

2. False Synthesis

Much of scientific research is called basic, but the search for the ultimate building blocks of matter is truly basic. Imagine an extraterrestrial being eavesdropping on a game of chess. His first problem is to figure out what are the rules of the game. Only then can he attempt to devise a winning strategy. Most of basic science is of the second category and consists of the working out of consequences of known rules. The discoveries of relativity theory and of quantum mechanics are of the first category. So also is the search for an understanding of the atomic nucleus.

The chemical properties of an atom are determined by the electrical charge of its nucleus. There are at least as many kinds of nuclei as there are elements. Again, there are simply too many nuclei for them to be truly elementary. Moreover, the nucleon charge is always precisely an integral multiple of the electron's charge. Today we think we have a good reason for why this is true, but this will come toward the end of our story.

The discovery of radioactivity was another clue to the structured nature of the nucleus. The alchemists' dream of atomic transmutation was fulfilled, but in a doubly disappointing way. No laboratory procedure would alter the natural processes of radioactive decay. Neither acid nor base nor heat nor cold could affect the spontaneous and relentless nuclear process that would change expensive radium into cheap lead.

Chlorine represented the first spectacular departure from Prout's law. Its atomic weight is as far from an integer as it could be. Isotopes were discovered, and it turned out that terrestrial chlorine consisted of 75 percent Cl^{35} and 25 percent Cl^{37}. Integers were in again, and a Proutian view of nuclear structure developed. The nucleus was to consist of protons and electrons somehow bound together: protons to supply the integer mass, and electrons to correct the nuclear charge. The electron was to play a dual role in atomic structure: "outsies" that obeyed the rules of quantum mechanics and explained the chemistry of atoms; and "insies" that were mysteriously confined to the interior of the nucleus. Improbable dynamical schemes to explain all this were proposed.

Next there was the case of nitrogen. The nitrogen nucleus was observed to behave like a boson and to satisfy symmetric statistics. Thus, it must have an even number of fermionic constituents. But, the nitrogen nucleus was supposed to consist of fourteen protons and seven electrons, and twenty-one is certainly odd. (Years later, an analogous "statistics problem" was to worry the early proponents of quarks.) The proton-electron model of the nucleus was in desperate trouble.

Enter the neutron! With it, a rational model of nuclear structure was at hand. Nuclei were simply made up of protons and neutrons. But what was to hold them together? The necessary existence of a third fundamental force was recognized. Yukawa postulated that it, like electromagnetism, would be mediated by fundamental new particles — mesons. And, just as he predicted, pions were found.

Of the three known types of radioactivity, two could be quite well understood in terms of the strong (nuclear) and electromagnetic forces. The third process, β-decay, presented a number of peculiar puzzles. A new particle, the neutrino, had to be invented, and a fourth force as well.

By the early 1950s a commonly accepted picture of the microworld had evolved. There were just four fundamental forces. Gravitation plays no role in atomic or subnuclear physics; it is simply too weak. Strong interactions, mediated by fundamental pions, are the glue holding protons and neutrons together in nuclei. Electromagnetic interactions, mediated by photons, are much weaker and play a central role in atomic physics and a lesser role in nuclear physics. Weak interactions were a necessary evil to account for nuclear beta decay. Perhaps they were mediated by heavier mesons that were not yet observed. This prediction of Yukawa is still unfulfilled.

The subnuclear world was quite simple: Nuclei were made of protons and neutrons held together by pions. The neutrino was soon to be more or less directly observed. Although mysterious and elusive, it was seen to play an essential role in those processes that make the sun shine. It was seriously put forward that we had reached the end in our search for the ultimate constituents of matter.

It was the development of large particle accelerators that put a decisive end to this view. But, early indications of trouble were to come from the heavens above, from cosmic ray experiments. *How do muons and strange particles fit in?* Things were to get a lot more complicated before they got simple again. Only now, when we have become convinced of the structured nature of protons, neutrons, and pions does this question begin to strike terror into our hearts.

3. New Synthesis

Protons and neutrons are not at all the fundamental pointlike particles that they were first imagined to be. They have all the properties one expects of structured composite systems. All but one, that is: The conjectured quarks and gluons that make up a nucleon have not been, and perhaps cannot be, seen as isolated components.

Atoms and nuclei show their structure by their spectra. Their stable ground states are merely the lightest of an infinite number of quantum states. Higher excitations are generally characterized by higher angular momenta. Energy-level diagrams depicting the first dozen or so states of a typical atom and a typical nucleus are similar, except for a difference in scale by a factor of a million. So it is with the nucleon.

Fermi discovered in 1954 what was to become known as the first pion-nucleon resonance. Today there are more than two dozen observed states that may be regarded as excited states of the nucleon. Spins of at least 11/2 have been identified. The scale of the energy-level diagram is about 100 times larger than that of a nuclear system. The excited nucleon decays by strong interactions rather than by electromagnetism, and by the emission of mesons rather than photons. Aside from this, the pattern is a familiar one. Nucleons, like nuclei and atoms, must be composites.

Bound systems cannot be pointlike. The diameter of the proton was determined by careful measurements of elastic electron-proton scattering, which were interpreted in terms of a spread-out distribution of the proton's electric charge. By its very nature elastic scattering cannot reveal anything more about the inner structure of a proton. With higher energy electrons available at Stanford Linear Accelerator Center, experimenters were able to study inelastic electron scattering and find truly startling results. The electric charge of the proton seems to be concentrated on pointlike particles within. Fifty years after Rutherford had discovered the small heavy constituents of atoms, modern-day workers in an analogous way, had exposed the basic constituents of nucleons. At first these constituents were called partons.

Quarks had been postulated several years before partons were "seen." The nucleon was to be composed of exactly three conjectured fermionic constituents. With just two possible values of the quark charge, we could understand why the nucleons and their excited states always display one of four electric charges: $-1, 0, +1$, or $+2$. The up quark had to have $Q = 2/3$, the down quark $Q = -1/3$. There were indeed just two basic constituents to nuclei, but these were not the proton and neutron themselves. The parton is a quark, after all.

Mesons too are to be made up of quarks. They consist of one quark and one antiquark rather than three quarks. The rules of valency are these:

$$qqq = \text{baryon}$$
$$\bar{q}q = \text{meson}$$
$$\bar{q}\bar{q}\bar{q} = \text{antibaryon}$$

No other combination of quarks is to correspond to an observable hadron. (Rules of this kind often have exceptions with no fundamental import. Compounds of the inert gases have been found without overthrowing the foundations of chemistry. So it is that we have seen or may be on the verge of seeing hadrons made up of six quarks, or of two quarks and two antiquarks.)

If mesons are made up of quarks, then they are not the fundamental nuclear glue. Yukawa was wrong. It may be useful to regard the nuclear force as a consequence of meson exchange, but all this must be explained in terms of more basic concepts.

The nuclear force must be a secondary consequence of the fundamental force that binds quarks into nucleons, just as the chemical force between neutral atoms is a secondary consequence of the Coulomb force that binds the atom. No longer can we regard nuclear physics as a truly fundamental discipline. Einstein's apparent indifference to the field is now justified. But, what is the basic force between quarks that is at the root of all strong and nuclear interactions?

It is amazing to me that we finally seem to have the answer to this question. The theory underlying quark interactions is known as quantum chromodynamics (QCD), and its quanta are known as gluons. It is an unbroken gauge theory based upon the local symmetry group $SU(3)$. It acts upon a truly hidden variable known as color. Each flavor of quark (up, down, and so forth) is to come in three colors. Although the color degree of freedom is unobservable, it makes itself felt indirectly. The antisymmetry of the 3-quark wave function that is the proton is relegated to the color degree of freedom. This is why quarks often behave as if they satisfied Bose-Einstein statistics. The number of colors reveals itself more or less directly in the observed rate of π^0 decay and in the annihilation of leptons into hadrons.

The successes of QCD are many, and so are its challenges. Most physicists agree that it is likely to be the correct theory of strong interactions. Nucleons, nuclei, and mesons — all hadronic matter — are now believed to be made up of quarks held together by gluons. Neither can be seen in isolation from the hadrons of which they are part.

4. Perverse Directions

As invented, the quark model needed three flavors of quarks, not just two. We have put off the sad story of strange particles, and their even stranger kin. The classical hadrons were known to display three additive quantum numbers: baryon number, electric charge, and strangeness. Baryon number is superceded by quark number: baryons are made of three quarks. Just as they can have four possible charges, they may have four values of strangeness. Plots of strangeness versus charge show simple geometric forms. Hadrons make up supermultiplets in the unitary symmetry scheme: singlets, octets, and decimets of baryons; singlets and octets of mesons. Oddly, no higher representations showed up. Like the periodic table of elements, the scheme was predictive. It anticipated the existence and the properties of the Ω^-; it predicted the correct value for the magnetic moment of the Λ; and much more. All this was explained by the quark hypothesis. Two quarks were all that were needed to describe the nuclear particles; a third strange quark was the characteristic ingredient of the strange particles. QCD, our first true theory of strong interactions, could be phrased only in terms of the quark model. But this is a terrible weakness as well as a strength. QCD does not tell us how many flavors of quark there must be.

As long ago as 1961, strangeness posed a puzzle to those of us who favored a gauge theory of weak interactions: Strangeness changing neutral currents would upset the successes of such a theory. This resulted from the asymmetry of a world with four leptons and only three quarks. How much nicer things would look if there were four quarks as well! The existence of a fourth quark would remove all phenomenological obstacles to a proper theory of weak interactions. But, where were the predicted charmed hadrons? The J/Ψ, discovered in 1974, was the first particle seen that contained charmed quarks. Soon, there were to be many more.

For a brief time, it seemed that nature used just four quarks and four leptons. But, this was not to be. Now we know of a third charged lepton, tau. The upsilon particle is generally supposed to contain a fifth flavor of quark. These new states fit into a symmetrical pattern involving six leptons and six quarks. The sixth quark has not yet been found. Nor is there any convincing argument as to what its mass should be. Perhaps these are all the quarks and leptons there are. Then again, perhaps not: such a simplistic view has failed once before. In any case, it is a clearly central and unanswered question. How many quarks and leptons are there? And, why are there that many? Are there other basic particles lurking whose properties we can only guess?

Our discipline has taken a new and disturbing tack. No longer are we delving into the structure of an atom. It is made of up and down quarks and electrons. Muons, tau leptons, and heavy quarks are simply not needed as elemental building blocks. They are simply there, for no known reason. The superfluous fermions do not seem to indicate the existence of a deeper layer of structure. They are not excited states of one another. Quark energy-level diagrams are not akin to anything we have seen before.

No higher spins have appeared, nor are there any departures from quark and lepton pointlikeness. Perhaps we will never solve the mystery of the superfluous fermions. Muons are now forty years old, and we still do not know who ordered them. And after all, the biologists have come a long way without hope of learning why there are rabbits.

5. Groups of Forces; Families of Fermions

Strong interactions are supposed to result from a gauge theory based on the group $SU(3)$. Electromagnetic interactions, as has long been known, also result from a gauge theory. Since the 1950s it was suspected that weak interactions might come from a gauge theory too. Such a theory could not be developed until the chaos of strong interaction physics gave way to the underlying simplicity of quarks.

Weak and electromagnetic interactions result from a gauge group acting upon flavor rather than color. It is the group $SU(2) \times U(1)$ rather than $SU(3)$, and it is spontaneously broken rather than exact. Its predictions are quantitatively confirmed.

Thus it is that today weak, strong, and electromagnetic interactions are all recognized to be gauge theories. Just why did nature do this? Is there something particularly compelling about gauge theories, among all other renormalizable quantum field theories? The interactions of particle physics depend upon the existence of a local twelve-parameter Lie group, $SU(3) \times SU(2) \times U(1)$. We have the freedom to change these twelve parameters freely at each space-time point. Who asked for such a dubious privilege? Perhaps the reason that each of the three types of interaction is described by a gauge theory is that somehow the three are really one.

This is the idea behind unification. Perhaps the three-component phenomenological group $SU(3) \times SU(2) \times U(1)$ can be fit into a simple unifying group with just one component. We are led directly to the unified $SU(5)$ theory or to one of its elaborations. It seems that weak, strong, and electromagnetic interactions must first be put together before we may implement Einstein's dream of marriage to gravity.

Given the phenomenological gauge group, the observed fermions may be classified. They transform as fundamental or as trivial representations of $SU(3)$: quarks and leptons. They transform as fundamental or as trivial representations of weak $SU(2)$ as well, depending simply upon their handedness. And, they fit into (three) characteristic and boringly repetitive families.

The relevant quarks and leptons, those that play an essential role in the running of Spaceship Earth are these: the up and down quarks, electrons, and electron neutrinos. They form the first family of fundamental fermions. They comprise fifteen chiral fields that form a fifteen-dimensional reducible representation of $SU(3) \times SU(2) \times U(1)$. In the simplest unified schemes ($SU(5)$, $SU(6)$ or $SO(10)$) this family retains its identity as a representation of the unifying group.

There is something special about the fifteen-member family. It is the smallest subset of fields that is free of triangle anomalies. One family would be sufficient to yield a renormalizable gauge theory. Leave out a lepton or a quark and disaster is sure to follow. It is also the smallest subset of fields for which the electric charge sums to zero. Thus, it is the simplest unit that may form the basis of a unified theory.

Unification, renormalizability, and phenomenology are all satisfied with only a single family of fermions. Had muons and strangeness and all that not existed at all, we should have all been quite happy. Why, then, does nature present us with two additional fifteen-member families, each equivalent to the first family? The extra families are simply heavier, phenomenologically irrelevant supercargo in our search for the one and true theory.

Suppose that there are in toto just three families of fermions and nothing more; so argue astrophysicists who put limits on the number of massless neutrinos that exist. Where has the search for simple beginnings led us? I call it the scenario of the seventeen parameters.

The semisimple gauge group has three components, corresponding to three dimensionless coupling constants. The Higgs domain requires at least two parameters, say G_F and the Higgs meson mass. There are nine quark and lepton masses, and four Cabibbo-like angles. Delete one for a measuring rod, and we are left with seventeen. Unless we have the courage to buy unification, this seems to be the end of the road: seventeen unknowable dimensionless parameters, all on a par with the fine structure constant. Surely this is a totally unacceptable situation. But we are on the horns of a dilemma. To unify means to escape (perhaps) from a seventeen-parameter universe. The price is high; we must believe in all sorts of interesting physics, new particles, and new forces, at very high energies. These energies are so high that we cannot even dream of a day when they will be accessible to direct experimental study.

6. Unity

The gauge group $H = SU(3) \times SU(2) \times U(1)$ is the basis to an entirely satisfactory theory of strong, weak, and electromagnetic interactions. That is to say, it offers a description of elementary particle phenomena that is apparently correct and in principle complete. It is not, however, a unified theory: it is a seventeen-parameter theory. And, it leaves a number of questions unanswered:

Why are strong interactions stronger than the others? Why are there three separate but similar fermion families? Why is there Cabibbo mixing and CP violation? Why are all electric charges commensurate? Are there no relations among the fermion masses? Can we understand the five-decade spread between the electron and intermediate vector boson masses? Will there be surprises at higher energies?

We could hope to answer some of these questions if the phenomenological group H were a subgroup of a larger unifying simple group G. Immediately we would understand the fact of electric charge quantization. The three independent gauge-coupling constants are reduced to only one. We must still explain the empirical disparities in the strengths of weak, strong, and electromagnetic interactions. This may be done if the group G is very badly broken. In particular, all the generators of G that are not generators of H must correspond to exceedingly heavy mesons, with masses of order 10^{14} GeV. Renormalization group corrections will then produce the observed interaction strengths. Furthermore, if H is the whole of the phenomenological gauge group, there is some confirming evidence for this picture of unification. The weak mixing parameter $\sin^2\theta$ may be directly calculated, and the result is in good agreement with experimental determinations. This would not be the case if, for example, $SU(3) \times SU(2) \times SU(2) \times U(1)$ were the phenomenological gauge group.

Unification demands the existence of a rich physical sructure — new particles and new interactions — lying at energies that will be forever inaccessible to direct experimental search. Will there be no new structure between the known and accessible domain of energies and the superheavy domain? May we look forward only to the discovery of the t quark, the intermediate vector bosons, and a Higgs boson? This real possibility is cause for despair. It could foreshadow the imminent end of the experimental frontier. We may hope that the vast desert between 100 GeV and 10^{14} GeV will bloom after all, but there is now no indication that it will. Fortunately, there is at least one way by which we may learn something of the superheavy regime.

Most unified theories, but not all of them, require that protons and neutrons decay. Diamonds are *not* forever. The superheavy particles mediate a fifth force that need not conserve baryon number. Indeed, it may be possible that unification will expunge all of the apparent exact global symmetries: baryon number, muon number, electon number, and so on. Pehaps all symmetry is local symmetry.

In the simplest unified theory, based on $SU(5)$, the proton does decay, with a lifetime estimated to be 10^{32} years. This is a mere two decades removed from the current experimental lower limit. Experimental vindication of this result would be a strong hint that we are on the right track. Nonetheless, we must admit the possibility that unification may be correct, but that the proton is absolutely stable. An example of this kind uses the gauge group $SU(3) \times SU(3) \times SU(3)$ with a permutation symmetry. It has no great appeal, except to those who find proton decay unpalatable.

How the proton shall decay, at what rate and according to what selection rules, depends upon the theory. In the simplest version of $SU(5)$, the decay modes are as follows:

$$(\text{Nucleon}) \rightarrow (\text{antilepton}) + (\text{mesons with } S = 0 \text{ or } 1) \ .$$

Decays involving a positively charged lepton or into $S = -1$ hadrons or into several leptons are forbidden. For the correct theory, these selection rules may or may not be satisfied. Only experiments can tell. Even in the naive model, there is much to be measured, since final states with six experimentally distinguishable values of lepton number and strangeness are anticipated.

Minor modifications of the theory, still within the context of $SU(5)$, can radically change the decay selection rules. Naive $SU(5)$ preserves conservation of baryon number minus lepton number. A richer Higgs system can destroy this symmetry. In that case, decays into negatively charged leptons and hadrons could compete effectively. One may even arrange for a substantial $\Delta B = 2$ interaction, in which case two nucleons in a nucleus could annihilate into mesons releasing 2 GeV of hadronic energy. This interaction could also reveal itself by producing a small baryon-number-violating mass term connecting the neutron to the antineutron. Experiments with ultraslow neutrons could reveal such a coupling with a strength as small as 10^{-24} eV.

Although $SU(5)$ is the simplest model, it is not at all clear that it is the correct model. Indeed, the simplest version of $SU(5)$ is an evident failure. Minimal $SU(5)$ employs Higgs bosons forming an adjoint representation and a quintet. The following renormalization-group corrected formulas for quark-lepton mass ratios follow directly:

$$\frac{m_d}{m_e} = \frac{m_s}{m_\mu} = \frac{m_b}{m_\tau} \approx 3 \ .$$

Only one of these three results is true. There simply must exist a Higgs 45-plet in addition to a quintet. The Higgs sector cannot be as simple as it might have been.

Another argument points in a similar direction. It concerns the origin of the net baryon surplus in the universe. Yoshimura has suggested that the existence of at least six quark flavors, of baryon number violation, and of CP violation together present a framework from which the cosmic baryon abundance may be explained. The right answer can indeed emerge — but again, it is necessary that the Higgs sector is more than minimally complex.

One theoretical obstacle does stand in the way of unification — the so-called gauge hierarchy problem. At least one Higgs boson must survive as an observable particle with a normal mass. Its colored counterparts, which mediate proton decay, must be superheavy. Coupling constants of the Higgs sector must be rigged to preposterous accuracy to accomplish this feat. We have no principle that can ensure this happening, and no faith in twenty-five-decimal place coincidence. Something is wrong, or something is missing from our philosophy.

Let me conclude this paper with some mention of the problem of flavor: how to understand the superfluous replication of fermions. There have been a number of suggestions, but none of them work out well or seem right.

Substructure. Perhaps quarks and leptons are made up of more fundamental things: "maons" or whatever. Why have no higher spin entities shown up? Can one really make do with fewer maons than we already have chiral fields? How can a massless neutrino be regarded as a composite system?

Horizontal Symmetry. Symmetries connecting the three or more families of fermions may be discrete or local. Discrete symmetries are the subject of K. Gatto's chapter in this volume. They do not seem to go anywhere. Local symmetries, unless they are very badly broken, only serve to induce unobserved interactions. And, they are not easily incorporated into a unified theory. Elaborate theories involving global symmetries broken by terms of lower dimension have been proposed. So do Georgi and Nanopoulos compute the t quark mass to be 15 GeV, but it is hard to believe that nature has chosen so inelegant a way.

Oblique Embedding. Perhaps the three fifteen-member families are the residue of a large representation of a large group. Particles that can obtain a mass that is invariant under H do, and this mass is superheavy. Georgi has invented a model based on $SU(11)$ in which a 561-dimensional anomaly-free representation leaves precisely three families at normal mass. What a sacrifice of elegance for expediency!

Unification at Low Mass. Perhaps there is a unifying group, but at an accessible rather than a superheavy mass. In such a theory, the proton is presumably stable. A wondrous variety of new particles remains to be discovered at the next generation of accelerators. It is possible to invent such theories, but not nice ones. Georgi (again) has invented one based on $SU(11)$ (again). Perhaps it is right.

A Larger Phenomenological Group. It could be that a fifth force lurks at TeV energies. Susskind has called it technicolor. Perhaps it, not Higgs bosons, is ultimately responsible for the masses of intermediate vector bosons. Then what produces the masses of quarks and leptons? Thus is born the subject of "extended technicolor," about which I have nothing to say.

Orthodox strong, weak, and electromagnetic interactions have been explained in terms of a gauge theory. The theory begs to be unified, and some such scheme is no doubt correct. We now recognize that we live at a very fortuitous time. In youth and in old age, our universe is symmetric between matter and antimater. We live in the Age of Nuclei. Perhaps we will some day succeed in putting gravity back into our unified field theory. At present, the prospect seems dim. Our accelerators are about to explore the great desert in which no new particles will be found except those that have already been predicted. Want to bet?

Acknowledgment

This research was supported in part by the National Science Foundation under Grant No. PHY77-22864.

Discussion

Following the Paper by S. L. GLASHOW

Y. Achiman (U. of Wuppertal): I want to point out that if one wants only one representation for all the fermions, there is a more elegant way to do it. You have to remark that the two 27-dimensional representations of $E(6)$ are actually related to each other by a kind of parity operation, and if you add parity invariance to the game, then you have actually only one representation. So the $E(6)$ model practically gives also an answer to the question of how many families there are: there is only one family. And one more remark that is related: All the good results of $E(6)$ can be obtained from a model that is $SU(3) \times SU(3) \times SU(3)$ that is globally embedded in $E(6)$. My previous remark is also true for this group.

Glashow: I agree with your remark. However, there are some difficulties with every $E(6)$ theory I have ever seen. There are phenomenological questions with $E(6)$ though there are $SU(3) \times SU(3) \times SU(3)$ theories that make phenomenological sense.

13. QUARKS AND LEPTONS: THE GENERATION PUZZLE

Haim Harari

1. **Introduction: "Standard Wisdom"**

A well-defined "standard view" of the world of quarks and leptons now exists. Much of it has been brilliantly confirmed by experiment. Some of it is yet to be confirmed, but most of us believe that this is just a matter of time. Beyond the "standard" picture, we face a long list of crucial questions, about which we know very little. The exciting physics of the next decade will probably focus on these questions. I devote this paper to a discussion of some of them.

The first generation of quarks and leptons is, undoubtedly, the best studied. We know that the left-handed (u,d) quarks and (ν_e, e) leptons form doublets of the electroweak $SU(2) \times U(1)$ gauge group[1] and that their right-handed counterparts are in singlets of the same group.[2] We know that the quarks come in three colors and believe that they interact with gluons, presumably according to the rules of quantum chromodynamics (QCD). We believe that W^+, W^-, and Z are the gauge bosons of the weak interactions and that the weak and electromagnetic couplings are related by the parameter $\sin^2\theta_W \sim 0.23$.

We also know that a second and, probably, a third generation of quarks and leptons exist. All their known properties are consistent with those of the first generation, but many experimental facts are yet to be confirmed. The t-quark is still to be discovered, the b-quark and τ-neutrino are only indirectly "observed," the electroweak properties of b are not known and even the right-handed c, s, μ, and τ are not fully investigated.[3] Nevertheless, it is very likely that they will all turn out to reproduce the properties of the first-generation fermions.

The "standard" description then consists of three generations of fermions (each containing two quarks and two leptons), and twelve gauge bosons: eight $SU(3)$ gluons and four $SU(2) \times U(1)$ electroweak vector bosons. All their interactions are specified by the overall $SU(3) \times SU(2) \times U(1)$ gauge group. The free parameters of the theory include the masses of all quarks and leptons, the generalized Cabibbo angles, and the coupling constants of the three gauge groups or, alternatively: α, α_s, $\sin^2\theta_W$.

So much for the "standard wisdom."

2. The Generation Pattern: Unlikely Alternatives

The presently accepted pattern of generations has two *independent* striking features:

1. Within each generation, the pattern of quarks is very similar to the pattern of leptons.
2. Each generation is similar to the other generations.

Each of these features leads to interesting implications. The first suggests a profound connection between quarks and leptons. The second indicates that the old e-μ puzzle is now generalized into a puzzle of apparently redundant generations of both leptons and quarks which, like e and μ, differ from each other only by their masses.

It is still possible, however, that the correct pattern is different. For instance, we cannot completely exclude the possibility that each generation actually contains, say, three quarks of charges 2/3, −1/3, −1/3 (for example, u,d,b; c,s,h). This would break the quark-lepton similarity, unless there are two charged leptons for each neutrino.

It is also possible that future experiments will show that different generations have different structures. Higher generations may contain more fermions or they may involve right-handed $SU(2) \times U(1)$ doublets et cetera. An even wilder possibility is the existence of "exotic" quarks and leptons. These might include doubly charged leptons and/or quarks with charges 5/3 or −4/3, and/or spin 3/2 quarks and leptons and/or color sextets.

There is no experimental evidence or theoretical need for any of the foregoing suggestions. However, we should constantly keep an open eye for any hints in such unconventional directions. Although we do not understand the pattern of identical generations, we at least have a well-defined puzzle. Any deviations from the standard pattern will radically change our puzzle.

3. The Electroweak Group: Interesting Extensions of $SU(2) \times U(1)$

The experimental evidence for the validity of $SU(2) \times U(1)$ as the correct gauge theory for electroweak interactions is quite impressive. We have no reason to doubt it. However, it is entirely possible that some higher gauge group G provides a full description of the electroweak interactions, and contains $SU(2) \times U(1)$ as a subgroup. Those gauge bosons of G that lie outside $SU(2) \times U(1)$, must be heavier than W^+, W^-, and Z. All the present phenomenological studies of $SU(2) \times U(1)$ could then remain essentially unchanged.

Do we have good theoretical reasons to go beyond $SU(2) \times U(1)$? There are at least three such reasons and they are related to the three obvious open questions of $SU(2) \times U(1)$:

1. How (or why) is parity violated, leading to a very different $SU(2) \times U(1)$ classification of left-handed and right-handed fermions?
2. What determines the value of $\sin^2\theta_W$ which, in $SU(2) \times U(1)$, remains a free parameter?
3. A third possible motivation might be to include different generations in one large gauge multiplet. We will return to it in Sec. 8.

The question of parity is extremely interesting. There are two rather simple "orthogonal" views. One possibility is that parity is fundamentally broken at all momenta and distances. There is no energy scale in which the electroweak interactions conserve parity, and there is always a difference between the response of left-handed and right-handed fermions to the electroweak bosons.

This view does not explain how or why parity is violated. It fits well with the apparent masslessness of the neutrinos (without explaining it, of course). If this approach is correct, there is no need to extend the electroweak group beyond $SU(2) \times U(1)$.

The opposite view is that at very short distances and large momenta, parity is actually conserved. The full electroweak group conserves parity in the symmetry limit and is, therefore, larger than $SU(2) \times U(1)$. Parity is spontaneously broken and its observed violation at present energies results from the fact that left-handed and right-handed fermions couple (with *identical* couplings!) to gauge bosons of *different* masses. The violation of parity is introduced via the mass spectrum of the guage bosons, and is triggered by the same mechanism that produces the fermion and boson masses and the Cabibbo angles. The simplest group[4] that may accomplish this task is $SU(2)_L \times SU(2)_R \times U(1)$. Its gauge bosons are $W_L^\pm$, Z_L, $W_R^\pm$, Z_R, γ. If we identify $W_L^\pm$ and Z_L with the "usual" $W^\pm$ and Z, and if $W_R^\pm$, Z_R are substantially heavier, the entire $SU(2) \times U(1)$ phenomenology remains unchanged except for small corrections. At the same time we have a parity-conserving theory of electroweak interactions, at energies well above the masses of $W_R^\pm$ and Z_R. Present data place the lower limit on these masses around 300 GeV or so. It should be interesting to improve the accuracy of the "old" β-decay and μ-decay parameters in order to increase these limits (or to discover right-handed charged currents). In such left-right symmetric theories, a massless neutrino is extremely mysterious and somewhat unlikely. However — small neutrino masses cannot be experimentally excluded.

I believe that $SU(2)_L \times SU(2)_R \times U(1)$ is an attractive possibility and that its theoretical and experimental implications should be further studied.

The second motivation for extending the electroweak group beyond $SU(2) \times U(1)$ is the desire to calculate $\sin^2\theta_W$. Putting it more bluntly: $SU(2) \times U(1)$ is not a true unified theory of electromagnetic and weak interactions because it still has two independent coupling constants. The simple solution to this problem would be to embed $SU(2) \times U(1)$ in a larger simple Lie group that has only one coupling constant and, therefore, determines $\sin^2\theta_W$. We refer to such a theory as "simple unification" (as opposed to "grand unification" on one hand, and to $SU(2) \times U(1)$ on the other hand).

4. Simple Unification: An Attractive Idea That Does Not Work

In order to find a simple unification scheme, we must seek an electroweak gauge algebra G that has the following properties:

1. G is either a simple Lie algebra or a direct product of isomorphic Lie algebras having identical coupling constants. In both cases *all couplings are defined in terms of one overall parameter*.
2. G contains $SU(2) \times U(1)$. The gauge bosons of G that are outside $SU(2) \times U(1)$ are necessarily heavier than W^+, W^-, and Z. However, there is no reason to believe that they are superheavy (say, 10^{15} GeV) and we can assume that their masses are, at most, a few orders of magnitude above 100 GeV (say, less than 100 TeV). In such a case, coupling-constant relations predicted by the gauge symmetry are likely to remain essentially unchanged when tested at present energies. (This would not be the case if the additional bosons were superheavy, as they are in grand unification theories.)
3. Since G is "only" unifying the electromagnetic and weak interactions, G commutes with color $SU(3)$. It therefore cannot relate quarks to leptons.

The idea of simple unification is very attractive. It would provide for a true and complete electroweak unification and would uniquely determine $\sin^2\theta_W$. This could then be the starting point for attempts to connect quarks and leptons or for schemes of unifying strong and electroweak interactions.

Unfortunately, simple unification does not work. It has been shown[5] that, if all quarks have charges 2/3 and −1/3, simple unification leads to $\sin^2\theta_W = 3/4$ or 3/8, in clear disagreement with experiment. Simple unification also necessitates a pattern of quarks that is completely different from that of leptons and leads to unpleasant flavor-changing neutral currents. Simple unification could conceivably be made to work if the extra gauge bosons are superheavy. However, I do not know of any reason to make such an assumption (as long as the strong interactions remain unrelated).

The failure of simple unification teaches us an extremely important lesson. It seems that a step-by-step approach may not work, whereas a "catch-all" solution is more successful. It is likely that complete unification of electromagnetic and weak interactions is more difficult than strong-electromagnetic-weak unification. It also appears that our ability to calculate $\sin^2\theta_W$ may depend in a peculiar way on the existence and properties of the strong interactions.

5. The Strong Group: Unlikely Alternatives to $SU(3)$

There is rather convincing (although indirect) evidence for the existence of three colors of quarks. We may be on the verge of obtaining evidence for the existence of gluons (more precisely, gluon jets). Quantum chromodynamics is far from being confirmed experimentally, in spite of many unsubstantiated claims, and is even further away from being fully understood theoretically. However, it is a beautiful theory, essentially without competition. The features of perturbative QCD at high momenta are very attractive and are in *qualitative* agreement with observations. The relevance of nonperturbative effects to the question of quark and gluon confinement is less certain. In any event color $SU(3)$ as the gauge theory of the strong interactions appears to be an extremely good bet.

Two unlikely, but interesting, variations should be kept in mind:

1. Color $SU(3)$ may not be exact. It may be slightly broken, perhaps by the standard Higgs mechanism. This might provide mass to gluons, with or without affecting their alleged confinement. There is no experimental reason to suggest that color $SU(3)$ is broken. On the other hand, the notion of a quantum number that can never be detected is perhaps somewhat chilling.
2. It is also possible that color $SU(3)$ is the exact gauge subsymmetry of a larger, broken gauge group. One candidate for the larger group is $SU(3)_L \times SU(3)_R$ with ordinary color $SU(3)$ as the "diagonal" subgroup.[6] There is some appealing analogy between this "chiral color" and the analogous left-right symmetric electroweak group $SU(2)_L \times SU(2)_R \times U(1)$. However, the overall case for chiral color is not very convincing, in my opinion. I will return to it in Sec. 7.

6. Grand Unification: A Possible Quark-Lepton Connection

The analogy between quarks and leptons in each generation indicates that they must be somehow related. There are at least two attractive approaches to this problem:

1. Quarks and leptons may be composite states of the *same* set of fundamental entities.[7]
2. Quarks and leptons belong to the same multiplet of a large gauge group that necessarily contains color $SU(3)$ and therefore unifies the strong and electroweak interactions.[8]

These two possibilities are not mutually exclusive. I find the ideas of composite quarks and leptons to be very attractive and I discuss it in a separate publication.[9] Here I comment on the more popular approach of grand unification of electroweak and strong interactions.

The various motivations, the competing models and the resulting predictions and theoretical problems have all been discussed extensively[10] and will not be repeated here. I wish only to emphasize a few points:

1. The choice of a grand unification scheme depends, among other things, on the choice of the full gauge groups for electroweak interactions and for strong interactions. Thus, if $SU(5)$ is the "final word," $SU(2)_L \times SU(2)_R \times U(1)$ is excluded and parity remains violated at very short distances. Similarly, if color $SU(3)$ is a subgroup of "chiral color," $SU(3)_L \times SU(3)_R$, most popular models are excluded. $SU(5)$ and $SO(10)$ are, respectively, the most natural and simple candidates corresponding to the two views of parity violation discussed in Sec. 3.
2. An attractive feature of grand unification theories is the fact that the same "superheavy" mass scale is independently calculated on the basis of two arguments. It can be estimated both from renormalization considerations of the various coupling constants and from the present experimental limit on the proton lifetime. An unattractive, unexplained feature is the emergence of two radically different mass scales (10^2 and 10^{15} GeV) for the masses of gauge bosons. Even more unattractive and very unlikely is the notion that no new physics arises between 10^2 and 10^{15} GeV and that one can freely extrapolate over so many orders of magnitude. (The same ratio exists between the sizes of a proton and a billiard ball. Lots of things happen there.)
3. Most grand unification schemes do not address the pattern of generations. No known scheme can accommodate the "standard" three generations in one multiplet. This is disappointing. It may indicate, however, that the reason for generation duplication is different from the reason for a quark-lepton connection. Grand unification may be the answer to the quark-lepton similarity within a generation. It certainly does not explain the pattern of repeating generations.
4. It may be possible[11] to achieve grand unification at energies well below 10^{15} GeV. However, the predicted values of $\sin^2\theta_W$ seem to be too high, as shown in the next section.
5. It is customary to assign u, d, v_e, e to one generation and to one multiplet of $SU(5)$ or $SO(10)$. However, in the same way that the "partner" of u is really $d' = d \cos\theta_c + s \sin\theta_c$, we may ask what are the partners of e^- in $SU(5)$. In general we should define new generation-mixing angles that define the combination of u, c, t residing in the same

multiplet with e^-, etc. Such angles are presumably small, but they might influence $SU(5)$ predictions of mass relations and provide additional decay modes such as $\mu^+ + \pi^0$. These angles become additional parameters of the theory, on equal footing with the generalized Cabibbo angles.

7. Calculating $\sin^2\theta_W$: The Only Available Test of Grand Unification

The only numerical prediction of grand unification models that can be presently compared with experiment and that does not depend on detailed assumptions of the Higgs structure is the calculation of $\sin^2\theta_W$. Any model that follows the "standard wisdom" (see Sec. 1) without adding new quarks and leptons, gives $\sin^2\theta_W = 0.375$ at the grand unification mass. This prediction suffers substantial renormalization when we try to apply it to presently available energies, which are thirteen orders of magnitude away. The calculation of these renormalization effects is, by now, standard.[12] However, the final result depends on the relatively low-energy subgroup of the specific grand unification scheme. The phrase "relatively low energy" refers to energies that may be a few orders of magnitude above 10^2 GeV, but are far below 10^{15} GeV. Consequently, the renormalization of $\sin^2\theta_W$ is essentially accomplished between the grand unification mass and the relatively low energy of the nonunified subgroup.

We may wish to consider the electroweak subgroups of $SU(2) \times U(1)$ or $SU(2)_L \times SU(2)_R \times U(1)$ and the strong subgroups of $SU(3)$ or $SU(3)_L \times SU(3)_R$. There are four combinations, leading to four expressions for $\sin^2\theta_W$:

(A) $G_{EW} \equiv SU(2) \times U(1)$; $G_S \equiv SU(3)$. This is the case for $SU(5)$, but also for larger groups provided they break down to $SU(5)$. The obtained expression is

$$\sin^2\theta_W = \frac{1}{6} + \frac{5}{9}\frac{\alpha}{\alpha_s}.$$

For $\alpha/\alpha_s \sim 0.05$ we get $\sin^2\theta_W \sim 0.195$,

(B) $G_{EW} \equiv SU(2)_L \times SU(2)_R \times U(1)$; $G_S \equiv SU(3)$. This is the case for $SO(10)$, provided that $W_R^\pm$ and Z_R have masses of order, say, TeV rather than 10^{15} GeV. In this case:

$$\sin^2\theta_W = \frac{1}{4} + \frac{1}{3}\frac{\alpha}{\alpha_s} \sim 0.27 .$$

Note that an $SO(10)$ scheme with superheavy W_R and Z_R gives the result in (A).

(C) $G_{EW} \equiv SU(2) \times U(1)$; $G_S \equiv SU(3)_L \times SU(3)_R$. This corresponds to an extremely unlikely situation in which we have chiral color but no left-right symmetry in the electroweak interactions. We find:

$$\sin^2\theta_W = \frac{2}{7} + \frac{10}{21}\frac{\alpha}{\alpha_s} \sim 0.31 .$$

(D) $G_{EW} \equiv SU(2)_L \times SU(2)_R \times U(1)$; $G_S \equiv SU(3)_L \times SU(3)_R$. This is the case for the $[SU(4)]^4$ scheme.[13] We obtain:

$$\sin^2\theta_W = \frac{1}{3} + \frac{2}{9}\frac{\alpha}{\alpha_s} \sim 0.34 \; .$$

It is clear that with the presently accepted value of $\sin^2\theta_W \sim 0.23$, cases (C) and (D) are excluded. It is the failure of this prediction that makes chiral color and early grand unification unattractive. Cases (A) and (B) are both acceptable.

I must repeat, however, my general reluctance to rely heavily on calculations that are based on extrapolations covering thirteen orders of magnitude.

8. What Identifies a Generation?

We do not know the reason for the existence of three similar generations of quarks and leptons. The fermions in each generation respond in an identical way to the gauge bosons of $SU(3) \times SU(2) \times U(1)$. They differ from each other by their masses and (not independently) by their couplings to the Higgs bosons. What is the secret behind the existence of generations? What defines them? Is there a quantum number that labels the generations?

One possibility is that each generation is, in some sense, an excited state of the first generation. If quarks and leptons are composite, the first-generation fermions may represent the ground state of some composite system while the next generations represent higher excitations. However, these are not spin or angular momentum excitations and they cannot be radial excitations because of the relatively small mass differences between generations, as compared with the necessary large mass scale corresponding to the small dimensions of quarks and leptons. The excitations must therefore be of something else, and we do not know anything about it.

Another possibility is to suggest that there is a discrete "phase" symmetry or a $U(1)$ symmetry that acts differently on different generations. This is a completely arbitrary hypothesis that explains nothing and is not motivated by any theoretical idea. However, such an assumption, together with simple constraints on the Higgs particles, leads to interesting relations between quark masses and Cabibbo angles that are discussed in Sec. 10.

One may imagine that there is a "horizontal" gauge symmetry[14] among the generations. The overall gauge symmetry would then be $SU(3) \times SU(2) \times U(1) \times H$, where H acts on the generations and its quantum numbers label the generations. Such a scheme leads naturally to a duplication of generations. The number of similar generations is the dimensionality of the multiplet of H.

All horizontal models must yield flavor-changing neutral currents. The gauge bosons of H are, of course, neutral, and they do change flavor. Hence, they must be heavy. A particularly interesting experimental quantity related to such masses is the width for $K_L^0 \to \mu^- e^+$. If we assign quantum numbers a_1, a_2, a_3 to the three generations, respectively, the simplest process that conserves this quantum number and that involves flavor-changing neutral currents is $K_L^0 \to \mu^- e^+$. The present upper limit on the rate yields a lower limit of 30 TeV for the mass of the gauge boson of H that connects the first generation to the second generation.

An interesting problem in horizontal gauge symmetries relates to the hierarchy of generations. Can every generation transform to every other generation by a gauge boson in H, or is there a hierarchy (for example, only I ↔ II and II ↔ III transitions are induced to lowest order)? The simplest examples of these two options would be $H \equiv SU(3)$ and $H \equiv SU(2)$, respectively. The second possibility is more attractive, in our opinion, because of the apparent smaller Cabibbo mixing of "distant" generations (I and III). The $SU(2)$-group has another advantage: it has no anomalies. However, no completely satisfactory horizontal model has been proposed so far.

An even more ambitious approach would be to embed $SU(3) \times SU(2) \times U(1) \times H$ (or $G \times H$, where G is a grand unification group) in an even larger group, such that all fermions belong to one multiplet. This seems to be impossible if all generations have identical structure. However, there may be some clever ways around this difficulty.

A last tool that might prove useful is the permutation symmetry among generations, which is automatically contained in the Lagrangian of the full QCD and electroweak theory, except for its Higgs sector. This can shed no light on the generation pattern but may be useful in discussing the connection between quark masses and Cabibbo angles.

All in all, the generations puzzle is well defined but no solution is in sight.

9. Quark Masses and Cabibbo Angles: The Framework

The standard electroweak gauge model envisages two logical stages of development: In the symmetry limit all quarks and leptons are massless. There is no difference between u and c, e and μ, d and s. Cabibbo angles are meaningless. All generations are equivalent.

The complete symmetry is then spontaneously broken, presumably via the Higgs mechanism. Quark and lepton mass matrices appear. If we know all the properties of all Higgs particles (their number, their representations, their vacuum expectation values, their couplings) we obtain mass terms of the form:

$$(\bar{u}_0 \ \bar{c}_0 \ \bar{t}_0)_L \ M_U^0 \begin{pmatrix} u_0 \\ c_0 \\ t_0 \end{pmatrix}_R + \text{h.c.}$$

and similar expressions for d,s,b and so on. The matrix M_U^0 is the mass matrix in an arbitrarily chosen basis u_0, c_0, t_0. It need not be Hermitian and it can be diagonalized by a biunitary transformation, yielding the "physical" quark masses:

$$M_U = L_U^{-1} M_U^0 R_U ,$$

$$M_D = L_D^{-1} M_D^0 R_D .$$

M_U, M_D are diagonal matrices with eigenvalues m_u, m_c, m_t and m_d, m_s, m_b respectively. L_U, R_U, L_D, R_D are unitary matrices. The standard generalized Cabibbo angles are contained in the matrix:

$$C = L_U^{-1} L_D .$$

A complete knowledge of the mass matrices M_U^0, M_D^0 determines all quark masses and all Cabibbo angles (including the CP-violating phase[15]). A complete understanding of the Higgs sector of the theory (or of whatever is the responsible mechanism for generating the masses) is necessary for a complete knowledge of the mass matrices.

In the absence of a convincing theoretical description of the physics behind the mass matrices, we are reduced to simple "games" with mass and angle parameters. If the correct number of generations is three, we have six quark masses and four angle parameters.[16] Hence, if the mass matrices M_U^0 and M_D^0 can be expressed in terms of fewer than ten parameters, relations among masses and angles must follow.

Note that if we perform the *same* unitary transformations on M_U^0 and M_D^0, no physical parameters change. This would only amount to a redefinition of our original arbitrary quark basis. Consequently, the number of physically meaningful parameters in M_U^0 and M_D^0 is smaller than would originally appear.

But why should we believe that there are mass-angle relations? A complete theory should probably enable us to calculate all quark masses as well as all angles. However, even if all quark masses are accepted as God-given parameters, one may argue that the angles should be expressed in terms of the masses. The argument runs as follows: Certain low-energy quantities such as the $K_S^0 - K_L^0$ mass difference and certain other weak amplitudes, are *increasing* functions of the masses of their intermediate quark lines (for example, ΔM_{K^0} has a term proportional to m_t^2, and so on). It is extremely unlikely that such low-energy quantities would dramatically change if the mass of the heaviest quark is changed. There is only one way of avoiding this, and it is physically very attractive. If the squared Cabibbo-like angle connecting a heavy quark of mass m_Q to the lightest quarks is inversely proportional to m_Q, the contributions of m_Q to, say, ΔM_{K^0}, would always remain small, regardless of the value of m_Q. Although I cannot express this argument in a general and rigorous way, I believe that it is essentially correct. It leads to two interesting conclusions: the elements of the generalized Cabibbo matrix must depend on the quark masses, and the off-diagonal matrix elements should be small (actually if there are many generations, elements of the Cabibbo matrix that are further away from the main diagonal must be smaller).

These considerations have led people to suspect that relations among quark masses and Cabibbo angles may be derived by making relatively naive assumptions, even without a profound understanding of the generation structure. Many such attempts have been published.[17,18] An interesting exercise of this nature is discussed next.

10. An Interesting Quark Mass-Matrix

An amusing exercise may teach us several interesting lessons concerning the quark mass matrices. Let us assume that the electroweak group is $SU(2)_L \times SU(2)_R \times U(1)$ and that the quark mass matrices M_U^0, M_D^0 are real. (The latter assumption is made only for the sake of simplicity and will be relaxed later.) It is clear that the full gauge-invariant Lagrangian (excluding the Higgs sector) is invariant under permutations among the different generations. In the case of n generations, we have a discrete $S_{nL} \times S_{nR}$ symmetry. Now allow a completely arbitrary "phase symmetry," which may be discrete or continuous, such that each generation of quarks transforms into itself, times a phase factor.

$$q_i \to e^{i n_i} q_i ,$$

where i is the generation number and η_i is arbitrary. Such an arbitrary symmetry may or may not distinguish between some or all of the generations. Each Higgs field presumably has well-defined properties under our arbitrary "phase symmetry":

$$\phi_j \to e^{i\chi_j} \phi_j ,$$

where, again, χ_j is arbitrary. Yukawa couplings will, of course, be allowed, only if they are invariant under the "phase symmetry." If two or more generations remain indistinguishable by their η_i, we may assume that their Yukawa couplings possess the residual permutation symmetry among them. Finally, let us assume that the total number of Higgs multiplets that couple to quarks is, at most, two.

This set of assumptions is, of course, quite elaborate and arbitrary. It represents, however, a "phenomenological" approach to the question of identifying the generations. It is quite general in the sense that many published models[19] are specific cases of our exercise.

On the basis of the foregoing assumptions, we may now try to construct all possible mass matrices. It is clear that in each case either we have vanishing matrix elements (because of the "phase symmetry") or we have relations among matrix elements (because of the permutation symmetry). A careful study of all possible cases shows that there is a surprisingly small number of solutions. If we ignore "trivial" solutions (those in which at least two quark masses or at least one angle or the trace of the mass matrix vanish) we can prove[20] that there is an essentially unique form of the mass matrix.

In the case of two generations, M_U^0 and M_D^0 must have the forms:

$$\begin{pmatrix} 0 & x_U \\ x_U & y_U \end{pmatrix} ; \begin{pmatrix} 0 & x_D \\ x_D & y_D \end{pmatrix}$$

whereas in the case of three generations there are two solutions. The first solution is:

$$M_U^0 = \begin{pmatrix} 0 & A_U & 0 \\ A_U & 0 & B_U \\ 0 & B_U & C_U \end{pmatrix} ; M_D^0 = \begin{pmatrix} 0 & A_D & 0 \\ A_D & 0 & B_D \\ 0 & B_D & C_D \end{pmatrix}$$

with

$$\frac{A_U}{C_U} = \frac{A_D}{C_D} .$$

The second solution is:

$$M_U^0 = \begin{pmatrix} W_U & X_U & 0 \\ X_U & 0 & Y_U \\ 0 & Y_U & Z_U \end{pmatrix} \quad M_D^0 = \begin{pmatrix} W_D & X_D & 0 \\ X_D & 0 & Y_D \\ 0 & Y_D & Z_D \end{pmatrix}$$

with

$$\frac{W_U}{Y_U} = \frac{W_D}{Y_D} \; ; \; \frac{X_U}{Z_U} = \frac{X_D}{Z_D} \; .$$

If we relax the assumption of real mass matrices, the only added complications are some arbitrary phases in the mass matrices. In the case of two generations we obtain one prediction[22]:

$$\theta_c = \left| \text{arc tan} \sqrt{\frac{d}{s}} + e^{i\delta} \text{arc tan} \sqrt{\frac{u}{c}} \right|$$

where δ is an arbitrary phase and the quark labels denote their masses. Using the standard "current-quark" masses, this yields

$$9° \leq \theta_c \leq 16°$$

in good agreement with experiment ($\theta_c = 13°$). In the case of three generations we get from the first solution:

$$\frac{uct}{(u - c + t)^3} = \frac{dsb}{(d - s + b)^3}$$

$$\theta_1 \equiv \theta_c \sim \left| \sqrt{\frac{d}{s}} + e^{i\delta} \sqrt{\frac{u}{c}} \right|$$

$$\theta_2 \sim \frac{1}{\theta_1} \sqrt{\frac{d}{s}} \left| \sqrt{\frac{c}{t}} + e^{i\zeta} \sqrt{\frac{s}{b}} \right|$$

$$\theta_3 \sim \frac{1}{\theta_1} \sqrt{\frac{u}{c}} \left| \sqrt{\frac{c}{t}} + e^{i\zeta} \sqrt{\frac{s}{b}} \right|$$

where δ, ζ are arbitrary phases and we assume $u \ll c \ll t$, $d \ll s \ll b$. These expressions give

$$m_t \sim 13 \text{ GeV}, \quad 8° < \theta_2 < 28°, \quad 2° < \theta_3 < 8°.$$

All of these predictions are consistent with the present bounds on the relevant parameters.[23] The second solution in the case of three generations predicts that in the limit of m_b, $m_t \to \infty$, the Cabibbo angle $\theta_1 = \theta_c$ vanishes, while for m_u, $m_d \to 0$, the mixing between the second and third generations vanishes. This is an extremely unattractive feature. In contrast, the first solution gives a value of θ_c that does not depend at all on m_t, m_b, and the mixing between the two higher generations is unaffected if m_u, $m_d \to 0$. I therefore discard the second solution and suggest that both for two generations and for three generations there is a unique solution. The solution has several attractive features:

1. Each angle is inversely related to the mass of the heavy quark that it mixes.
2. The three-generations solution joins smoothly with the two-generations solution, both for m_b, $m_t \to \infty$ and for m_u, $m_d \to 0$.
3. The form of the mass matrix can easily be generalized to the case of an arbitrary number of generations while the preceding two features are kept intact.

Although the assumptions that led to the derivation of these mass matrices are arbitrary and unsatisfactory, it is entirely possible that the matrices themselves are approximately correct. In fact, many authors,[24] starting from many different (and always arbitrary) sets of assumptions, have "derived" the same forms of matrices. The more general derivation given here explains why such different starting points always lead to the same conclusions. However, so far, neither I nor anyone else has shed any light on the question of identifying the physical differences among the generations. We may have some correct relations between quark masses and Cabibbo angles, but we are far from understanding the generation structure.

11. Some Open Questions and Some Prejudices

Some of the central open questions of the world of quarks and leptons are the following:

Are quarks and leptons related to each other?
Are the higher generations some kind of excitations of the first generation?
Is parity conserved at very short distances?
Are quarks and leptons composite?
Are there relations among quark masses and Cabibbo angles?

I suspect that the answers to all of these questions are in the affirmative, but we are far from fully understanding any of them.

Other open questions involve the number of generations; the possible existence of "exotic" quarks and leptons; the absolute conservation of quantum numbers such as baryon number, lepton number, and color; the calculation of $\sin^2\theta_W$; and, last but not least, the confinement of quarks and gluons. We have a full agenda for the next few years (or decades!).

Acknowledgment

This work was supported by the Department of Energy under contract number DE-AC03-76SF00515.

NOTES

1. S. Weinberg, Phys. Rev. Lett. **19**, 1264 (1967); A. Salam in *Proceedings of the 8th Nobel Symposium* (Stockholm, 1968).
2. See, e.g., L. F. Abbott and R. M. Barnett, Phys. Rev. **D18**, 3214 (1978).
3. For a review see, e.g., H. Harari, Phys. Rep. **42C**, 235 (1978).
4. See, e.g., S. Weinberg, Phys. Rev. Lett. **29**, 388 (1972); J. C. Pati and A. Salam, Phys. Rev. **D10**, 275 (1974); R. N. Mohapatra and J. C. Pati, Phys. Rev. **D11**, 566 (1975).
5. Harari, op. cit. in n. 3; S. Okubo, Hadronic J. **1**, 77 (1978).
6. V. Elias, J. C. Pati, and A. Salam, Phys. Rev. Lett. **40**, 920 (1978).
7. See, e.g., Pati and Salam, op. cit. in n. 4.
8. J. C. Pati and A. Salam, Phys. Rev. **D8**, 1240 (1973); H. Georgi and S. L. Glashow, Phys. Rev. Lett. **32**, 438 (1974).
9. H. Harari, Phys. Lett. **86B**, 83 (1979).
10. See the references cited in notes 3 and 8.
11. Elias, Pati, and Salam, op. cit. in n. 6.
12. H. Georgi. H. R. Quinn, and S. Weinberg, Phys. Rev. Lett. **33**, 451 (1974).
13. Elias, Pati, and Salam, op. cit. in n. 6.
14. See, e.g., F. Wilczek and A. Zee, Phys. Rev. Lett. **42**, 421 (1979).
15. M. Kobayashi and K. Maskawa, Prog. Theoret. Phys. **49**, 652 (1973).
16. Harari, op. cit. in n. 3.
17. See, e.g., S. Weinberg, Trans. New York Acad. Sci. **38** (1977); H. Fritzch, Phys. Lett. **70B**, 436 (1977); F. Wilczek and A. Zee, Phys. Lett. **70B**, 418 (1977); H. Fritzsch, Phys. Lett. **73B**, 317 (1978); T. Kitazoe and K. Tanaka, Phys. Rev. **D18**, 3476 (1978); M. A. de Crombrugghe, Phys. Lett. **80B**, 365 (1979).
18. A. de Rujula, H. Georgi, and S. L. Glashow, Ann. Phys. **109**, 258 (1977); T. Hagiwara, et al., Phys. Lett. **76B**, 602 (1978); H. Harari, H. Haut, and J. Weyers, Phys. Lett. **78B**, 459 (1978); G. Segrè, H. A. Weldon, and J. Weyers, Phys. Lett. **83B**, 351 (1979); E. Derman, Phys. Lett. **78B**, 497 (1978); S. Pakvasa and H. Sugawara, Phys. Lett. **73B**, 61 (1978); A. Ebrahim, Phys. Lett. **76B**, 605 (1978), and **72B**, 457 (1978); H. Georgi and D. V. Nanopoulos, Harvard preprint, 1979; R. Barbieri, R. Gatto, and P. Strocchi, Phys. Lett. **74B**, 344 (1978).
19. Op. cit. in n. 17.
20. H. Harari, unpublished.
21. Harari, op. cit. in n. 20.
22. Op. cit. in n. 17.
23. Harari, op. cit. in n. 3.
24. Op. cit. in n. 17.

Discussion

Following the Paper by H. HARARI

G. 't Hooft (U. of Utrecht):   I have a brief comment in connection with your ruling out "simple" unification as you call it. I think I have one argument to add to that: namely, as soon as you have a simple compact gauge group, then, necessarily, electric charges are also quantized, and you have a problem in finding representations that have only integer-charged leptons and fractional-charged quarks, with the appropriate restrictions of color and fractional charge. I guess that is practically impossible.

Harari: No. I'm sorry: if you are unifying only the weak and electromagnetic interactions, you will not put quarks and leptons in the same multiplet. So, it is true that you can never find one such representation, but there is no problem in claiming that the triplet that has noninteger charge, contains the quark, and for instance, the octet, which have integer charges contains the leptons.

't Hooft: I am not saying that. I am just saying we have to find one representation by itself, with integer charges for the leptons, and one completely different representation by itself, with only fractional charges for the quarks. Now, the charge operator in the theory must be the same for the leptons and the quarks, and I suggest this is extremely artificial, but not impossible to find.

Harari: You must be thinking about something different than what I understood, because if you take $SU(3)$ for instance as a simple unification group, you have a triplet of quarks, with noninteger charges, and an octet of leptons. Their charge operator is the same charge operator.

't Hooft: Yes, there is nothing wrong with that, except that I cannot imagine a simple-looking representation that has all these properties. There is no compelling reason anymore for quarks to have fractional charges.

Y. Achiman (U. of Wuppertal):   I am sorry that I am the only one here protecting $SU(3)$ models of flavor. There are models built that have all these properties that you mentioned and actually there is a kind of symmetry between leptons and quarks that is even nicer than in $SU(2) \times U(1)$ models (or $SU(5)$ models) and for simple reasons: if you add color to the game, then you have three doublets of quarks to each doublet of leptons. But if you add color to the game for $SU(3) \times SU(3)$, it means you have $SU(3) \times SU(3) \times SU(3)$ (not grand-unification, only the group). Then you have actually three representations: $3 \times \bar{3} \times 1, \bar{3} \times 1 \times 3$, and $1 \times \bar{3} \times 3$. And there is a permutation symmetry that, I think, is at least as nice as the relation between three doublets and one doublet. There is a relation between flavor and color, which is also a kind of symmetry. So on the grounds of symmetry there is no way to avoid such models and there are models that are consistent with all the details of the experimental observations.

14. THE MASS MATRIX

R. Gatto

1. Symmetry and Quark Masses

The logical beauty of unified theories (such as the unification of electromagnetism and weak interactions,[1] or further unification with strong interactions and possibly with gravitation[2] is achieved at the price of postulating complex (as opposite to uniform) behaviors at high energies (for instance massive resonances, such as the W and Z^0 poles in unified weak and electromagnetic theory, and additional vector particles in more unified theories). Since large mass parameters are already present in the low-energy description even before any attempt at unification (such as for instance the inverse square root of the Fermi coupling G_F), high-energy complexity is in fact to be expected. Uniformity, on the other hand, would have certainly encountered difficulties (for instance the unitarity problem with a four-linear Fermi coupling). The particular aspect of such theories that leaves one not completely satisfied has rather to do with the explicit Higgs mechanism, with all its as yet unsettled arbitrariness. Proposals have been made to reduce such arbitrariness, such as the inclusion of the Higgs field into supersymmetric multiplets. Also, radiative corrections contribute to the potential in a calculable way[3] and consistency of the theory puts limits on the arbitrary parameters.[4] One might speculate, unfortunately without any mathematical support so far, that the Higgs picture is a phenomenological description of some dynamical mechanism. In any case, whatever will be the final formulation of the theory, it is clear that one would welcome any attempt to reduce the great arbitrariness involved in the description of the breaking via Higgs multiplets, which are coupled to fermions and to themselves in a way that is not dictated by the gauge principle.

The current frame at low energy, that is much before "grand unification," starts from the direct product of the color group $SU(3)_c$ with the flavor group $SU(2) \times U(1)$ (or, maybe $SU(2)_L \times SU(2)_R \times U(1)$). The gauge bosons associated to $SU(3)_c$ are the gluons, those associated to the flavor group are the photon and the intermediate weak bosons. Two additional groups of multiplets are introduced: those describing quarks and leptons, and the scalar multiplets, transforming nontrivially under flavor. Their vacuum expectation values determine, in the classical picture, the pattern of flavor symmetry breaking and lead to masses for the intermediate weak bosons and for some components of the scalar multiplet itself. Through Yukawa couplings among fermions and

---

Yuval Ne'eman (ed.), To Fulfill A Vision: Jerusalem Einstein Centennial Symposium on Gauge Theories and Unification of Physical Forces
ISBN 0-201-05289-X
Copyright © 1981 by Addison-Wesley Publishing Company, Inc., Advanced Book Program.
All rights reserved. No part of this publication may be reproduced, stored in a retrieval system, or transmitted, in any form or by any means, electronic, mechanical photocopying, recording, or otherwise, without the prior permission of the publisher.

scalars, the quarks and the leptons acquire masses, in terms of the vacuum expectation values. Also some global symmetries that were present in the original massless theory now become broken explicitly; some of them in addition may have undergone spontaneous breaking. The latter situation occurs particularly with global symmetries such as for instance strong chiral $SU(2) \times SU(2)$, broken by the quark masses m_u, m_d with the pion as a quasi-Goldstone particle.[5] In the larger chiral $SU(3) \times SU(3)$, broken by the three quark masses m_u, m_d, m_s, the K-meson is also a quasi-Goldstone particle. In principle the symmetry is entirely broken at the end but isospin-$SU(2)$ remains a good working symmetry for certain reasons: the smallness of m_u and m_d, and some difficulties in observing the expected $SU(2)$ violations. (In pionic processes one expects only electromagnetic violations, and other processes — $\pi^0 N$, πK scattering length, and so on — are difficult to verify; see Weinberg.[6]) From the assumption of partial conservation of the axial currents (with π and K quantum numbers) and approximate $SU(3)$ for the vacuum expectation values of the quark bilinears one has

$$\frac{m_d}{m_u} = \frac{\pi^+ + \Delta_K}{\pi^+ - \Delta_K - 2\Delta_\pi} = 1.8 , \quad (14.1)$$

$$\frac{m_s}{m_d} = \frac{K^0 + \Delta_{K\pi}}{K^0 - \Delta_{K\pi}} = 20 , \quad (14.2)$$

where particle symbols stand for the corresponding squared mass and $\Delta_K = K^0 - K^+$, $\Delta_\pi = \pi^+ - \pi^0$, $\Delta_{K\pi} = K^+ - \pi^+$. The large deviations from unity in the foregoing ratios are essentially due to the smallness of the mass of the pion, a quasi-Goldstone particle. In particular in (14.2) the denominator is very small, the term K^0 almost compensating the term $\Delta_{K\pi} = K^+ - \pi^+ \cong K^+$.

2. Naturality Groups

In an important paper Glashow and Weinberg have discussed "natural" flavor-conservation in neutral weak currents.[7] Within $SU(2) \times U(1)$ the main implication of such a natural conservation is that quarks of the same helicity and charge have the same weak isospin (that is, the same weak T_3 and T^2). The condition on T_3 follows directly from the form of the weak neutral charges

$$Y_{L \atop R} = T_{3L \atop R} - xQ \qquad (x = \sin^2\theta_w) ,$$

where Q is the electric charge. Under unitary transformations of the right- and left-quark fields, which preserve electric charge, the charges Y remain diagonal if T_3 is a function of Q, and, vice versa, if they must remain naturally diagonal, T_3 has to be a function of Q. The condition on T^2 follows from the analysis of second-order processes ($\sim \alpha G$) and limiting to small masses and momenta (with respect to m_w). However, Higgs exchange might also give rise to flavor-nonconserving processes. Glashow and Weinberg give a sufficient condition to avoid such flavor nonconservation through Higgs exchange. The condition is that quarks of the same charge receive mass from a single Higgs field. In such a case, in fact, the quark mass is clearly diagonal simultaneously with the quark couplings for neutral Higgs exchange.

The problem that poses itself is that of finding the necessary and sufficient condition for natural flavor-conservation through Higgs exchange. Because of the naturality requirement this problem will have to be formulated in group-theoretical terms. That is, one replaces the single effective Higgs field of the Glashow-Weinberg sufficient condition with a general Higgs structure and assigns quarks and scalars to representations of some additional symmetry (that is, beyond the flavor group) in such a way as to naturally guarantee flavor conservation through Higgs exchange. This problem I shall call "searching for naturality groups." There is, however, a second stage in dealing with these problems, once such naturality groups have been found, namely that of looking for ensuing conditions on masses and mixing angles. Such relations, if they exist, are again natural — that is, not spoiled by renormalization. In this paper I shall, among other things, summarize the (by now well-defined) conclusions on such a line of research, to which have contributed, besides myself, G. Morchio and F. Strocchi, G. Sartori, and at an earlier stage, R. Barbieri.[8]

Writing the Higgs coupling, in compact notation, as $\bar{q}_L \Gamma \phi q_R + \bar{q}_R \Gamma^\dagger \phi^\dagger q_L$, the mass term is $\bar{q}_L M q_R + \bar{q}_R M^\dagger q_L$, with $M = \Gamma \langle \phi \rangle_0$, $\langle \phi \rangle_0$ denoting the scalar vacuum expectation values. Under a transformation of the quark basis, $q'_L = U_L q_L$, $q'_R = U_R q_R$, the matrix M becomes $M' = U_L M U_R^\dagger$. Such a biunitary diagonalization problem requires unitary diagonalization for the matrices (Hermitian and non-negative)

$$\mathcal{M} = MM^\dagger , \qquad \overline{\mathcal{M}} = M^\dagger M , \qquad (14.3)$$

which transform as

$$U_L \mathcal{M} U_L^\dagger , \qquad U_R \overline{\mathcal{M}} U_R^\dagger . \qquad (14.4)$$

With quark charges 2/3 and −1/3, we shall speak of an up-sector (charges 2/3) and a down-sector (charges −1/3). Correspondingly the matrices U_L, U_R, $\mathcal{M}$, $\overline{\mathcal{M}}$ are block-diagonal with blocks $U_{L,u}$, $U_{L,d}$, and so on. Thus $U_{L,u}$ diagonalizes $\mathcal{M}_u$, $U_{L,d}$ diagonalizes $\mathcal{M}_d$, and similarly for $U_{R,u}$, $U_{R,d}$ relatively to $\overline{\mathcal{M}}_u$, $\overline{\mathcal{M}}_d$.

I shall call K-symmetry the additional symmetry, beyond the flavor group. Rather than constructing all sorts of possible different models I would like to concentrate on general results. So, for the group K I shall not assume any particular restriction. It will be a discrete group or a continuous compact group — in the latter case, however, unwanted Goldstone modes may appear. As far as the classification is concerned, it will be based on sequential classification with separate up and down structure (standard $SU(2) \times U(1)$ is the typical case, or sequential $SU(5)$). I shall comment on left-right symmetric groups such as $SU(2)_L \times SU(2)_R \times U(1)$, where the up-sector and down-sector are not separated (u_R and d_R are in the same flavor multiplet).

The quark-scalars Lagrangian can be written, in compact notation,

$$\mathcal{L}_{qH} = \bar{\psi}_{iL} \varphi_\alpha^u (\Gamma_{ij}^\alpha)^u \psi_{jR}^u + \bar{\psi}_{iL} \varphi_\alpha^d (\Gamma_{ij}^\alpha)^d \psi_{jR}^d + \text{h.c.} , \qquad (14.5)$$

where i, j are generation indices $(i,j = 1,2, \ldots n)$, α labels the scalar doublets $(\alpha = 1,2, \ldots n_H)$, and $SU(2)$ indices are omitted ($\bar{\psi}_L$ are $SU(2)$-doublets, ψ_R singlets, φ doublets). For any element g of the group K the statement of K-symmetry for Eq. (14.5) is

$$K_L^\dagger(g) \, (\Gamma^\alpha)^u \, K_R^u(g) = (\mathcal{D}_{\alpha\beta}^*(g))^u \, (\Gamma^\beta)^u \tag{14.6}$$

same for d in place of u ,

where the matrices (on generation space) $K(g)$ and $\mathcal{D}(g)$ represent the transformation generated by g on the quarks and on the scalars

$$\psi_L \to K_L(g)\psi_L$$

$$\psi_R^u \to K_R^u(g)\psi_R^u \qquad \varphi_\alpha^u \to \mathcal{D}_{\alpha\beta}^u(g)\varphi_\beta^u \tag{14.7}$$

$$\psi_R^d \to K_R^d(g)\psi_R^d \qquad \varphi_\alpha^d \to \mathcal{D}_{\alpha\beta}^d(g)\varphi_\beta^d \ .$$

The requirement of natural flavor-conservation can most simply be analyzed by assuming irreducible K-representations. The extension to reducible representations formally complicates our work but does not essentially change the picture. One then finds the following (necessity and sufficiency) characterization of naturality groups and representations:

1. The representations K_L, $K_R^{u,d}$ must be unitarily equivalent to the matrix representations,

$$K_L(g)_{ij} = \delta_{i,P_g(j)} \, e^{i\alpha_i(g)} \qquad (i,j = 1, \ldots, n)$$

$$K_R(g)_{ij} = \delta_{i,P_g(j)} \, e^{i\beta_i(g)} \quad \text{(both for } u, d\text{)} , \tag{14.8}$$

where $\alpha_i(g)$, $\beta_i(g)$ are real phases and P_g permutations on n indices uniquely associated to each element g ($P_g(j)$ is the action of P_g on the index j). In other words, there is a homorphism of K into a (transitive) subgroup of the permutation group on the n quark generations. The K-symmetry is then a group-theoretical extension of such a transitive subgroup of S_n via an Abelian group K_0, the latter providing for the phases $\alpha(g)$, $\beta(g)$. Equation (14.8) shows that the matrices K have only one nonzero element in each row and in each column, such an element being of modulus 1.

2. The Higgs representation $\mathcal{D}_{\alpha\beta}$ must be equivalent to a subrepresentation of the n-dimensional representation D:

$$D(g)_{ab} = \delta_{a,P_g(b)} \, e^{-i[\beta_a(g) - \alpha_a(g)]} \tag{14.9}$$

(both for u and d) .

In other words $\mathcal{D}^*$ must appear in the reduction of $K_L^\dagger \times K_R^T$ only as a subrepresentation of D.

3. Each irreducible subrepresentation of $\mathcal{D}$ must occur the same number of times in D as in the decomposition of the tensor product $(K_L^\dagger \times K_R^T)^*$.

In proving these results one assumes a nondegenerate mass matrix. The result under (1) follows from the unitary mapping effected by K_L of $\Gamma\Gamma^\dagger$ into $(\mathscr{D}*\Gamma)(\mathscr{D}*\Gamma)^\dagger$ (Γ is some nondegenerate linear combination of the coupling matrices Γ^a), as well as from the biunitary mapping $K_L^\dagger \Gamma K_R = \mathscr{D}*\Gamma$, which implies the identity of $P(g)$ for both left- and right-transformations. The condition under (2) expresses the biunitary action of the pair $(K_L(g), K_R(g))$ on the linear space of the $n \times n$ diagonal matrices.

The multiplicity condition (2) eliminates the possibility of some Γ^a being nondiagonal and thus violating flavor.

It is possible from the foregoing characterization to construct naturality groups and representations. For instance, for three generations, $n = 3$, one starts by noting that there are two transitive subgroups of S_3: the alternating A_3, and S_3 itself. The choice of the Abelian group gives possible extensions such as: $V_4 \boxtimes A_3 \equiv A_4$ (the symbol $\boxtimes$ denotes semidirect product) with representations $\underline{3}$ for L and R and reducible $\underline{1}_0 + \underline{1}_1 + \underline{1}_2$ for D; $C_7 \boxtimes A_3$ (C_7 is the cyclic group of order 7) with various possible choices of representations; $\mathscr{A}_3(1,1) \boxtimes A_3 (\mathscr{A}_3(1,1)$ is the Abelian group of order $3^{1+1} = 9$ for the partition $2 = 1 + 1$); $C_{13} \boxtimes A_3$; $V_4 \boxtimes A_3 \equiv S_4$ with $L = R = \underline{3}$, $D = \underline{2} + \underline{1}$ or $L = \underline{3}_1$, $R = \underline{3}_2$, $D = \underline{2} + \underline{1}$. One can thus enumerate for any n the possible naturality groups and representations. The preceding examples are, for three generations, perhaps the most interesting, and I would rather at this point go over to a discussion of the charged currents (the quark coupling matrix).

3. The Quark Coupling Matrix

For a charged current of the form $\bar{q}_L \gamma_\mu T^+ q_L$, where T^+ is the charged generator, the transformation to the physical basis $q_L' = U_L q_L$ gives $\bar{u}_L' \gamma_\mu U d_L'$

$$U = U_L^u (U_L^d)^\dagger , \qquad (14.10)$$

where U is the coupling matrix. It acts between the physical vector $(\bar{u}_L, \bar{c}_L, \bar{t}_L, \ldots)$ and the vector obtained by transposing $(d_L, s_L, b_L, \ldots)$ in the usual notation for quarks. The parametrization of U requires $(n-1)^2$ independent real parameters, of which $\frac{1}{2}(n^2 - 3n + 2)$ are phases, for n generations. In fact, a unitary matrix has n^2 independent real parameters, but $2n - 1$ phases can be absorbed in defining the $2n$ quark fields. Of the remaining $(n-1)^2$ real parameters, $\frac{1}{2}n(n-1)$ would be needed to describe a rotation and the remaining $\frac{1}{2}(n^2 - 3n + 2)$ are phases. For $n = 2$ one has the Cabibbo matrix[9]:

$$U = \begin{vmatrix} \cos\theta_c & \sin\theta_c \\ -\sin\theta_c & \cos\theta_c \end{vmatrix} \qquad (14.11)$$

with no phases (no CP violation by the current-couplings); for $n = 3$ one has the Kobayashi–Maskawa matrix[10]:

$$U = \begin{vmatrix} c_1 & s_1 c_3 & s_1 s_3 \\ -s_1 c_2 & c_1 c_2 c_3 - s_2 s_3 e^{i\delta} & c_1 c_2 s_3 + s_2 c_3 e^{i\delta} \\ -s_1 s_2 & c_1 s_2 c_3 + c_2 s_3 e^{i\delta} & c_1 s_2 s_3 - c_2 c_3 e^{i\delta} \end{vmatrix} \quad (14.12)$$

with the abbreviations $c_i = \cos\theta_i$, $s_i = \sin\theta_i$ and the phase angle δ (a possible origin of CP violation). The angle θ_1 can be identified with the Cabibbo angle, but now $c_1^2 + s_1^2 c_3^2 = 1 - s_1^2 s_3^2$, so that s_3 will have to be small. A recent discussion on the bounds on the matrix elements of (14.12) has been given by Schrock, Treiman, and Wang.[11] For the coupling to $\bar{u}d$, from $0^+ - 0^+$ Fermi β-decay in nuclei and μ-decay one obtains

$$c_1^2 = .948 \pm .005 , \quad (14.13)$$

whereas semileptonic strange-particle decay gives

$$s_1^2 c_3^2 = .048 \pm .005 . \quad (14.14)$$

Shrock, Treiman, and Wang suggest

$$|s_3| = 0.28 \, {}^{+0.21}_{-0.28} .$$

Whereas essentially no dynamical assumptions are needed to obtain (14.13) and (14.14) (apart from inclusion of radiative corrections, and $SU(3)$ breaking in strange decays), to go further, the standard procedure uses the dynamical one-loop estimate of the $\bar{s}d \to \bar{d}s$ transition. This transition, through additional dynamical assumptions, determines the observables of the $K^0 - \bar{K}^0$ system, in particular the mass difference $m(K_L^0) - m(K_S^0)$ and the projection of K_L^0 into K_S^0, $\langle K_S^0 | K_L^0 \rangle$. The estimates for θ_2 and δ depend on some choices of signs and mass values. To give an example, choosing all angles θ_1, θ_2, θ_3, δ between $0°$ and $90°$, and taking $m_t = 15$ GeV, Shrock, Treiman, and Wang find, for $|s_3| = 0.3$, the values $|s_2| = 0.2$ and $|\sin\delta| \sim 10^{-2}$. Such values give in turn a full set of predictions for decay of charm-, bottom-, and top-quarks, so that one can hope in the near future to learn much more on the elements of the Kobayashi–Maskawa coupling matrix for charged currents, Eq. (14.12).

It is clearly important to speculate about a possible theory for quark mixing and masses that could lead to a determination of such important phenomenological parameters. The problem is essentially identical to the general problem of symmetry breaking of the underlying gauge theory, which I consider essentially unsettled, as I pointed out in the introduction of this paper.

The smallness of the masses of e, u, and d (the first generation) may lead one to the speculation that they are zero at first order, in a natural way, and become different from zero from radiative corrections. This point has been investigated by Weinberg and has to face some difficulties. The masses of e, u, d would have to come through virtual gauge boson effects, giving the following[12]:

$$\delta m \sim \sum_N \frac{g^2}{4\pi^2} m_0 \log \mu_N^2 , \qquad (14.15)$$

where N labels the gauge boson of mass μ_N and m_0 is the mass matrix. Equation (14.15) does not hold between massive eigenstates of m_0. The gauge bosons of $SU(2)_L \times U(1)$ cannot give mass to the initially massless particles. There must be very heavy gauge bosons that couple such massless particles to the massive ones; they must be $SU(2)$ singlets since the right particles are assigned to singlets and they must couple to them. Within the first generations the relevant transitions due to emission and reabsorption of such singlets are $e_L \to \mu \to e_R$, $u_L \to c \to u_R$, $d_L \to s \to d_R$ and correspondingly, from Eq. (14.15):

$$m_e \sim \left(\frac{g^2}{4\pi} \Sigma \log \mu_N^2\right) m_\mu, \text{ etc. },$$

and in particular $m_e/m_\mu \sim m_u/m_c \sim m_d/m_s$. One expects $g \sim e$. Whereas $m_u/m_c = 0.0037$ and $m_e/m_\mu = 0.0048$ do not differ very much from $e^2/4\pi^2 = 0.0023$, $m_d/m_s \sim 0.05$ is much larger. Besides this difficulty, the nondiagonal transitions $d \to s$ would be $\sim g^2 m_s/4\pi^2$ (and $(u \to c) \sim g^2 m_c/4\pi^2$), leading to a small Cabibbo angle of the order 10^{-3}, rather than 10^{-1}. The speculation that all e, u, d masses be naturally zero at lowest order and arise from radiative corrections seems therefore to face some difficulties. This conclusion has led Weinberg, De Rujula, Georgi and Glashow, Fritzsch, Wilczek and Zee, and other authors to speculate on the possibility that all masses and mixings are already present in lowest order.[13] A simplest example for such a possibility is given by Weinberg.[14] For d and s take, in some particular basis, the mass matrix, in the notations we are using, in the form:

$$M = \begin{vmatrix} 0 & A \\ B & C \end{vmatrix} . \qquad (14.16)$$

If, in that basis, one has, for some reason, parity conservation, so that $|A| = |B|$, the unitary transformation to the physical quark basis gives for the Cabibbo angle

$$\theta_c \cong \sqrt{\frac{m_d}{m_s}} . \qquad (14.17)$$

This relation coincides with an old conjecture.[15] In $SU(2)_L \times SU(2)_R \times U(1)$ it is not difficult to reproduce the mass matrix (14.16). The two left-quark multiplets $\psi_L^{(i)}$ ($i = 1,2$) transform as ($\frac{1}{2}$,0), the right ones ψ_R^i as $(0,\tfrac{1}{2})$, the scalars as $(\tfrac{1}{2},\tfrac{1}{2})$. One takes two such scalars Φ and Ω (2×2 matrices) and imposes a discrete symmetry

$$\psi_L^{(1)} \to i\psi_L^{(1)}, \psi_R^{(1)} \to -i\psi_R^{(1)}, \psi_{L\atop R}^{(2)} \to \psi_{L\atop R}^{(2)},$$

$$\Phi \to \Phi, \Omega \to i\Omega.$$

In addition one assumes space inversion: $L \to R$, $\Phi \to \Phi^\dagger$, $\Omega \to \Omega^\dagger$. The scalar-quark interaction satisfying such requisites gives the mass matrix (14.16) with $|A| = |B|$. Such derivation is, of course, natural, in the sense that it is stable under renormalization. Unfortunately natural flavor-conservation is not satisfied in the simplest realistic $SU(2)_L \times SU(2)_R \times U(1)$ models.[16] For three generations the extension is straightforward[17]: one follows the previous pattern exactly, with the only change that it is now $\psi^{(3)}$ that stays invariant, $\psi^{(1)}$ transforms as before, and $\psi^{(2)}$ transforms like $\psi^{(1)}$ apart for changing i into $-i$. Instead of (14.11) (with $|A| = |B|$) one has

$$M = \begin{vmatrix} 0 & A & 0 \\ A^* & 0 & B \\ 0 & B^* & C \end{vmatrix} \qquad (14.18)$$

for the up-sector and analogously for the down-sector. Most of the models proposed are indeed for $SU(2)_L \times SU(2)_R \times U(1)$, as in the example just discussed. Let us therefore deal first with this case and then come back to sequential models of the Weinberg–Salam type (that is, with separate up- and down-sectors).

4. A Theorem for $SU(2)_L \times SU(2)_R \times U(1)$

Nonvanishing mixing angles cannot arise in lowest order in any flavor-conserving model based on $SU(2)_L \times SU(2)_R \times U(1)$. The proof is trivial. For natural flavor-conservation left-quarks transform as $(2,1)$, right-quarks as $(1,2)$; the coupled scalars must transform as $(2,2)$. The certainly allowed alternative of right-quarks transforming as $(1,1)$ is like $SU(2) \times U(1)$, and need not be discussed. The scalar-quark Lagrangian is

$$\sum_\alpha \bar\psi_L \phi_\alpha \Gamma^\alpha \psi_R + \text{h.c.},$$

and the vaccum expectation values are of the form

$$\langle \phi_\alpha \rangle = \begin{vmatrix} V_\alpha & 0 \\ 0 & V'_\alpha \end{vmatrix}. \qquad (14.19)$$

This gives for up and down, respectively,

$$M_u = \Sigma v_a \Gamma^a , \quad M_d = \Sigma v_a' \Gamma^a . \tag{14.20}$$

Demanding that

$$U_L^\dagger \Gamma^a U_R \tag{14.21}$$

be simultaneously diagonal implies, from Eq. (14.20), that all mixing angles vanish.

I do not consider this result negative. Looking at the matrix (14.18) one can imagine that indeed the nondiagonal elements arise from radiative corrections. A similar possibility is indeed envisaged also by Fritzsch on the basis of the relative magnitudes of the matrix elements.[18]

The foregoing general and extremely elementary argument solves for any added symmetry (and assignment to its representations) our problem of searching for a connection between the mass matrix and flavor conservation in the case of $SU(2)_L \times SU(2)_R \times U(1)$, or in any similar situation where the up-sector and down-sector are connected to the same set of Higgs couplings. Let us now go back to the extended $SU(2) \times U(1)$ model (and analogous situations) where the separation of the two sectors renders the mathematical problem slightly more complicated.

5. Implications of Flavor Conservation in the Extended $SU(2) \times U(1)$ Model

The simplest solution for natural flavor-conservation in Higgs exchange in the extended $SU(2) \times U(1)$ model is that suggested by Glashow and Weinberg,[19] as discussed in Sec. 2: quarks of same charge receive mass from a single Higgs. The study of the necessary and sufficient condition for natural flavor-conservation has led my colleagues and me, however, to classify more general situations, that is, to the study of what we had called the naturality groups.

For the case of singlet Higgs (singlet, of course, with respect to the additional symmetry) one can demonstrate in a few steps the following result, for any additional symmetry and for whatever number and representations of quarks.

> THEOREM: In sequential $SU(2) \times U(1)$ for any K-symmetry and for any number of quarks the Cabibbo angle is either zero or $\pi/2$ or indeterminate within the standard Higgs structure (two doublets).

The proof is so simple that I shall reproduce it here. On the other hand I shall omit the proofs for the more general situation for an arbitrary naturality group, which requires more cumbersome mathematics.

Let us suppose any number of quarks, transforming, in general, according to some reducible representation of the additional symmetry group K. In each of the two sectors there is, however, only one Higgs $SU(2) \times U(1)$ doublet. Invariance under the symmetry K requires

$$K_L^\dagger(g) (\Gamma^a)^u K_R^u(g) = e^{i\chi^u(g)}(\Gamma^a)^u$$
$$K_L^\dagger(g) (\Gamma^a)^d K_R^d(g) = e^{i\chi^d(g)}(\Gamma^a)^d , \tag{14.22}$$

where $K_L(g)$, $K_R^u(g)$, $K_R^d(g)$ represent the group element g within the left-doublets, the up-singlets, and the down-singlets, respectively. These equations imply for the matrices $\mathcal{M}^u = M^u M^{u\dagger}$ and $\mathcal{M}^d = M^d M^{d\dagger}$:

$$[K_L(g), \mathcal{M}^u] = 0 \tag{14.23}$$

$$[K_L(g), \mathcal{M}^d] = 0 \ . \tag{14.24}$$

And now, just trivially, observe that in the basis where $\mathcal{M}^u$ is diagonal, $K_L(g)$ must be diagonal because of (14.23). Then (14.24) implies that $\mathcal{M}^d$ be block-diagonal, each block being left unconstrained. The quark coupling matrix, in this basis, is simply $(U_L^d)^\dagger$. Therefore the mixing angles can only be zero or 90° or indeterminate. No K-symmetry exists to further constrain U_L^d, which diagonalizes $\mathcal{M}^d$.

The preceding proof is almost a triviality. The negative result leads to the much more complex study of the implications of the infinite possible naturality groups. Here, the key equation, which allows one to solve the problem, is the following. Suppose the quarks transform irreducibly. We use a basis where all Γ^u are diagonal and obtain the representations $K_L(g)_u$, $K_R^u(g)_u$, where the lower index u denotes such a basis. Similarly in a basis where all Γ^d are diagonal we have $K_L(g)_d$, $K_R^d(g)_d$. The quark coupling matrix (generalized Cabibbo matrix) $U = U_L^u (U_L^d)^\dagger$ is exactly that matrix that transforms one basis into the other:

$$U^\dagger K_L(g)_u U = K_L(g)_d \tag{14.25}$$

for each group element g. Then U is uniquely determined by Schur's lemma. Because of the particular forms of K_L, within naturality groups, Eq. (14.25) implies that a matrix element U_{kl} of U has the same modulus as a matrix element

$$U_{P_g^u(k), P_g^d(l)} ,$$

where permutations P_g^u and P_g^d have been applied to the indices k, l, respectively. The equivalence implied by such permutations, for all group elements g, extends, by Schur's lemma, to all pairs of indices. Thus all elements have the same modulus, and this contradicts experience ($|\sin\theta_c|$ does not equal $|\cos\theta_c|$). Next, in the case of reducibility, an argument too long to be reproduced here shows that a similar negative result holds, in the sense that one only has in addition some well-specified indeterminacies.

I believe that the idea of insisting on natural flavor-conservation is in any case basically useful. The idea, as well known, originated by the experimental success of the Glashow-Iliopoulos-Maiani mechanism. It was suitably phrased by Glashow and Weinberg, who examined its implications for the currents coupled to the intermediate bosons and gave the simplest sufficient condition for the couplings of the Higgs. I have reported here the result of the search for the necessary and sufficient condition. Natural flavor-conservation, in its more general formulation, whenever more than a single Higgs field contributes to the masses of quarks of given charge, requires an additional symmetry group, the naturality group, and particular choices of its representations. These groups can be classified. They are extensions of the symmetric group acting on the generations and of an Abelian group. The second question was to look for the implications of such groups on the mixing angles and possible mass relations. The mass matrix was found to be still indeterminate or with unacceptable determinations (for instance geometric mixing angles, often 0° or 90°). The following possibility then suggests itself: One starts with a solution leading for instance to zero mixing angles, a naturally guaranteed solution. Radiative corrections may then lead to small nonzero mixings. The phenomenological pattern of quark masses and mixing is not a priori in contradiction with such a possibility. At the same time flavor violation is induced by these radiative corrections, something one has always been prepared to accept.

An alternative possibility, which seems less attractive to me, but certainly not rejectable is to have already flavor violations before radiative corrections. These violations must be small, at least for the lowest quarks, but this requisite is not in itself a very stringent one. In particular an interesting possibility in this direction is suggested by Georgi and Nanopoulos.[20] They suggest ascribing flavor violation to very massive Higgs fields. The originally introduced scalars can be linearly redefined so as to leave all of them but one without vacuum expectation value. Except for this one, the others can be given very large masses. It thus seems that some well-defined possibilities exist to solve within the gauge theories the problem of the quark mass matrix. Further work should be carried out along the various possible directions.

NOTES

1. S. Weinberg, Phys. Rev. Lett. **19**, 1264 (1967); A. Salam, in *Elementary Particle Theory*, edited by N. Svartholm (Wiley, New York, 1969).
   See other papers in this volume, in particular the chapters by S. Glashow, S. Weinberg, M. Gell-Mann, J. Pati, F. Gürsey, Y. Ne'eman and D. Z. Freedman.
2. Ibid.
3. S. Coleman and E. Weinberg, Phys. Rev. **D**, 1889 (1973).
4. S. Weinberg, HUTP-78/A060, 1978.
5. S. L. Glashow and S. Weinberg, Phys. Rev. Lett. **20**, 224 (1968); M. Gell-Mann, R. J. Oakes, and B. Renner, Phys. Rev. **175**, 2195 (1968).
6. S. Weinberg, HUTP-77/A057, published in *Rabi Festschrift*, (New York Academy of Sciences, 1978).
7. S. L. Glashow and S. Weinberg, Phys. Rev. **D15**, 1950 (1977).
8. R. Barbieri, R. Gatto, and F. Strocchi, Phys. Lett. **74B**, 344 (1978); R. Gatto, G. Morchio, and F. Strocchi, Phys. Lett. **80B**, 265 (1979), and SNS 2/78; G. Sartori, UGVA-1978/12-187, Geneva University, 1978.
9. N. Cabibbo, Phys. Rev. Lett. **10**, 531 (1963).
10. M. Kobayashi and T. Maskawa, Prog. Theor. Phys. **49**, 652 (1973).
11. R. E. Shrock, S. B. Treiman, and L. Wang; for previous work see J. Ellis, M. Gaillard, and D. V. Nanopoulos, Nucl. Phys. **B109**, 213 (1976); H. Harari, Phys. Rep. **42C**, 235 (1978).
12. Weinberg, op. cit. in n. 6.
13. Ibid.; A. De Rujula, H. Georgi, and S. L. Glashow, Ann. Phys. (New York) **109**, 258 (1977); H. Fritzsch, Phys. Lett. **70B**, 436 (1977), and **73B**, 317 (1978); F. Wilczek and A. Zee, Phys. Lett. **70B**, 418 (1977); S. Pakvasa and H. Sugawara, Phys. Lett. **73B**, 61 (1978); A. Ebrahim, Phys. Lett. **72B**, 457 (1978), and **76B**, 605 (1978); R. N. Mohapatra and G. Senjanovic, Phys. Lett. **73**, 176 (1978); G. C. Branco, Phys. Lett. **76B**, 70 (1978); H. Harari, H. Haut, and J. Weyers, Phys. Lett. **78B**, 459 (1978); J. M. Frere, Phys. Lett. **80B**, 369 (1979); F. Wilczek and A. Zee, Phys. Rev. Lett. **42**, 421 (1979); H. Hagiwara et al., Phys. Lett. **76B**, 602 (1978); T. Kitazoe and K. Tanaka, Phys. Rev. **D18**, 3476 (1978); D. Wyler, Phys. Rev. **19D**, 333 (1979); M. A. De Crombrugghe, Phys. Lett. **80B**, 365 (1979); G. Segrè and H. A. Weldon, UPR-0113T, University of Pennsylvania.
14. Weinberg, op. cit. in n. 6.

15. R. Gatto, G. Sartori, and M. Tonin, Phys. Lett. **28B**, 128 (1968); N. Cabibbo and L. Maiani, Phys. Lett. **28B**, 131 (1968); R. Jackiw and H. J. Schnitzer, Phys. Rev. **D5**, 2008 (1972); H. Pagels, Phys. Rev. **D11**, 1213 (1978).
16. R. N. Mohapatra and D. P. Sidhu, Phys. Rev. **D16**, 2843 (1977); Q. Shafi and C. Wetterich, Phys. Lett. **69B**, 464 (1977); R. N. Mohapatra, in *New Frontiers in High-Energy Physics,* edited by B. Kursunoglu (Plenum, New York 1978).
17. Fritzsch, op. cit. in n. 13.
18. Ibid.
19. Glashow and Weinberg, op. cit. in n. 7.
20. H. Georgi and D. V. Nanopoulos, Phys. Lett. **82B**, 95 (1979).

PART VI: STRONG–WEAK–ELECTROMAGNETIC UNIFICATION IDEAS

Sheldon Glashow's chapter in Part V can be read as a partial introduction to this final part describing the most ambitious attempt at unification in our recent research. Murray Gell-Mann's summary discusses the main features of these theories and can be regarded as an introduction as well as a summary.

Steven Weinberg's chapter discusses several aspects and applications of such theories with exact color conservation. Jogesh Pati's presentation discusses the alternative view in which the color subgroup is broken.

Yuval Ne'eman (ed.), To Fulfill A Vision: Jerusalem Einstein Centennial Symposium on Gauge Theories and Unification of Physical Forces
ISBN 0-201-05289-X

Copyright © 1981 by Addison-Wesley Publishing Company, Inc., Advanced Book Program.
All rights reserved. No part of this publication may be reproduced, stored in a retrieval system, or transmitted, in any form or by any means, electronic, mechanical photocopying, recording, or otherwise, without the prior permission of the publisher.

Chairman's Remarks: Einstein in Paris (Some Personal Reminiscences)

ALFRED KASTLER

It is for me a great honor as an official delegate of the French Academy of Sciences in Paris to transmit to the organizing committee and to the Israel Academy of Sciences the heartiest greetings of our Academy, and the best congratulations for this centennial symposium.

Albert Einstein came to Paris in 1922, invited to the College de France by Paul Langevin, who lectured on relativity. I was at this time a young student at the Ecole Normale Supérieure attending Langevin's lectures. I remember with great pleasure those days, I have written my souvenirs of that time in a recent issue of Technion-Informations. I am happy to transmit this paper to the organizing committee:

Souvenirs sur Albert Einstein*

Le Technion, sur le Mont Carmel qui domine Haïfa, est le symbole de la vitalité d'Israël, de sa foi en l'avenir de la jeune nation, symbole aussi de l'union harmonieuse et féconde entre science, technologie, et humanisme.

J'ai appris avec plaisir que son Institut de Physique porte le nom d'Albert Einstein. Les scientifiques du monde entier s'apprêtent à célébrer le centenaire de la naissance de ce grand physicien. Pour le grand public, c'est l'homme de la Relativité, pour le scientifique c'est aussi le physicien dont l'oeuvre, associée à celle de Max Planck, a inauguré la Physique du XXe siècle, la physique des quanta. C'est l'introduction en 1905 du concept de quantum d'énergie lumineuse, de "photon" comme nous disons aujourd'hui, qui a été le motif pour l'attribution à Einstein du prix Nobel en 1921. Sa contribution ne s'est pas arrêtée la: le concept d'émission stimulée de lumière énoncé par lui en 1917 à conduit a l'invention du laser, la statistique de Bose-Einstein édifée en 1924 a fourni la clef de la compréhension des effets quantiques macroscopiques: suprafluidité et supraconductivité.

Permettez-moi d'évoquer un souvenir de jeunesse: j'étais élève en première année a l'Ecole Normale Supérieure de la rue d'Ulm lorsque Paul Langevin eut l'idée hardie, en 1922, d'inviter Einstein à Paris. L'entreprise à cette date n'était pas sans danger, car Einstein était alors professeur à Berlin et l'on pouvait craindre des manifestations anti-allemandes. Mais tout s'est bien passé. Einstein a fait d'abord pour le grand public un exposé sur l'idée de relativité dans la grande salle du Collège de France qui était bondée. Il s'efforçait de parler en un francais teinté d'accent germanique. Lorsqu'un mot lui manquait, il s'adressait à Langevin assis à sa droite qui lui soufflait l'expression française. L'exposé était simple et clair et le public était ravi. C'était l'époque où la théorie de la relativité commençait à déborder le cercle des scientifiques et à intéresser l'opinion publique.

Cet exposé d'introduction fut suivi, le lendemain et le surlendemain, de séances de discussions réservées a des initiés. Comme je faisais partie de la douzaine d'auditeurs qui suivaient au Collège de France le cours de Langevin, j'avais droit à une carte d'entrée. J'ai gardé de ces séances un souvenir inoubliable. Tous les grands physiciens, mathématiciens, philosophes étaient là, réunis autour d'Einstein et de Langevin. Je me souviens d'un professeur Guillaume venu de Suisse Romande qui avait couvert le tableau noir de panneaux sur lesquels s'entrecroisaient des cercles et des ellipses et qui terminait un long exposé par ces mots prononcés avec onction: "Et voilà la plus grave objection qu'on peut faire à la théorie de la relativité.'' Langevin se tourna vers Einstein: "Qu'avez-vous à répondre à cette critique?" Einstein leva les bras et dit d'un ton bon enfant: "Je regrette, je n'ai rien compris." Toute la salle partit d'un éclat de rire.

Mais il y eut aussi des moments dramatiques. Ainsi, lorsque le vieux Sagnac, inventeur d'un interféromètre ingénieux, partit dans une explosion de colère contre la théorie de la relativité qu'il accusa de tous les maux. Il n'y avait qu'à laisser passer l'orage. Et aussi des moments difficiles: lorsque le grand mathématicien Paul Painlevé, évoquant l'aventure des deux amis — celui resté sur place, et celui qui était parti en train et revenu — s'obstinait à ne pas comprende porquoi ce dernier était resté plus jeune que le premier. Il faut bien dire que l'interprétation triviale qu'on donne de cet effet n'est pas satisfaisante car on oublie de faire remarquer que le voyageur, en se retournant, subit une accélération considérable et que le problème, pour être traité à fond, doit être étudié du point de vue de la relativité générale.

Mon camarade, Théophile Aron, en 3e année d'école, qui avait été invité à diner chez Langevin avec Einstein, me raconta que celui-ci se comportait en société comme un grand enfant timide embarrassé de ses bras et de ses mains. Nous savons par ailleurs qu'il était bon violoniste. Un de ses plaisirs préférés était de jouer, accompagné de Planck au piano, une sonate de Beethoven. Nous serons heureux d'évoquer bientôt, en Israël et ailleurs, la figure de ce grand physicien qui a été aussi un grand pacifiste et qui s'est voulu citoyen du monde.

*Reprinted from Technion-Informations **11** (December 1978).

The "Secrétaire perpétuel" of our Paris Academy was during long years Louis de Broglie, whose work on wave mechanics owed much to Einstein. He started from the Lorentz–Einstein transformations and showed that a standing wave associated to a particle in its rest frame was transformed into a propagating wave in a traveling frame in which the particle had velocity v. The wavelength was given by the relation $\lambda = h/mv$ where h is Planck's constant, and de Broglie showed the v could be interpreted as the group velocity of the waves associated to the particle. It was through Einstein that de Broglie's work became known to Schrödinger and gave him the idea to establish his wave equation. Louis de Broglie retired from his position at our Academy five years ago and now lives at the age of eighty-seven a monastic life. Like Einstein, he rejects the Copenhagen interpretation of quantum mechanics, and he may be happy to learn about the opinion Professor Dirac expressed during this symposium concerning future developments.

15. GRAND UNIFICATION

Steven Weinberg

1. Introduction

The program of "grand unification" aims at the combination of the gauge theories of electroweak and strong interactions, and the combination of quarks and leptons in the same multiplets. As such, it is to be distinguished from the more ambitious program of "superunification," in which one tries to put particles of different spin together in larger multiplets. I will treat here only grand unification, not superunification.

In grand unified theories, it is assumed that the electroweak gauge group $SU(2) \times U(1)$ and the strong gauge group $SU(3)$ of quantum chromodynamics are subgroups of a larger "grand" gauge group G that connects quarks and leptons. (There are a number of specific grand unified models, based on various grand gauge groups G; a list of some leading examples is given in note 1.) Just as $SU(2) \times U(1)$ is spontaneously broken to the $U(1)$ of electromagnetism, giving masses 80–90 GeV to $W^{\pm}$ and Z^0, so also the grand gauge group must be more strongly broken down to $SU(3) \times SU(2) \times U(1)$, giving larger masses to other vector bosons.

It is naturally attractive to suppose that the gauge coupling constants of the strong and electroweak interaction are related by their common origin so that they are of comparable magnitude. The simplest possibility is that the grand gauge is simple, either in the narrow sense of having a simple Lie algebra, or in the somewhat extended sense of being a direct product of isomorphic simple gauge groups together with a discrete group that connects these factors irreducibly. The distinction is not an important one physically; in either case one has only one free gauge coupling constant for all interactions. The ratios of the squares of the strong and electroweak couplings are then rational numbers of order unity. But any theory that makes the $SU(3)$ and $SU(2) \times U(1)$ gauge couplings comparable is in apparent disagreement with the fact that the strong interactions are strong and the electroweak interactions are not.

The property of quantum chromodynamics that makes it possible to have comparable $SU(3)$ and $SU(2) \times U(1)$ couplings is its asymptotic freedom, discovered by Gross and Wilczek and

---

Yuval Ne'eman (ed.), To Fulfill A Vision: Jerusalem Einstein Centennial Symposium on Gauge Theories and Unification of Physical Forces
ISBN 0-201-05289-X

Copyright © 1981 by Addison-Wesley Publishing Company, Inc., Advanced Book Program.
All rights reserved. No part of this publication may be reproduced, stored in a retrieval system, or transmitted, in any form or by any means, electronic, mechanical photocopying, recording, or otherwise, without the prior permission of the publisher.

Politzer. The strong interaction coupling constant is a decreasing function of the energy $\varkappa$ at which it is measured

$$g_{QCD}^2(\varkappa) \simeq \frac{8\pi^2}{11 - 2n/3} \frac{1}{\ln(\varkappa/\Lambda)}, \tag{15.1}$$

where n is the number of quark flavors and $\Lambda \simeq 500$ MeV. At energies $\varkappa \geqslant 100$ GeV, above the $W^\pm$ and Z^0 masses, the weak and electromagnetic couplings g, g' also vary with $\varkappa$, but more slowly. Thus there is at least a chance that all couplings will come together at some very large scale. This is the scale that characterizes the spontaneous breakdown of the grand gauge group G, and at which the couplings g_s, g, and g' have whatever ratios are dictated by G. (Note that from this point of view the strong interactions must be asymptotically free — otherwise they would not be strong!)

This approach to grand unification was introduced five years ago by Georgi, Quinn, and myself (GQW).[2] It has become more popular lately because of experimental developments, to which I will return in the next section.

The particular grand unified models that have been proposed so far have not yet succeeded in accomplishing all the things that one would have wanted from the unification of strong with weak and electromagnetic interactions. In particular, for the most part they leave the generations of fermions separate, different aspects of different specific models are somewhat contrived, and gravity is left out. It seems to me useful therefore to try to explore those aspects of grand unified gauge theories that do not depend on the details of any specific model. Here I will follow the approach of GQW[3] and avoid any commitment to any specific grand unified model.

However, there is one important special assumption on which I will rely. I will assume (as in GQW) that the spontaneous breakdown of the grand gauge group into $SU(3) \times SU(2) \times U(1)$ takes place in only one stage, so that the only gauge bosons with mass far below the grand unification scale M that are non-neutral under $SU(3) \times SU(2) \times U(1)$ are the eight gluons of $SU(3)$ and the $W^\pm$, Z^0, and γ of $SU(2) \times U(1)$.

Of course, there are many other possibilities. Harari elsewhere in this volume gave a survey of them. For instance, it is possible that the $SU(3)$ gauge group of the strong interactions is a remnant of the spontaneous breakdown of a larger strong-interaction gauge group, which is broken at a mass scale intermediate between that of the W and Z masses and the final scale M at which the grand gauge group itself breaks down. In the same way, it is possible that the $SU(2) \times U(1)$ group of the weak and electromagnetic interactions is the heir of a larger flavor group, which is broken at a scale much larger than $m_{Z,W}$ but less than M.

Generally speaking, the introduction of such intermediate scales of spontaneous symmetry-breaking has the effect of lowering the final scale at which the grand unified gauge group was spontaneously broken. This is because the effective gauge group at energies larger than the intermediate scale would generally involve factors larger than $SU(3)$ or $SU(2)$, and the larger the group, the faster the change in the coupling, and hence the less far up you have to go in energy before the individual coupling constants can approach each other.

The alternative, which I will be exploring here, is that nothing happens to $SU(3) \times SU(2) \times U(1)$ between the scale of the Z and W masses and the scale M of the grand unification breakdown. This makes the scale M enormously large, because $SU(3)$ is not a very large group, so that the QCD coupling drops very slowly as the energy increases, and hence one must go to enormously high energies before grand unification is achieved. I think that some physicists are repelled by the idea of a vast desert between the Z and W masses and the grand unification scale. I cannot say that I am particularly attracted to it myself. Certainly it is reasonable to explore the possibility of various kinds of intermediate mass scales. But it is also reasonable to explore the simpler alternative, that $SU(3) \times SU(2) \times U(1)$ does not merge into any larger group until the grand unification scale is reached.

Of course, one naturally hopes that the next energy level at which really new physics appears will lie within the range of practicable accelerators. But this hope cannot be used as a basis for physical theories. Einstein is often quoted to the effect that God is subtle but not malicious. I do not think that this should be interpreted to mean that the laws of nature were designed with attention to the capabilities of our accelerators.

2. Calculation of M and $\sin^2\theta$

This section will present a slightly updated review of the calculation of the grand unification scale M and the $Z^0 - \gamma$ mixing parameter $\sin^2\theta$ in GQW.[4]

We must first consider the relations that are imposed by any simple grand unified gauge group on the coupling constants g_s, g, g' of $SU(3)$ and $SU(2) \times U(1)$. The general rule is very simple: At the grand unification scale, the square of any coupling times the trace of the square of the corresponding matrix generator is the same for all couplings. This rule must be satisfied for all representations, so all we need to know in order to calculate the ratios of the couplings g_s, g, g' is the $SU(3) \times SU(2) \times U(1)$ content of any one representation of the grand unified gauge group.

This would be perfectly straightforward if we knew what the grand gauge group actually was. Since we do not know that, we have to guess: we can look at the particles that we see in nature that have some definite helicity, and assume that they form at least a good sample of some representation of the grand gauge group. In particular, we know of plenty of left-handed spin 1/2 fermions. They all seem to fall into "generations":

$$l_{aL} = \begin{bmatrix} \nu_a \\ e_a^- \end{bmatrix}_L \qquad \overline{e_{aR}^-}$$

$$q_{a\alpha L} = \begin{bmatrix} u_{a\alpha} \\ d_{a\alpha} \end{bmatrix}_L \qquad \overline{u_{a\alpha R}} \qquad \overline{d'_{a\alpha R}} \tag{15.2}$$

with α a color index, and a a generation index:

$$e_1 = e \qquad e_2 = \mu \qquad e_3 = \tau \qquad \ldots$$

$$u_1 = u \qquad u_2 = c \qquad u_3 = t\ (?) \qquad \ldots$$

$$d_1 = d \qquad d_2 = s \qquad d_3 = b \qquad \ldots\ ,$$

It is at least plausible either that all such generations of left-handed fermions form a complete representation (reducible or irreducible) of the grand gauge group or that any additional fermions that are needed to complete the representation are neutral under $SU(3) \times SU(2) \times U(1)$. ($SU(5)$ gives an example of the first possibility; $SU(4)^4$ and $SO(10)$ of the second.) In either case, we can compute the traces of the squares of the generators of $SU(3)$ and $SU(2) \times U(1)$ by simply adding up the squares of the $SU(3)$ and $SU(2) \times U(1)$ quantum numbers for all fermions in these generations. For instance, the $SU(3)$ generator λ_3 has eigenvalues $+1/2$ for red quarks, $-1/2$ for green quarks, and 0 for blue quarks, so including quarks and antiquarks of charge $+2/3$ and $-1/3$, we have

$$Tr\, \lambda_3^2 = (2 \times 2) \times (\tfrac{1}{4} + \tfrac{1}{4}) N = 2N \,, \tag{15.3}$$

where N is the number of fermion generations. The $SU(2)$ generator t_3 has eigenvalues $+1/2$ for v_a and $u_{a\alpha L}$ and $-1/2$ for e_a and $d_{a\alpha L}$, so

$$Tr\, t_3^2 = (1 + 3) \times (\tfrac{1}{4} + \tfrac{1}{4}) N = 2N \,. \tag{15.4}$$

The $U(1)$ generator $y \equiv t_3 - q$ has eigenvalues $+1/2$ for v_{aL} and e_{aL}; -1 for e_{aR}; $-1/6$ for $u_{a\alpha L}$ and $d_{a\alpha L}$; $2/3$ for $u_{a\alpha R}$; and $-1/3$ for $\overline{d_{a\alpha R}}$; so

$$Tr\, y^2 = \left(\frac{1}{4} + \frac{1}{4} + 1 + \frac{6}{36} + \frac{3 \times 4}{9} + \frac{3 \times 1}{9}\right) N = \frac{10}{3} N \,. \tag{15.5}$$

As we have said, the coupling constants g_s, g, g' that are associated with λ_3, t_3, and y must be related by

$$g_s^2(M)\, Tr\, \lambda_3^2 = g^2(M)\, Tr\, t_3^2 = g'^2(M)\, Tr\, y^2 \tag{15.6}$$

or, using (15.3)–(15.5),

$$g_s^2(M) = g^2(M) = \tfrac{5}{3} g'^2(M) \,. \tag{15.7}$$

These relations are clearly independent of the number of fermion generations. Furthermore, even if there were some extra fermions that are non-neutral under $SU(3) \times SU(2) \times U(1)$ and did not fall into any generation, the relations (15.7) would continue to be approximately valid, provided there were only a few of these extra fermions.

The relations (15.7) apply at the grand unification mass scale M. What do they tell us about the coupling constants at ordinary energies? This can be answered entirely in the context of the known $SU(3) \times SU(2) \times U(1)$ interactions, with no further reference to grand unification. We must integrate the Gell-Mann–Low renormalization group equations

$$\kappa \frac{d}{d\kappa} g_i(\kappa) = b_i g_i^3(\kappa) + \ldots \tag{15.8}$$

using Eq. (15.7) as an initial condition at $\varkappa = M$. The calculation of the b_i is now taught in ordinary graduate school courses on quantum field theory; the results are

$$16\pi^2 \, b_{SU(3)} = -11 + \frac{4N}{3} \tag{15.9}$$

$$16\pi^2 \, b_{SU(2)} = -\frac{22}{3} + \frac{4N}{3} \tag{15.10}$$

$$16\pi^2 \, b_{U(1)} = \frac{20N}{9} \,. \tag{15.11}$$

(Small scalar particle contributions are omitted here.)

Here N is, strictly speaking, the number of generations of fermions with masses $\lesssim \varkappa$, and it is assumed that the fermions within any one representation do not have enormously different masses. All N-dependence is eliminated in two combinations of these equations,

$$16\pi^2 \, \varkappa \, \frac{d}{d\varkappa} \left[\frac{1}{g_s^2(\varkappa)} - \frac{1}{g^2(\varkappa)} \right] = \frac{22}{3} \,, \tag{15.12}$$

$$16\pi^2 \, \varkappa \, \frac{d}{d\varkappa} \left[\frac{1}{g^2(\varkappa)} - \frac{3}{5g'^2(\varkappa)} \right] = \frac{44}{3} \,. \tag{15.13}$$

According to Eq. (15.7), the quantities in brackets vanish at the grand unification scale $\varkappa = M$, so the solutions are

$$\frac{1}{g_s^2(\varkappa)} - \frac{1}{g^2(\varkappa)} = \frac{11}{24\pi^2} \ln \frac{\varkappa}{M} \,, \tag{15.14}$$

$$\frac{1}{g^2(\varkappa)} - \frac{3}{5g'^2(\varkappa)} = \frac{11}{12\pi^2} \ln \frac{\varkappa}{M} \,. \tag{15.15}$$

These equations are valid down to $\varkappa \simeq 100$ GeV, where the spontaneous breakdown of $SU(2) \times U(1)$ begins to affect the right-hand sides of Eqs. (15.12) and (15.13). Below this energy, $g^2(\varkappa)$ and $g'^2(\varkappa)$ vary more slowly, so we can use their observed values as the values at 100 GeV:

$$g^2(100 \text{ GeV}) = e^2/\sin^2\theta \quad g'^2(100 \text{ GeV}) = e^2/\cos^2\theta \,. \tag{15.16}$$

Equations (15.14) and (15.15) can then be used to give formulas for the $Z^0 - \gamma$ mixing angle θ and the grand unified scale M in terms of e^2 and g_s^2 (100 GeV):

$$\ln\left(\frac{M}{100 \text{ GeV}}\right) = \frac{4\pi^2}{11}\left(\frac{1}{e^2} - \frac{8}{3g_s^2(100 \text{ GeV})}\right), \qquad (15.17)$$

$$\sin^2\theta = \frac{1}{6}\left(1 + \frac{10e^2}{3g_s^2(100 \text{ GeV})}\right). \qquad (15.18)$$

(I should perhaps emphasize again that these results are independent of the number of fermion generations.) Taking g_s^2 (100 GeV)/4π in the range of 0.12 to 0.14, one finds

$$M = 2 \times 10^{16} \text{ GeV} \quad \text{to} \quad 4 \times 10^{16} \text{ GeV} \qquad (15.19)$$

$$\sin^2\theta = 0.20 \quad . \qquad (15.20)$$

Goldman and Ross and Marciano[5] have refined this analysis, including mass-dependent and two-loop terms in the renormalization group equations. They find a 10 percent correction in the logarithm, so that the masses of the superheavy gauge bosons are about 50 times smaller than the nominal grand unified mass scale (15.19)

$$M_v \simeq 4 \times 10^{14} \text{ GeV} \quad \text{to} \quad 8 \times 10^{14} \text{ GeV} \quad .$$

However, $\sin^2\theta$ is shifted only very slightly, to a value $\simeq 0.21$.

When Eqs. (15.17) and (15.18) were first obtained by Georgi, Quinn, and myself[6] about five years ago, experimentalists were quoting for $\sin^2\theta$ a value of about 0.3 or 0.35, so our prediction that $\sin^2\theta = 0.2$ was not regarded as a great triumph of the grand unification program. Since that time, as more and more neutral current data has accumulated, $\sin^2\theta$ has become much more accurately known, and its value has dropped. The most precise measurements today are

$$\sin^2\theta = 0.24 \pm 0.02 \qquad (vN \rightarrow vX, \text{ CDHS})$$

$$\sin^2\theta = 0.224 \pm 0.02 \qquad (eN \rightarrow eX, \text{ SLAC-Yale}) \quad .$$

It is entirely possible that the actual value of $\sin^2\theta$ will turn out to agree with the grand unification prediction, but to settle this it will be necessary to push measurements of $\sin^2\theta$ to even higher precision and perhaps also make further improvements in the calculations of radiative corrections. However, the present fair agreement between predicted and observed values of $\sin^2\theta$, together with the fact that the prediction $\sin^2\theta = 0.20$–0.21 applies for such a broad range of grand unified theories, has led many physicists to take a fresh look at the implications of these theories.

3. Stages of Symmetry-Breaking

In the previous section, we have been led to contemplate mass scale ratios $M/m_{Z,W}$ as large as 10^{13} to 10^{14}. Such huge ratios are inevitable once we impose the condition that the slowly decreasing $SU(3)$ coupling drops to a strength comparable to the $SU(2) \times U(1)$ couplings at a scale M. Also, as I will discuss shortly, these large values of M fit in well with considerations of baryon nonconservation. But this still leaves us with the question: Why should there be such very different scales of symmetry-breaking?[7] The answer is that we do not know, and for the present we must just do our best to guess.

First, let me be a little more precise about what is meant by having two stages of spontaneous symmetry-breaking. We can define a potential $V(\phi)$ as the free energy of the vacuum for a given set of scalar field vacuum expectation values ϕ_i. If the grand gauge group G were simply broken down to $SU(3) \times SU(2) \times U(1)$, then $V(\phi)$ would have a minimum at some point $\phi = \lambda$ that is $SU(3) \times SU(2) \times U(1)$-invariant, but not G-invariant. In order for there to be a second stage of symmetry-breaking, in which $SU(2) \times U(1)$ is broken more weakly than G, we have to suppose that there is a second stationary point of the potential, invariant under $SU(3)$ and electromagnetic gauge invariance but not under $SU(2) \times U(1)$, which is near the first stationary point, but which represents a true minimum of the potential.

So that is a way of talking about stages of symmetry-breaking, but it does not answer the question of *why* there should be a minimum of the potential close to a different stationary point. In exploring the neighborhood of the first stationary point $\phi = \lambda$, we can use an effective field theory, in which the fields are just those that correspond to degrees of freedom that did not get superlarge masses from the first stage: the gauge bosons of $SU(3) \times SU(2) \times U(1)$, plus those quarks and leptons that belong to nonreal representations of $SU(3) \times SU(2) \times U(1)$, plus any scalars that for one reason or another did not get large masses from the first stage. To be more specific, these scalars correspond to those eigenvectors of the matrix $[\partial^2 V(\phi)/\partial \phi_i \partial \phi_j]_{\phi=\lambda}$ with eigenvalues $m^2 \ll M^2$. It is the presence of such light scalars that gives the effective field theory the seeds of a second stage of symmetry-breaking.

The simplest case is that the small eigenvalues m^2 actually vanish. The second stage of spontaneous symmetry-breaking is then of the type described by Coleman and E. Weinberg,[8] in which radiative corrections play an essential part in establishing a nontrivial minimum of the potential. With arbitrary members of quartic couplings, the minimum of the potential was shown by Gildener and myself[9] to be at a point $\phi_i = \lambda_i + n_i x_0$, where x_0 is the renormalization scale at which the quartic couplings $f_{ijxl}(x)$ of the effective field theory satisfy the constraint

$$\min_{U_i U_i = 1} \{f_{ijxl}(x_0) U_i U_j U_x U_l\} = 0 ,$$

$$f_{ijxl}(x) \equiv \left[\frac{\partial^4 V(\phi)}{\partial \phi_i \partial \phi_j \partial \phi_x \partial \phi_l}\right]_{|\phi - \lambda| \sim x} ,$$

(15.21)

and n_i is the value of the unit vector U_i at which the minimum in Eq. (15.21) is attained. To find the value of x_0 at which this constraint is satisfied, we must integrate the renormalization group equations for the quartic couplings. If the minimum of $f_{ijkl}(\varkappa)U_i U_j U_k U_l$ on the unit sphere $|U| = 1$ is of order e^2 for $\varkappa$ of the order of the grand unification scale M, then Eq. (15.21) will be satisfied only if we integrate out to very large values of $|ln(\varkappa/M)|$ — that is, to $\varkappa \ll M$. It is for this reason that the true minimum $\lambda_i + \varkappa_0 n_i$ is likely to be very close to the stationary point $\phi_i = \lambda_i$, so that symmetries that are broken at $\phi_i = \lambda_i + \varkappa_0 n_i$ but not at $\phi_i = \lambda_i$ are very much more weakly broken than those broken at $\phi_i = \lambda_i$.

To see what sort of numerical results are to be expected, take the effective $SU(3) \times SU(2) \times U(1)$ field theory to have six flavors of quarks and of leptons, and one light scalar doublet (ϕ^+, ϕ^0). There is in this case just one quartic coupling $f(\varkappa)$, which satisfies the renormalization group equations

$$16\pi^2 \varkappa \frac{df}{d\varkappa} = 4f^2 - (9g^2 + 3g'^2)f$$

$$+ \tfrac{9}{4}(3g^4 + 2g^2 g'^2 + g'^4)$$

with $SU(2) \times U(1)$ gauge couplings satisfying the equations

$$16\pi^2 \varkappa \frac{dg}{d\varkappa} = -\frac{13}{4} g^3$$

$$16\pi^2 \varkappa \frac{dg'}{d\varkappa} = +\frac{27}{4} g'^3 .$$

The constraint (15.21) just gives

$$f(\varkappa_0) = 0 .$$

This value of $\varkappa_0$ sets the scale of the breakdown of $SU(2) \times U(1)$ — that is, of m_W and m_Z — so it is to be taken of order 100 GeV. The "initial" conditions for these differential equations can then be completed with

$$g(\varkappa_0) = e/\sin\theta \qquad g'(\varkappa_0) = e/\cos\theta .$$

A numerical integration shows that $M/\varkappa_0$ has a value of order 10^{13} if the quartic coupling at the grand unification scale M is given the value $f(M) = 1.7 e^2$, certainly not an implausible coupling.

Incidentally, you can if you like play with the possibility that the scalar mass m^2 is not zero but only very small. In this way, the ratio $M/\varkappa_0$ can be made even larger, but only by an additional factor $e^{1/4} = 1.28$. If you try to increase m^2 past this point, a first-order phase transition is encountered, in which the first-stage point $\phi = \lambda_i$ becomes the minimum of the potential.

We do not know why the matrix $\partial^2 V(\phi)/\partial \phi_i \partial \phi_j$ should have eigenvalues m^2 at $\phi = \lambda$ that are zero or much less than M^2. All I can offer here is to note that the point λ is characterized by a high degree of symmetry, at least $SU(3) \times SU(2) \times U(1)$, so perhaps there is some other symmetry, whose nature we do not yet understand, that enforces a condition that some of these eigenvalues vanish. But it should be stressed that this condition, of a small or zero mass eigenvalue, is all that is needed to bring about a huge ratio of symmetry-breaking scales.

In any case, the hypothesis that the small eigenvalues m^2 of the scalar mass matrix vanish has an interesting experimental consequence: the curvature of the potential $V(\phi)$ along the direction n_i is due only to radiative corrections, so the corresponding Higgs boson has a relatively small but calculable mass:

$$m_S^2 = \frac{\sqrt{2}G_F}{8\pi^2} [6m_W^4 + 3m_Z^4 - 4\sum_F m_F^4 + \sum_H m_H^4] ,$$

the sums running over all fermions and all other Higgs bosons. Gildener and I called this particle the "scalon," because it is the pseudo-Goldstone boson associated with broken scale invariance. If we assume that there are not any fermions or scalars as heavy as W and Z, and take $\sin^2\theta$ between 0.25 and 0.20, then m_S turns out to be between 9 GeV and 11 GeV. It will be hard to find the S, because its couplings conserve all internal quantum numbers: charge, strangeness, C, P, T, et cetera. But if it is at all possible, this would certainly be something worth looking for.

There is another possibility, that no scalars escape getting masses from the spontaneous breakdown of the grand gauge group, and that the spontaneous breakdown of $SU(2) \times U(1)$ to $U(1)_{em}$ is "dynamical," caused by unobserved extra strong interactions,[10] whose gauge couplings become of order unity at scales of order 300 GeV. Such theories have many attractive aspects, but it is difficult to construct models in which ordinary quarks and leptons get realistic masses, and in which there are no unacceptable pseudo-Goldstone bosons of small or zero mass.

4. Observable Hyperweak Interactions

Let us face the prospect of a theory that contains superheavy particles — gauge bosons and perhaps others as well — with masses of the order of 10^{15} GeV. Clearly our accelerators are never going to get up to energies that high, at least not until someone figures out how to concentrate macroscopic amounts of energy on a single particle. What then can we observe at ordinary energies? In a sense we live among the debris of the first stage of spontaneous symmetry-breaking, and we must root about among this debris to find clues to the original theory.

Of what does this debris consist? There are of course the renormalizable interactions among those particles — leptons, quarks, $W^\pm$, Z^0, γ, gluons, and low-mass scalars — that for one reason or another did not get masses comparable to the grand unification scale M from the first stage of symmetry-breaking. No more needs to be said about these now. In addition, when we "integrate out" the degrees of freedom of the superheavy particles, we find all kinds of further effective induced couplings, with coupling strengths of the order of various negative powers of the grand unification mass scale M. At ordinary energies, such induced couplings would be enormously suppressed, by powers of about 10^{15}. I can think of only two types of such "hyperweak" interactions that might ever be observed.

First, there is gravitation. The Newton coupling constant can be written as $(1.22 \times 10^{19}$ GeV$)^{-2}$, so it is hyperweak in the same sense as before, except that the characteristic mass $M \approx 10^{15}$ to 10^{16} GeV is replaced with 10^{19} GeV. The reason that we can observe gravitation at all is that it has the unique[11] feature of long range and coherence: the gravitational field of a macroscopic body is given by adding up the fields produced by each particle within it. We would never even know of the existence of gravitation if this feature did not make it possible to observe gravitation macroscopically.

In the same way, there may be a great number of other interactions roughly as weak as gravitation, the existence of which we do not suspect because they can not be observed macroscopically. (Indeed, we are severely limited even in our ability to observe gravitational interactions. When Einstein developed general relativity, in effect he restricted the gravitational action to just a single term, the famous $\sqrt{g}R/16\pi G$. There is an unlimited number of other possible terms, proportional to $\sqrt{g}R^2$, $\sqrt{g}R_{\mu\nu}R^{\mu\nu}$, $\sqrt{g}R^3$, and so on. Einstein ruled out these terms, on the grounds that the field equations should be at most second-order differential equations. This certainly worked very well, but it leaves us wondering why nature should care about whether the field equations are of second or higher order. It seems to me far more attractive to suppose that all the higher interactions are really there in the gravitational Lagrangian, but with coupling constants involving higher and higher powers of M^{-2} or $G = (1.2 \times 10^{19}\text{ GeV})^{-2}$, as we add more and more derivatives or powers of the curvature tensor. In this case, the only term that could be observed macroscopically, and hence the only term about which we could have any direct knowledge, would be the Einstein term $\sqrt{g}R/16\pi G$.)

It is striking that the Planck mass 1.2×10^{19} GeV is not so very different from the grand unification mass scale $M \simeq 10^{15}$ to 10^{16} GeV. Among other things, this suggests that when we talk about grand unification in terms of ordinary Yang-Mills gauge groups, and leave gravity out, we may be being a little too naive. It is in part for this reason that I said earlier that I did not want to tie what I was saying here to any specific grand unified model, but would instead concentrate on those aspects that are common to a wide variety of models and hence that have a chance of being found to remain valid when we understand grand unification in a wider context.

Apart from gravitation, the only other kinds of hyperweak interactions that I can imagine could be detected with present techniques are interactions that violate cherished global conservation laws, laws that are believed otherwise to be exactly valid. The obvious possibilities are baryon and lepton number nonconservation. Baryon and lepton conservation are good candidates for violation, because unlike charge and color conservation, they appear to be unrelated to any unbroken local gauge symmetry.[12] Indeed, because baryon and lepton conservation are not gauge symmetries, their currents can have Adler-Bell-Jackiw anomalies, and such anomalies actually produce a low rate of nuclear decay[13] in the modern theory of weak and electromagnetic interactions. Also, the existence of a non-vanishing cosmic baryon density suggests that baryon conservation may have been violated in the very early universe.[14] Pati and Salam, in their early work on grand unification,[15] noted that baryon and lepton number were not conserved in their model. As they remarked, baryon and lepton nonconservation is possible (though not inevitable) in grand unified theories, because quarks and leptons are in the same multiplet of the grand gauge group, so that there exist superheavy gauge bosons that turn quarks into leptons. Baryon and lepton nonconservation has been found also to occur in all the other leading models[16]: $SU(5)$, $SO(10)$, E_6, E_7, and so on. But apart from this, the possibility of baryon or lepton nonconservation provides what may be the unique means of learning about the physics of extremely high energies.

5. Mechanisms for Baryon Nonconservation

This section will present a general analysis[17] of the possible mechanisms by which baryon and lepton conservation may be violated. I will make no direct use of grand unified models here, but will simply assume that the primary source of the violation is in the coupling of very heavy ($\gg$ 100 GeV) vector or scalar "X bosons" to pairs of fermions and/or antifermions of the familiar types (2). I will also assume that the gauge group $SU(3)$ of QCD is unbroken, and hence the analysis presented here will differ from some of that of Pati and Salam,[18] but it applies to any theory that incorporates QCD.

The tool for this analysis is the familiar $SU(3) \times SU(2) \times U(1)$ gauge symmetry of strong and electroweak interactions. This is an exact symmetry of the Lagrangian, and although $SU(2) \times U(1)$ is spontaneously broken at energies $\lesssim$ 100 GeV, at the mass of our X bosons this symmetry-breaking can be ignored.

All we need then to do in classifying all bosons that can interact with pairs of the familiar type of fermions and/or antifermions is to add up the $SU(3) \times SU(2) \times U(1)$ quantum numbers of all such pairs. We take the ordinary fermion masses to be negligible compared with the superheavy boson masses; in consequence for vector bosons the pair consists of one left- and one right-handed fermion and/or antifermion; for scalar bosons there are two left- or two right-handed fermions and/or antifermions. (The analysis presented here would apply even for nonrenormalizable boson-fermion interactions, except for the identification of the bosons as scalars or vectors. However, effects of nonrenormalizable couplings are suppressed at ordinary energies by extra powers of superheavy masses.) In this way, it is an elementary exercise to compile a complete list of all vector or scalar bosons that can interact at all in an $SU(3) \times SU(2) \times U(1)$-invariant way with a fermion and/or antifermion pair. Such a list is presented in Tables 15.1 and 15.2, with vector and scalar bosons cataloged according to their $SU(3) \times SU(2) \times U(1)$ quantum numbers.

A striking fact emerges when we inspect this catalog. Almost all of the vector and scalar bosons couple to channels with only a single value of the baryon and the lepton number. Such bosons can be assigned a baryon and a lepton number, in such a way that baryon and lepton numbers are conserved in boson-fermion interactions. Thus for instance the (1,1,0) vector boson [the B of $SU(2) \times U(1)$] has $B = L = 0$; the (1,2,3/2) vector bosons have $B = 0, L = 2$; the (6,2, $-5/6$) and (6,2,1/6) vector bosons have $B = 2, L = 0$; and so forth. There are just five possible kinds of boson that couple to channels of nonunique baryon or lepton number. These are the vector bosons labeled X_V and X_V' in Table 15.1, and the scalar bosons labeled X_S, X_S', X_S'' in Table 15.2.

Note also that all bosons, including those with baryon- or lepton-violating interactions, couple only to channels having a common value of baryon number minus lepton number. Thus we can assign values of $B - L$ to all bosons, in such a way that $B - L$ is conserved in all boson-fermion interactions. (This is not obvious when one starts this analysis; $SU(3)$ alone would only tell us that a boson that decays into quark-lepton or quark-antilepton channels would have to belong to the $\underline{3}$ representation of $SU(3)$, so that it could also decay into antiquark plus antiquark but not quark plus antiquark or quark plus quark. In order to see that a boson that decays into two antiquarks can not also decay into quark plus antilepton, but only quark plus lepton, one needs to use $SU(2) \times U(1)$ as an unbroken symmetry as well.) An important practical consequence of the conservation of $B - L$ is that protons and neutrons can decay into positrons and mesons (for example, $p \to e^+\pi^0$, $n \to e^+\pi^-$) but not into electrons and mesons (for example, $p \to e^-\pi^+\pi^+$, $n \to e^-\pi^+$).

	$SU(3)$	$SU(2)$	Y	Charges	Decay Modes
B	1	1	0	0	$\bar{l}_L l_L, \bar{e}_R e_R, \bar{q}_L q_L, \bar{u}_R u_R, \bar{d}_R d_R$
	1	1	-1	$+1$	$\bar{d}_R u_R$
	1	2	3/2	$-1, -2$	$l_L e_R$
A	1	3	0	$+1, 0, -1$	$\bar{l}_L l_L, \bar{q}_L q_L$
	3	1	$-2/3$	$+2/3$	$\bar{e}_R d_R, \bar{l}_L q_L$
	3	1	$-5/3$	$+5/3$	$\bar{e}_R u_R$
X_V	3	2	5/6	$-1/3, -4/3$	$l_L d_R, e_R q_L, \bar{q}_L \bar{u}_R$
X'_V	3	2	$-1/6$	$2/3, -1/3$	$l_L u_R, \bar{q}_L \bar{d}_R$
	3	3	$-2/3$	$4/3, 2/3, -1/3$	$\bar{l}_L q_L$
	6	2	$-5/6$	$4/3, 1/3$	$q_L u_R$
	6	2	1/6	$1/6, -5/6$	$q_L d_R$
Gluons	8	1	0	0	$\bar{q}_L q_L, \bar{u}_R u_R, \bar{d}_R d_R$
	8	1	-1	$+1$	$\bar{d}_R u_R$
	8	3	0	$+1, 0, -1$	$\bar{q}_L q_L$

Table 15.1

Catalog of all vector bosons that can have $SU(3) \times SU(2) \times U(1)$-invariant couplings to a pair of ordinary light fermions and/or antifermions. (Corresponding antibosons are not explicitly shown.) Here the $SU(3)$ and $SU(2)$ columns give the multiplicities of the representation of each boson; Y is the $U(1)$ quantum number $T_3 - Q$; l_L denotes any of $(\nu_e e)_L, (\nu_\mu \mu)_L, (\nu_\tau \tau)_L \ldots$; e_R denotes any of $e_R, \mu_R, \tau_R, \ldots$; q_L denotes any of $(u,d)_L$, $(c,s)_L, (t,b)_L, \ldots$; u_R denotes any of $u_R, c_R, t_R, \ldots$; d_R denotes any of $d_R, s_R, b_R, \ldots$; a bar indicates an antiparticle; and L and R indicate multiplication with $\frac{1}{2}(1 \pm \gamma_5)$.

It is not clear whether the conservation of $B - L$ is an accidental property of the boson-fermion interaction, which could be weakly violated in the self-interactions of bosons with $B - L \neq 0$, or whether it is an exact symmetry principle. As Gell-Mann has emphasized, exact $B - L$ symmetry would have the important consequence that $\nu_L \leftrightarrow \bar{\nu}_R$ transitions would be forbidden; in the absence of right-handed neutrinos (as opposed to antineutrinos), neutrinos would have to be strictly massless.

Baryon decay has been under study by many theorists and experimentalists. In the theoretical framework described here, it is produced by exchange of an X boson between two quarks in the

	$SU(3)$	$SU(2)$	Y	Charges	Decay Modes
	1	1	-1	1	$\bar{l}_L\bar{l}'_L$
	1	1	2	-2	$e_R e_R$
(ϕ^+, ϕ^0)	1	2	$-1/2$	1,0	$\bar{e}_R l_L, \bar{q}_L u_R, \bar{d}_R q_L$
	1	3	-1	2,1,0	$\bar{l}_L \bar{l}_L$
	3	1	$-2/3$	2/3	$\bar{d}_R \bar{d}'_R$
X_S	3	1	1/3	$-1/3$	$l_L q_L, e_R u_R, \bar{d}_R \bar{u}_R, \bar{q}_L \bar{q}_L$
X'_S	3	1	4/3	$-4/3$	$d_R e_R, \bar{u}_R \bar{u}'_R$
	3	2	$-7/6$	5/3, 2/3	$\bar{l}_L u_R, \bar{e}_R q_L$
	3	2	$-1/6$	2/3, $-1/3$	$\bar{l}_L d_R$
X''_S	3	3	1/3	2/3, $-1/3$, $-4/3$	$l_L q_L, \bar{q}_L \bar{q}'_L$
	6	1	$-4/3$	4/3	$u_R u_R$
	6	1	$-1/3$	1/3	$u_R d_R$
	6	1	2/3	$-2/3$	$d_R d_R$
	6	3	$-1/3$	4/3, 1/3, $-2/3$	$q_L q_L$
	6	1	$-1/3$	1/3	$q_L q'_L$
	8	2	$-1/2$	1,0	$\bar{q}_L u_R, \bar{d}_R q_L$

Table 15.2

Catalog of all scalar bosons that can have $SU(3) \times SU(2) \times U(1)$-invariant couplings to a pair of ordinary light fermions and/or antifermions. (Corresponding antibosons are not explicitly shown.) Notation is same as for Table 15.1. A prime distinguishes cases where the fermions must be of different generations; in all other cases the fermions may be of the same or different generations.

baryon (either in the s or t channel) producing an antilepton and antiquark, with subsequent annihilation into mesons of the antiquark with the remaining quark. This gives a decay matrix element proportional to $1/m_X^2$ and hence a proton lifetime proportional to m_X^4. The proton lifetime τ_p was first estimated in this way in the paper by Georgi, Quinn, and myself;[19] we just guessed on dimensional grounds that τ_p would be of order m_X^4/m_p^5, or about 3×10^{32} years for $m_X = 10^{16}$ GeV. A more detailed estimate by Buras, Ellis, Gaillard, and Nanopoulos[20] gives a longer lifetime, due mainly to the presence of small coupling constant factors in the decay matrix element. But estimates of m_X have also decreased,[21] shortening the estimated lifetime to about 10^{31} years.

These estimates are of course very crude. Proton decay is a full-fledged strong-interaction process, and a precision calculation of its rate is no more possible than, say, for $\Lambda \to N\pi$ or $K \to \pi\pi$. Even so, rough as they are, these estimates of the proton lifetime suggest that we may be on the threshold of exciting developments. The best present limit on the proton lifetime is from an experiment of Reines et al.[22]; this gives a limit of 5×10^{-31}/year on the partial rate for $p \to \mu^+ +$ (neutrals). If we assume that $p \to \mu^+$ is not suppressed relative to other modes by a factor any smaller than $\sin^2\theta_c = 1/20$, we can conclude from this experiment that the proton lifetime must be longer than about 10^{29} years. Thus the prediction of τ_p is not contradicted by existing data, but it is close enough to present limits so that it can be tested if reasonable improvements are made in experimental sensitivity. I know of at least two groups in the United States, and I believe there are several in Europe, that are planning an assault on the next few orders of magnitude in the proton lifetime. I would bet that if these experiments can be made sensitive to proton lifetimes as long as 10^{32} years, then proton decay will be found.

6. Cosmological Baryon Production

The universe is not a strictly neutral soup of photons, neutrinos, and antineutrinos, but also contains a small seasoning of protons, neutrons, and electrons. One can express this a bit more quantitatively by comparing the baryon number density of the universe to its entropy density. Most of the entropy of the present universe is contained in the microwave radiation background and in the associated neutrino and antineutrino backgrounds. With a microwave background temperature of 2.9 °K, the photon entropy density is $s_\gamma \simeq 250k/\text{cm}^3$. The entropy density of each species of neutrino and antineutrino is calculated to be $(7/22)s_\gamma$. (The quantity s_γ/k is very roughly equal to the number density of background photons, but it is better to deal with entropy rather than photon number densities, because as long as the expansion of the universe is adiabatic, the total entropy in any co-moving volume remains fixed, while processes in the early universe like e^+e^- annihilation can affect the number of photons.) The baryon number density is not so well known, but limits on the deceleration of the cosmic expansion restrict n_B to be less than about $3 \times 10^{-6}/\text{cm}^3$, while the mass density of visible galaxies provides a lower bound on n_B (assuming that there is not much antimatter in the present universe[23]) of about 3×10^{-8} cm^{-3}. The ratio $n_B k/s$ of baryon number to entropy is thus in the range of 10^{-10} to 10^{-8}. This ratio remains constant throughout the history of the universe, as long as baryon number is conserved and the universe expands adiabatically.

Clearly, we would like to be able to understand why $n_B k/s$ does not vanish, and is yet so small. It is an old idea that the baryon excess in the universe may be a consequence of small violations of baryon conservation.[14] Recently this possibility has come under examination by a number of authors.[24]

One important point to come out of this work is that, even with nonconservation of baryon number, T, and CP, a cosmological production of baryon number can only occur when the particle distribution functions depart from their equilibrium form:

$$n_i(\mathbf{p}) = \left[\exp\left[\frac{\sqrt{\mathbf{p}^2 + m_i^2} - \mu_i}{kT} \right] \mp 1 \right]^{-1}. \tag{15.22}$$

(Here $n_i(p)$ = the phase space number density of particles of type i, μ_i is the sum of the chemical potentials for whatever conserved quantum numbers are carried by particle i, and the sign is -1 or $+1$ for bosons or fermions, respectively.) It is of course obvious that with baryon number not conserved, the net baryon number must vanish in a state of thermal equilibrium because TCP requires the masses of baryons and antibaryons to be equal, and there could not be any chemical potential in Eq. (15.22) that would distinguish baryons and antibaryons. However, it might be thought that weak collision processes that violate time-reversal invariance might drive the distribution functions slightly away from equilibrium, and if baryon number and CP are also violated, these non-equilibrium distributions might have a small baryon excess. But in fact, this cannot happen — even without time-reversal invariance, unitarity alone is enough to show that collisions cannot affect the form of particle distribution functions once thermal equilibrium has been established.[25] Some early calculations did seem to indicate that baryon number could be produced by collisions, starting with an equilibrium distribution of particles, but this was just because the calculations did not take account of all relevant diagrams, and therefore did not respect unitarity.

Fortunately, there is an external agency that can drive particle distributions out of the equilibrium form (15.22) — it is the expansion of the universe. A freely moving particle will have a momentum (as measured by local observers who move along with the universe) inversely proportional to the Robertson–Walker scale factor $R(t)$, while the phase-space volume element remains constant. For particles with negligible masses $m \ll kT$, the form of Eq. (15.22) is thus preserved for all $|\mathbf{p}| \leq kT$; the temperature and all chemical potentials simply scale like $1/R(t)$. On the other hand, if m is not negligible, a free expansion will change the form of the particle distribution functions. (These departures from equilibrium also produce small amounts of entropy through the effects of bulk viscosity,[26] a phenomenon known in the theory of imperfect fluids since the nineteenth century.) Hence departures from equilibrium can be produced if there is a period during which collision and decay rates are small compared with the cosmic expansion rate, and in which the temperature drops below the masses of one or more particles.

This is in fact what happens (as discussed in the papers by Toussaint et al., Dimopoulos and Susskind, and myself[27]). At the earliest times that we can consider at all, when the temperature was of the order of the Planck mass $m_P = G^{-1/2} = 1.22 \times 10^{19}$ GeV, gravitational interactions were strong, and all sorts of particles and antiparticles were being freely created and destroyed. Then as the universe expanded and cooled, gravitational interactions became ineffective. For a while, the expansion of the universe was faster than any particle's collision or decay rate. The cosmic expansion rate is

$$H \equiv \dot{R}/R = 1.66 \, N^{1/2} (kT)^2/m_P \tag{15.23}$$

(where N is the number of species of particles) while the decay rate of a particle X is of order

$$\Gamma_X \approx \alpha_X N m_X \quad , \tag{15.24}$$

where α_X is the "fine structure constant" describing the coupling of X to its decay modes. (For $kT > m_X$, there is an additional time-dilation factor of order m_X/kT.) We assume that $m_X \ll m_P$ and $\alpha_X N^{1/2} < 1$, so Γ_X was much less than H for kT comparable with m_P. Collision rates were even slower. Thus for a while after the temperature dropped below m_P, the universe passed through a period of essentially free expansion.

Nevertheless, if equilibrium distributions were established at $kT \approx m_P$, then as we have seen, they would have been maintained as long as the temperature remained above the masses of all particles. A particle X could begin to decay when the age $\approx H^{-1}$ of the universe became of the order of its lifetime Γ_X^{-1}; if kT was still above m_X at that time, then the unitarity relations among rates ensure that equilibrium was maintained through the era of decay, and little or no baryon number was produced. Specifically, this is because whatever baryon excess was produced in a decay $X \to A + B$ would have been destroyed in the inverse decay process $A + B \to X$. On the other hand, if kT dropped below m_X before H became of order Γ_X, then the distribution functions went out of equilibrium during the period of free expansion with $kT < m_X$; there was an overabundance of the X particles. These particles finally decayed when H dropped to order Γ_X, but inverse decay processes were then blocked by a Boltzmann factor $\exp(-m_X/kT)$, so the yield of baryon number was just whatever is produced when a thermal distribution (15.22) of particles decays.

The expansion rate $H \approx N^{1/2}(kT)^2/m_P$ becomes equal to the decay rate $\Gamma_X \approx \alpha_X N m_P$ when $kT \approx (\alpha_X N^{1/2} m_X m_P)^{1/2}$, so the condition that $kT < m_X$ at this time is satisfied if and only if

$$m_X \gtrsim \alpha_X N^{1/2} m_P \quad . \tag{15.25}$$

For the gauge bosons X_V, X_V' we expect α_X to be of order $\alpha = 1/137$ (or a little larger), so this would require that $m_X \gtrsim 10^{17} N^{1/2}$ GeV. As indicated in Sec. 2, we expect the superheavy gauge bosons to be several orders of magnitude lighter than this, so we can tentatively conclude that thermal equilibrium was preserved through their decay, and not much baryon number was then produced. (Indeed, even if thermal equilibrium distributions were not established at $kT \approx m_P$, and if there were an initial excess of baryons or antibaryons, the superheavy vector boson decay and inverse boson decay processes could have brought about thermal equilibrium, with no net baryon number remaining after they decayed.) However, the superheavy scalar bosons presumably have much weaker couplings to their decay channels, with $\alpha_X \approx G_F m^2 \approx 10^{-5}$, so they decayed later. Equation (15.25) is satisfied for these particles if $m_X \gtrsim 10^{14} N^{1/2}$ GeV, which is likely to be the case. We therefore are led to suppose a sequence of events in which particles and antiparticles are created at $kT \approx m_P$; then superheavy vector bosons decay while still relativistic, possibly wiping out any initial baryon excess or other departure from equilibrium; then the freely moving superheavy scalar bosons are red-shifted to nonrelativistic velocities, pulling the particle distributions for the first time out of equilibrium; and finally the superheavy scalar bosons decay, with no inverse decay processes available to cancel out the baryon number produced in these decays.

The net effect of this is that the universe acquires whatever baryon number is released when the superheavy X bosons, and most probably the superheavy X_S bosons, decay. As shown in Tables 15.1 and 15.2, the X boson decay modes are $X \to QL$ and $X \to \overline{QQ}$ (with Q and L any quark or lepton). Define the branching ratios as r and $1 - r$. The $\overline{X}$ bosons decay into the modes $\overline{X} \to \overline{QL}$ and $\overline{X} \to QQ$, with the same total rate, but with branching ratios $\overline{r}$ and $1 - \overline{r}$ that in general can be somewhat different from those for X decay. The average baryon number released when an X or an $\overline{X}$ decays is then

$$\Delta B = \tfrac{1}{2}[\tfrac{1}{3}r - \tfrac{2}{3}(1-r) - \tfrac{1}{3}\overline{r} + \tfrac{2}{3}(1-\overline{r})] = \tfrac{1}{2}(r - \overline{r}) \quad . \tag{15.26}$$

The number density of X or $\bar{X}$ bosons at very high temperatures is $n_X = \zeta(3)(kT)^3 N_X/\pi^2$, where N_X is the number of species of X or $\bar{X}$ bosons. But in a free expansion $T \propto 1/R$ and $n_X \propto 1/R^3$, so the same formula gives the number density just before the bosons decay. Also, the entropy density at temperature T is $s = 2\pi^2 k^4 T^3 N/45$. Putting this together, the baryon-entropy ratio just after the X bosons decay is the temperature-independent quantity

$$\frac{n_B k}{s} = \frac{n_X \Delta B k}{s} = \left[\frac{45\zeta(3)}{4\pi^4}\right]\left[\frac{N_X}{N}\right](r - \bar{r}) . \tag{15.27}$$

It is unlikely that there are any further baryon number-nonconserving reactions, so as long as the universe expands adiabatically, the baryon-entropy ratio will be given by Eq. (15.27).

The difficult problem is to calculate the branching ratio difference $r - \bar{r}$. Nanopoulos and I have been working on this lately.[28] It is immediately obvious that there are no nonvanishing contributions to $r - \bar{r}$ in the Born approximation, because the Born approximation is always time-reversal-invariant (if H' is Hermitian, then $|H'_{\beta\alpha}|^2$ is symmetric in β and α), and TCP invariance then implies CP invariance in Born approximation. More generally, the dispersive and absorptive parts of the decay matrix element are, respectively, Hermitian and anti-Hermitian, so $r - \bar{r}$ receives nonzero contributions only from the interference between dispersive and absorptive parts of the decay matrix element. Hence $r - \bar{r}$ will be rather small in any theory, so that $n_B k/s$ is inevitably much less than unity. It is a little less obvious, but also easy to prove,[29] that $r - \bar{r}$ receives no contribution from graphs of first order in baryon-nonconserving interactions, even if they involve an arbitrary number of other interactions. Therefore the lowest-order contributions to $r - \bar{r}$ come from graphs in which an X_S, X_V, or X'_V boson is exchanged between final particles in the decay of X_S, X_V, or X'_V. Even in the simpler grand unified models, there can be a nonvanishing contribution from graphs in which an X_S boson is exchanged between the final particles in the decay of a different species of X_S boson. These give a branching ratio difference of order

$$r - \bar{r} \approx G_F \overline{m^2} \mathcal{J} \varepsilon , \tag{15.28}$$

where $G_F \simeq 10^{-5}$ GeV^{-2}; $\overline{m^2} \simeq 1$ to 10 GeV^2 is the mean square quark or lepton mass (as measured at energies of order m_X); $|\mathcal{J}| \approx 10^{-3}$ to 10^{-2} is the one-loop integral; and ε is a phase, characterizing the CP violation in the X_S propagator and/or fermion interactions. On the basis of what we know of CP violation at ordinary energies, it seems reasonable to take $|\varepsilon|$ in the range of 10^{-2} to 1 radian, though this depends very much on how we think CP is violated. Then Eq. (15.28) gives $|r - \bar{r}| \approx 10^{-10}$ to 10^{-7}. Also, the number N of species of particles is at least 100, so the ratio N_X/N in Eq. (15.27) is probably in the range of 10^{-2} to 10^{-1}. Finally, the numerical coefficient $45\zeta(3)/4\pi^4$ in Eq. (15.27) has the value 0.13883. Putting this all together, Eq. (15.27) gives a baryon-entropy ratio

$$n_B k/s \approx 10^{-13} \text{ to } 10^{-9} . \tag{15.29}$$

This overlaps the range $n_B k/s \approx 10^{-8}$ to 10^{-10} of values allowed by astronomical observations, but these ranges of theoretical and observed values are clearly too broad for us to reach a definite conclusion that baryon nonconservation really is the source of the baryon excess of our universe.

7. Conclusion

In dealing with masses and temperatures of order 10^{15} GeV, one is working at the furthest reach of allowable speculation. We are dealing here with theories that are not yet well defined, and we are extending the domain of theoretical physics by many, many orders of magnitudes. And yet, I think it is fair to say that everything seems to hang together much more nicely than could have been anticipated. The mass of the superheavy scalar and vector bosons turns out to be not quite as large as the Planck mass 10^{19} GeV, so that it is not absurd to leave gravity out of these calculations, and yet they are somewhat larger than the lower bound 10^{14} GeV allowed (in this context) by the experimental lower bound on the lifetime of the proton. The mixing parameter $\sin^2\theta$ comes out about right. Furthermore, the mass of the superheavy scalar bosons is likely to be in the range, of $10^{14} N^{1/2}$ GeV to $10^{17} N^{1/2}$ GeV, for which there is a scenario leading to an appreciable cosmological baryon production. And the baryon-entropy ratio resulting from this scenario is quite compatible with what is observed astronomically.

It would be no great surprise to learn that the physics of the energy range from 10^3 GeV to 10^{19} GeV is infinitely more complicated than we now imagine, and that all the ideas that I have described here are therefore wrong. But these tentative ideas have so far been sufficiently successful that I think we would also not be too surprised to find that they are right.

NOTES

1. $SU(4)^n \times$ discrete symmetries: J. C. Pati and A. Salam, Phys. Rev. **D8**, 1240 (1973); Phys. Rev. **D10**, 275 (1974). $SU(5)$: H. Georgi and S. L. Glashow, Phys. Rev. Lett. **32**, 438 (1974). $SO(10)$: H. Georgi, in *Particles and Fields* (American Institute of Physics, New York, 1975); H. Fritzsch and P. Minkowski, Ann. Phys. **93**, 193 (1975); H. Georgi and D. V. Nanopoulos, Phys. Lett. **82B**, 392 (1979) and Harvard preprint HUTP-79/A001 (1979). E_6, E_7: F. Gürsey, P. Ramond, and P. Sikivie, Phys. Lett. **B60**, 177 (1975); F. Gürsey and P. Sikivie, Phys. Rev. Lett. **36**, 775 (1976); P. Ramond, Nucl. Phys., **B110**, 214 (1976); Q. Shafi, Phys. Lett. **79B**, 301 (1978). $SU(3) \times SU(3) \times SU(3) \times P$: Y. Achiman and B. Stech, Heidelberg preprint HU-THEP-78-20; and others.

2. H. Georgi, H. R. Quinn, and S. Weinberg, Phys. Rev. Lett. **33**, 451 (1974). Also see A. Buras, J. Ellis, M. K. Gaillard, and D. V. Nanopoulos, Nucl. Phys. **B135**, 66 (1978).

3. Georgi, Quinn, and Weinberg, op. cit. in n. 2.

4. Ibid.

5. D. Ross, Nucl. Phys. **B140**, 1 (1978); W. Marciano, Rockefeller University preprint COO-2232B-173, 1979; T. J. Goldman and D. A. Ross, CALT 68-704.

6. Op. cit. in n. 2.

7. The following discussion is based on that given by S. Weinberg, Phys. Rev. Lett. **82B**, 387 (1979). Other discussions are cited therein. Also see E. Gildener and S. Weinberg, Phys. Rev. **D13**, 3333 (1976).

8. S. Coleman and E. Weinberg, Phys. Rev. **D7**, 1888 (1973).

9. Gildener and Weinberg, op. cit. in n. 7.

10. S. Weinberg, Phys. Rev. **D13**, 974 (1976), **D19**, 1277 (1979); L. Susskind, Phys. Rev. **D20**, 2619 (1979).

11. S. Weinberg, Phys. Rev. Lett. **9**, 357 (1964), Phys. Rev. **B135**, 1049 (1964), and Phys. Rev. **B138**, 988 (1965); D. Boulware and S. Deser, Ann. Phys. **89**, 173 (1975). I understand that similar considerations were presented by R. Feynman in unpublished lectures at California Institute of Technology.

12. T. D. Lee and C. N. Yang, Phys. Rev. **98**, 101 (1955).

13. G. 't Hooft, Phys. Rev. Lett. **37**, 8 (1976).

14. See, e.g., S. Weinberg in *Lectures on Particles and Field Theory—Brandeis Summer Institute in Theoretical Physics 1964,* Vol. II, ed. by S. Deser and K. Ford (Prentice-Hall, Englewood Cliffs, N.J., 1965), p. 482; A. D. Sakharov, Zh. Eksp. Teor. Fiz. Pis'ma **5**, 32 (1967) [JETP Lett. **5**, 24 (1967)].
15. Pati and Salam, op. cit. in n. 1; and Phys. Rev. Lett. **31**, 661 (1973).
16. See references cited in n. 1.
17. Results of this analysis were reported by me in Phys. Rev. Lett. **42**, 850 (1979).
18. Pati and Salam, op. cit. in n. 1.
19. Op. cit. in n. 2.
20. Buras et al., op. cit. in n. 2. For more recent estimates of the proton and bound neutron lifetime see C. Jarlskog and F. J. Yndurain, to be published; M. Machacek, Harvard preprint HUTP-79/A021, 1979.
21. Marciano and Goldman and Ross, op. cit. in n. 5.
22. F. Reines and M. F. Crouch, Phys. Rev. Lett. **32**, 493 (1974). The first experiment that was specifically designed to set a limit on the proton lifetime was that of F. Reines, C. L. Cowan, Jr., and M. Goldhaber, Phys. Rev. **96**, 1157 (1954). For an early theoretical discussion, see G. Feinberg and M. Goldhaber, Proc. Nat. Acad. Sci. **45**, 1301 (1958).
23. For a discussion see G. Steigman, Ann. Rev. Astron. Astrophys. **14**, 339 (1976).
24. M. Yoshimura, Phys. Rev. Lett. **41**, 281 (1978), and **42**, 746(E) (1979); Tohoku University preprints TU/79/192, TU/79/193; S. Dimopoulos and L. Susskind, Phys. Rev. **D18**, 4500 (1978); Phys. Lett. **81B**, 416 (1979); A. Y. Ignatiev et al., Phys. Lett. **76B**, 436 (1976); B. Toussaint et al., Phys. Rev. **D19**, 1036 (1979); J. Ellis, M. K. Gaillard, and D. V. Nanopoulos, Phys. Lett. **80B**, 360 (1979), and **82B**, 444(E) (1979); Weinberg, op. cit. in n. 17; N. J. Papastamatiou and L. Parker, Phys. Rev. **D19**, 2283 (1979).
25. This result has been known (though perhaps not widely) for some time; a brief proof and references to earlier literature are given by me in the reference cited in n. 15. I first learned of this result in a special case from an early version of the paper by Toussaint et al., op. cit. in n. 24.
26. S. Weinberg, Astrophys. J. **168**, 175 (1971).
27. Toussaint et al., op. cit. in n. 22; Dimopoulos and Susskind, ibid.; Weinberg, op. cit. in n. 17.
28. D. V. Nanopoulos and S. Weinberg, Harvard preprint HUTP-79/A023. Also see S. Barr, G. Segrè, and H. A. Weldon (forthcoming).
29. Nanopoulos and Weinberg, op. cit. in n. 28.

Session Chairman's Remarks: On Einstein and on Anti-Semitism in Science
ALFRED KASTLER

Let me dedicate to Einstein this little poem on the Compton effect

COMPTON-EFFEKT

Zum Teufel! schrie das Elektron:
Wer hat mich angerempelt?
Verzeihung! weinte das Photon,
Mir hat ein Fusstritt von Compton
Die Flugkraft aufgestempelt.
Auch ich kriegt eine Beule,
Dass ich flugs seitwärts heule.

I think the Compton effect was an effect very dear to Einstein's heart as it showed that light quanta had not only energy but also momentum.

This game of billiards between photons and electrons was for Einstein a great triumph of physical reality.

When Einstein came to Paris in 1922, he was not invited to the Academy, not for scientific but for political reasons. He came from Berlin, Germany. At that time, after World War I, official relations with German scientists had not been reestablished. They were established two years later. The man who had made strong efforts in this direction was the general secretary of the International Union of Physics, Henri Abraham. He was the director of the Physics Laboratory of the Ecole Normale Supérieure and my teacher.

Hélas, how has he been thanked for it! His name indicates that he was a Jew. During the Second World War and the German occupation, he moved to the so-called unoccupied zone of France. There he was arrested at the age of seventy-five by the French police of the Vichy government and delivered to the Gestapo with one of his daughters who wanted to stay with him. We have never had news from him since. He disappeared probably in Auschwitz. His fate was also that of my beloved teacher, Eugène Bloch, his successor at the Ecole Normale, who had done so much to introduce quantum theory in France by his teaching and by his book, "L'ancienne et la nouvelle théorie des quanta." He was arrested by the Gestapo in 1943 and disappeared also in the Holocaust.

We very well understand Einstein's feelings after 1945 when these horrible things became known, his feelings against Germany and the Germans.

Nevertheless, I believe we should make the distinction between Germans and Nazis. The German Jews and the German Left-wing intellectuals were the first victims of the Nazi regime. Einstein himself took up friendly relations with Max Planck, whose son Erwin had been killed by the Nazis, and with Max von Laue. Probably he did not approve of his friend Max Born, who retiring from Edinburgh University returned to Germany, but this did not affect their friendly relations.

We know that during the Nazi period in Germany, two physicists, both Nobel laureates, supported the Nazi regime: Philipp Lenard and Johannes Stark. Psychoanalysis of Lenard might have shown that he was jealous of Einstein. He had worked on the photoelectric effect and had been unable to draw the conclusions that Einstein drew from his work. Stark, the discoverer of the Stark effect, was very angry also against Sommerfeld in spite of the fact that, in his wonderful book, *Atombau und Spektrallinien*, Sommerfeld had much appreciated Stark's work. As Sommerfeld stuck firmly to Einstein's theory of relativity, Stark called him "a Jew in mind" (*ein Geistesjude*). In Stark's mind this was an expression of depreciation. Let us take it as an expression of appreciation.

The attendants at this symposium are people from many countries — some of us are Jews, some of us are not. I am not a Jew, but I think we all feel together what unites us is that we are "Jews in mind."

The next chapter is by J. Pati, who worked with Dr. Abdus Salam, whom I had the pleasure to meet at the Trieste Institute, which he created. This is an international center working in training scientists for the developing world. I think this is also a task performed by the universities of Israel.

16. NEW PHYSICS WITH GRAND UNIFICATION

Jogesh C. Pati

1. Introduction

Einstein spent a good part of his later years in trying to unify electromagnetic and gravitational forces.[1] This particular ambition of Einstein remains to be fulfilled even today. Meanwhile a new set of ideas has developed over the recent past, carrying the same spirit of unification that Einstein put forth, that allows one to regard three of the basic forces of nature — weak, electromagnetic, as well as strong — as aspects of a single force.[2] Within the premises of these ideas there is no intrinsic asymmetry between quarks and leptons. They are regarded as members of one family (multiplet). Their "weak," "electromagnetic," as well as "strong" interactions are generated at once by a gauging of the symmetry group of this multiplet. The force thus generated is governed by a single coupling constant. This unified force divides itself into the observed weak, electromagnetic, and strong forces with effective coupling constants differing from each other at *low momenta* due to the spontaneous breakdown of the "grand" symmetry of quarks and leptons. The same symmetry-breaking deprives leptons from sharing the known strong interactions, even though it permits quarks and leptons to share the weak as well as the electromagnetic forces universally.

In short, the observed asymmetry between quarks and leptons as well as the asymmetries between the coupling constants of the strong, electromagnetic, and weak forces are interpreted within the premises of these ideas to be low-momentum phenomena. Such asymmetries must disappear and the "grand" symmetry of the basic particles (quarks *and* leptons) on the one hand and of the basic forces — the strong, electromagnetic, and weak — on the other must manifest itself at sufficiently large momenta exceeding the masses of the heaviest gauge particles in the theory; these are typically much heavier than the familiar $W^{\pm}$ and Z^0 particles, which carry masses of order 100 GeV.

The assumption that the basic particles as well their forces have *this* underlying unity is the so-called grand unification hypothesis. It possesses a few general characteristics as distinguished from more detailed features arising within a model.

Yuval Ne'eman (ed.), To Fulfill A Vision: Jerusalem Einstein Centennial Symposium on Gauge Theories and Unification of Physical Forces
ISBN 0-201-05289-X

Copyright © 1981 by Addison-Wesley Publishing Company, Inc., Advanced Book Program.
All rights reserved. No part of this publication may be reproduced, stored in a retrieval system, or transmitted, in any form or by any means, electronic, mechanical photocopying, recording, or otherwise, without the prior permission of the publisher.

1. The general characteristics
   a. Existence of a new class of interactions mediated by "exotic" gauge particles, which couple for example to quark-lepton currents and which typically carry masses much heavier than 100 Gev
   b. High-energy manifestations of absolute universality of quarks and leptons on the one hand as well as of weak, electromagnetic, and strong forces on the other
   c. Last but not least, violations of baryon and lepton number and in general also of fermion number, leading to instability of the proton.

These properties are direct consequences of the hypothesis of grand unification. As such they are common to alternative models of grand unification; they constitute the hallmarks of grand unification.

2. The important details
   a. The mass of the heaviest gauge particles in the theory
   b. The value of the weak angle
   c. The manner of violations of baryon and lepton number
   d. The nature of quark charges.

These details depend upon the choice of the symmetry group, the choice of the associated fermionic multiplet, and the pattern of spontaneous symmetry-breaking — in short, upon the model. The details are no doubt important for specific experimental signals and help distinguish between alternative models, but the general characteristics tied directly to the basic hypothesis of grand unification lie at a level much deeper than any particular model.

Before elaborating on these general and detailed characteristics, let me mention the motivations for grand unification. The grand unification hypothesis incorporates within it the hypothesis of electroweak unification.[3] This latter (sub)-unification is culminated by the now well-known Weinberg-Salam model,[4] which is in excellent agreement with all available data. The motivations for going beyond electroweak unification and for seeking an underlying unity of all matter and its forces are as follows. Within just electroweak unification one fails to answer a number of basic questions such as:

What is the value of the weak angle $\sin^2\theta_W$?
Why must there exist quarks *and* leptons?
Why must there exist weak, electromagnetic, *and* strong forces?
Why must low-energy weak interactions exhibit the same chiral projection (V-A) for the negatively charged electron and the positively charged proton rather than for the positively charged positron and the positively charged proton?
Last but not least, why must electric charges of quarks and leptons be so quantized?

It was observed in 1972 that one must go beyond electroweak unification to obtain an answer to these questions.[5] One must, despite their vast low-energy distinctions, assume that quarks *and* leptons are members of one family and that weak, electromagnetic, as well as strong forces are aspects of a single force. Within this picture there exist quarks *and* leptons, because they are members of one set. There exist weak, electromagnetic, *and* strong forces because they too are members of a single set. Without the existence of any one member the sets would be incomplete. Such a picture permitted the overall symmetry of quarks and leptons to be non-Abelian, which was not permissible realistically within the realm of just electroweak unification. This in turn accounted for quantization of electric charges. Simultaneously it was realized that at least within simple models only the negatively charged e^- (but not e^+) could belong to the same multiplet as ν_e and u and d quarks. This explained why e^- and proton (rather than e^+ and proton) exhibit the same chiral projection in weak interactions.

It was conjectured at that time[6] that the vast low-energy distinctions between the effective coupling constants of the strong versus electroweak interactions could be attributed to differing charge-renormalizations subsequent to spontaneous breakdown of the grand symmetry. The feasibility of this conjecture was demonstrated subsequently,[7] making essential use of the property of asymptotic freedom[8] of non-Abelian gauge interactions. The same analysis permitted a prediction for the value of the weak angle $\sin^2\theta_W$ at low energies. (Although the prediction depends upon the grand symmetry and the nature of spontaneous symmetry breakdown, the predicted value for a number of simple models agrees with the recent experimental value within the errors of theory and experiment.)

The motivations for the hypothesis of grand unification, as in 1972, are still primarily theoretical. However, because of the discovery of the weak neutral current force in 1973 and the subsequent agreement of the simple $SU(2)_L \times U(1)$ model of electroweak unification with experiments, the larger grand unification idea appears to be less of a heresy now than it did in 1972; it calls more urgently for an experimental search for its own signals. With a view to spelling out some of these signals, I now elaborate first on the general and then on the more detailed characteristics of grand unification.

2. The General Characteristics

Although the detailed model of interactions generated within the hypothesis of grand unification would depend upon the specific choice one makes of the symmetry group and the associated (quark + lepton) multiplet, it is good to reemphasize that there are three general characteristics that all such models would share[9]:

1. There must exist within such models a new class of interactions, mediated by exotic gauge particles X, which are coupled to quark-lepton currents carrying quark number $B_q = +1$, lepton number $L = -1$, and fermion number $F \equiv B_q + L = 0$. (By convention B_q here denotes quark number, which is $+1$ for quarks, $+3$ for protons, and 0 for leptons.) These are the so-called leptoquark gauge bosons. If the symmetry structure is extended to include fermions and antifermions in the same multiplet,[10] there must also exist gauge particles (Y, Y', and Y'') coupled to quark-antiquark, quark-antilepton, and lepton-antilepton currents carrying ($B_q = +2, L = 0$), ($B_q = +1, L = +1$) and ($B_q = 0, L = +2$), respectively; each of these currents possesses fermion number $F = +2$. See Fig. 16.1 for a listing of these gauge particles. These exotic gauge particles (X and Y) need in general to be much heavier than the weak $W^\pm$ and Z^0 bosons both because of the absence of (exotic) interactions generated by them at low energies and also because of the observed disparity between the strong versus the weak and electromagnetic coupling constants at such energies. Their precise masses will depend upon the model. It is these exotica and the interactions generated by them that are the *hallmarks* of grand unification in the same way as the Z^0 and the weak neutral-current interactions are the hallmarks of electroweak unification.

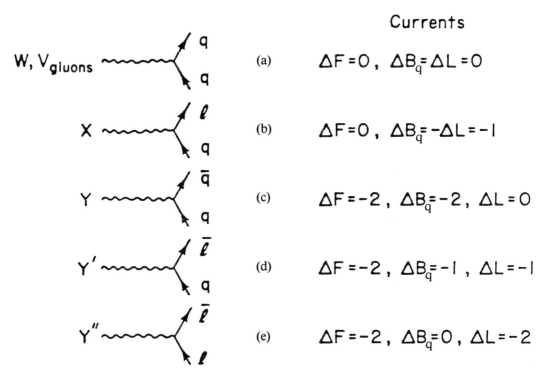

Fig. 16.1 Gauge particles within grand unification. Note that quark number B_q is related to baryon number B by $B_q = +3B$.

2. At center of mass energies exceeding the masses m_X, m_Y of the exotic gauge particles, quarks and leptons must interact universally in all respects. The distinctions between the strong, electromagnetic, and weak interactions must disappear simultaneously. At low energies the effective coupling constant of the strong interactions could differ from those of the electroweak interactions due to differing charge-renormalizations in the various sectors subsequent to spontaneous breakdown of the grand symmetry. As we will see, the "strong" interactions are stronger than the electroweak interactions at low energies simply because the former are generated by the symmetry structure $SU(3)$, while the non-Abelian components of the latter, as it turns out, are generated by $SU(2)$, which is smaller than $SU(3)$.[11] See Fig. 16.2 for a diagram of qualitative behavior of the effective coupling constants of the strong and electroweak forces as a function of the running momentum. The momentum scale M at which the differences between the various coupling constants disappear is the so-called grand unification mass. Typically it is about one order of magnitude larger than the heaviest gauge mass in the theory; its value would depend upon the model.

3. Finally with quarks and leptons in the same multiplet, baryon number B and lepton number L acquire a status, analogous to that of the third component of isospin I_3, hypercharge Y, and charm C within the basic Lagrangian. If the symmetry is extended to include fermions and antifermions in the same multiplet, fermion number F acquires a similar status as well. Even if these quantum numbers are conserved in the primary Lagrangian, given that no massless gauge

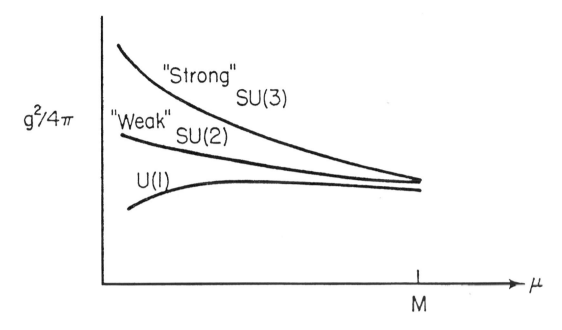

Fig. 16.2 Hierarchy of gauge coupling constants.

particle couples to B, L, or F,[12] there is no a priori reason why any of these quantum numbers should be conserved absolutely *after* spontaneous symmetry-breaking. We would in fact expect them to be violated after such a symmetry-breaking (subject only to the conservation of electric charge)[13] in a manner analogous to the violations of I_3, Y, and C. In turn, we would expect the proton to be ultimately unstable against decay into leptons. This is what had led Salam and myself to question the stability of the proton, notwithstanding its present high degree of stability.[14]

Before entering into the question of details let me discuss briefly the choice of the grand unification symmetry G. First, it is clear that G must either be a simple group or a semisimple group subject to discrete symmetries linking its various factors such that the interactions generated by G are governed by a single gauge-coupling constant in either case. There is one other criterion that G must satisfy to be in accord with reality. It must contain at least $SU(2) \times U(1) \times SU(3)$ as a subgroup, so that low-energy electroweak interactions may be generated by $SU(2) \times U(1)$ while low-energy strong interactions may be generated by $SU(3)$ (identified with $SU(3)$ color symmetry). That $SU(2) \times U(1) \times SU(3)$ is a desirable subgroup to generate low-energy electroweak and strong interactions, regardless of whether $SU(3)_{\text{color}}$ is ultimately broken spontaneously softly with quarks acquiring integer charges or $SU(3)_{\text{color}}$ is unbroken with quarks acquiring fractional charges, was noted in the references cited in note 2 and subsequently in the references cited in note 15.

Examples of semisimple unifying groups are:

$$[SU(4)]^4 \ , \ [SU(6)]^4 \ , \ \text{or} \ [SU(n)]^4 \quad (n > 4) \ .$$

For the smallest $[SU(4)]^4 = SU(4)_A \times SU(4)_B \times SU(4)_C \times SU(4)_D$, the basic fermions are given by a set of sixteen left-handed and a set of sixteen right-handed two-component fermions $F_{L,R}$:

$$F_{L,R} = \begin{bmatrix} u_r & u_y & u_b & \nu_e \\ d_r & d_y & d_b & e^- \\ s_r & s_y & s_b & \mu^- \\ c_r & c_y & c_b & \nu_\mu \end{bmatrix}_{L,R} \tag{16.1}$$

where $F_L \sim (4,1,\bar{4},1)$, $F_R \sim (1,4,1,\bar{4})$. These may thus be interpreted to carry four flavors and four colors, the fourth color being lepton number. These basic fermions must be supplemented by at least a parallel set of mirror fermions $F'_L = (1,4,1,\bar{4})$ and $F'_R = (4,1,4,1)$ to cancel triangle anomalies. Depending upon whether the discrete symmetry is maximal or minimal,[16] there will or will not exist additional fermions transforming as $(4,1,1,\bar{4})$, $(1,4,\bar{4},1)$, $(4,\bar{4},1,1)$, and $(1,1,4,\bar{4})$. These additional fermions, including the mirror fermions, provide new flavors and new heavy leptons, of which some are already seen.

One may enlarge the symmetry structures by putting fermions F_L and antifermions F_L^C (instead of F_R) into one multiplet. The maximal symmetry in the space of n left-handed fermions F_L and n left-handed antifermions F_L^C is $SU(2n)$. For $n = 16$ the maximal symmetry is $SU(32)$. One of course need not gauge the maximal symmetry $SU(2n)$ (or $SU(32)$ for $n = 16$) to put fermions and antifermions into one multiplet. For instance, one may gauge a suitable subgroup such as $SU(16)$ (as in unpublished work by Salam and myself) of which the left-handed (u_L and d_L) flavors with four colors and their charge conjugates (u_L^C and d_L^C) with four colors form a sixteenfold; likewise the (c and s) flavors form another sixteenfold. Similarly for the mirror fermions. Note that $SU(16)$ depending upon the pattern of SSB may break either via $SU(8) \times SU(8) \times U(1)_F$, or $SO(10)$. The former (in the space of basic fermions, that is, (u, d) and (c, s) flavors) can descend to

$$(SU(2)_L \times SU(3)_L^{\text{color}} \times U(1)_L) \times (SU(2)_R \times SU(3)_R^{\text{color}} \times U(1)_R) ,$$

while the latter $SO(10)$ could descend to $SU(2)_L \times SU(2)_R \times SU(3)_{L+R} \times U(1)_{L+R}$. Of course $SU(16)$ may break spontaneously directly through the heaviest mass-scale to the low-lying symmetries such as

$$SU(2)_L \times SU(2)_R \times SU(3)_{L+R}^{\text{color}} \times U(1)_{L+R} \quad \text{or} \quad SU(2)_L \times U(1) \times SU(3)_{L+R}^{\text{color}}$$

without passing through the intermediate steps of $[SU(8)]^2$ or $SO(10)$.

In all the foregoing examples there is one characteristic feature. Regardless of whether the group is semisimple (for example $[SU(n)]^4$) or simple (for example $SU(2n)$ or $SU(n)$), the basic quantum numbers such as fermion number F, baryon number B, and lepton number L are all conserved in the primary Lagrangian. They are violated only spontaneously. (Details of such spontaneous violation are discussed later.)

One may, however, choose to gauge still smaller subgroups of $SU(2n)$ or $SU(n)$ yet preserving the spirit of grand unification. Examples of such subgroups are:

$SU(5)$ (See Georgi and Glashow, cited in note 2.)
$SO(10)$ (See note 17.)
E_6, E_7 (See note 18.)

Note that all these subgroups are simple. Of these $SU(5)$ is indeed the smallest containing $SU(2)_L \times U(1) \times SU(3)_{\text{color}}$. $SU(5)$ provides different generations of fermions (that is, (u,d), (c,s), and (t,b) et cetera, together with associated leptons (ν_{eL}, e), $(\nu_{\mu L}, \mu)$, and $(\nu_{\tau L}, \tau)$) by postulating different sets of $(5 + \overline{10})$ representations. Each set by itself provides anomaly-free interactions. $SO(10)$ contains the bigger subgroup $SU(2)_L \times SU(2)_R \times SU(4)'_{L+R}$ as does $[SU(4)]^4$. These therefore provide the basis for a left-right symmetric parity-conserving Lagrangian. Parity is violated in these models as in $[SU(4)]^4$ spontaneously. $SO(10)$ introduces sets of 16-plets of fermions as does $[SU(4)]^4$ and preserves the interpretation of lepton number as the fourth color. E_7 a priori is an interesting candidate for grand unifying symmetry. While its size is bigger it introduces all two-component fermions (quarks + leptons) through a single basic representation $\underline{56}$ – 10 leptons + 18 quarks that transform as $(20,1)_L + (6,3)_L + (6^*,3^*)_L$ under the subgroup $SU(6) \times SU(3)_{\text{color}}$. A comparative summary of the different grand unifying symmetries is given later.

Now I discuss the question of details.

3. The Important Details

As indicated before, depending upon the choice of the symmetry group, the associated fermionic multiplet, and the pattern of spontaneous symmetry breaking, alternative models of grand unification can differ often drastically from each other in respect of 1) the mass of the heaviest gauge particles in the theory, 2) the value of the weak angle, 3) the manner of violations of B, L, and F, and 4) the nature of quark charges.

3.1 *The Heaviest Gauge Mass, the Grand Unifying Mass Scale, and the Weak Angle*

The question of the mass of the heaviest gauge particles $(X, Y, Y', \ldots)$ and thereby of the grand unifying mass M is of primary importance since it determines the energy scale at which the *new interactions* mediated by the exotic gauge particles X, Y, and Y' would begin to be visible. As I stressed before, it is these new exotic interactions between quarks and leptons that characterize the hallmarks of grand unification.

Georgi, Quinn, and Weinberg showed in 1974 that at least within a class of models the grand unifying mass scale M must be ultraheavy $\gtrsim 10^6$ GeV simply from the observed disparity between the electroweak versus the chromodynamic strong coupling constants at present energies.[19] They had assumed that 1) the grand unifying symmetry G, whatever it may be, breaks by some superheavy mass scale M_X to $G_{EW} \times SU(3)'_{\text{color}}$, 2) G_{EW} denoting electroweak symmetry may have the form $SU(2) \times U(1)$ (alternative forms such as $SU(2) \times SU(2) \times U(1)$ can easily be considered without making drastic changes), and 3) crucially the effective coupling constants g_W and g_c associated with the electroweak $SU(2)$ and strong $SU(3)$ symmetries are equal to each other in the grand unifying symmetric limit; that is, at high momenta:

$$g_W = g_c \, . \tag{16.2}$$

These assumptions, including the last one, are indeed satisfied for grand unifying symmetries such as $SU(5)$ and $SO(10)$.

Utilizing renormalization group equations *subject to condition* (16.2) and the property of asymptotic freedom of non-Abelian symmetries, one obtains:

$$\frac{\alpha_s(\mu)}{\alpha} = \frac{8}{3}\left[1 - \frac{11\alpha}{\pi}\ln\frac{M}{\mu}\right]^{-1}, \tag{16.3}$$

where $\alpha_s(\mu)$ denotes the effective chromodynamic strong coupling constant at the running momentum μ, $\alpha = 1/137$ and M is the grand unifying mass scale. To obtain $\alpha_s(\mu) \approx 0.2$ to 0.3 at present momenta ($\mu \sim 3$ GeV), it is necessary on the basis of (16.3) that M must be ultraheavy $\sim 10^{16}$ GeV. (Recall that for grand unifying symmetries such as $SU(5)$, $SO(10)$, E_6, and E_7, there is an independent reason that M_X and therefore M needs to be ultraheavy: proton decays in the second order of the basic gauge interactions in these models and therefore the exotic gauge particles must be sufficiently heavy $\gtrsim 10^{14}$ GeV to account for the known proton stability.)

The result that the grand unifying mass scale M is of order 10^{16} GeV and that the exotic interactions would show themselves only at energy and momentum scales of this order is rather disappointing. If this is indeed true, this would mean that the true hallmarks of grand unification characterized by the exotic quark-lepton interactions will have to remain unseen not only for our own generation but also for the conceivable future. Predictions such as the value of $\sin^2\theta_W$ and proton lifetime can no doubt be used to eliminate certain models of grand unification (see later); but they can hardly be used to provide tangible evidence for grand unification. One needs to see the exotic (X, Y, Y') interactions to obtain evidence for grand unification. Such a possibility would not arise if $M \sim 10^{16}$ GeV.

Furthermore, if M is nearly 10^{16} GeV, $M_{X,Y} \sim 10^{15}$ GeV and all other gauge masses lie below 100 to 1000 GeV (as in $SU(5)$), one would have a rather strange scenario insofar as discovery of new physics is concerned. This is depicted in Fig. 16.3. While the steps of eV, MeV, GeV, and 100 to 1000 GeV are all filled with interesting new phenomena appearing at each step, there would remain an unprecedented big gap (*desert*) in the region spanning between 10^2–10^3 GeV on one side and 10^{15} GeV on the other, where no truly *new* phenomenon would appear. This is to say the least an uncomfortable feeling. Be that as it may, there is a question of internal consistency of this scenario in that it is not clear whether the extraordinary gap (the desert) can sustain itself in spite of radiative corrections.

Must nature choose such a scenario? Elias, Salam, and I showed that "fortunately" that is not the case.[20] We found that assumptions (1)–(3) of Georgi, Quinn, and Weinberg are not general. They hold for $SU(5)$ and $SO(10)$, but they do not hold for a large class of grand unification models; specifically the assumption (3) does not hold for $[SU(4)]^4$, $SU(16)$, or the more general $[SU(n)]^4$ ($n \geq 4$) and its further extensions. We showed that for these later models the

eV	MeV	GeV	100 GeV	////////////// ← BIG DESERT →	10^{15} GeV
Atomic Physics	Nuclear Physics	Hadron Physics	Electroweak Physics $Z°, W^{\pm}$	NO NEW PHYSICS	Exotica of Grand Unification

Fig. 16.3 A scenario of new steps in physics for some models.

grand unification mass scale M need be no higher than 10^5–10^6 GeV, and correspondingly the exotic gauge masses M_X need be no higher than about 10^4–10^5 GeV.

I shall illustrate by just one example how this enormous reduction in M from 10^{16} to 10^5–10^6 GeV can come about. Consider the symmetry structure $[SU(4)]^4$ and its breaking chain:

$$G = [SU(4)_L \times SU(4)_R]_{\text{flavor}} \times [SU(4)'_L \times SU(4)'_R]_{\text{color}}$$

$$\downarrow$$

$$[SU(2)_L^{I+II} \times SU(2)_R^{I+II}]_{\text{flavor}} \times [SU(3)'_L \times SU(3)'_R \times U(1)'_L \times U(1)'_R]_{\text{color}} \quad (16.4)$$

$$\downarrow M' < m_W \sim 100 \text{ GeV}$$

$$SU(3)'_{L+R}$$

Here I have for convenience used the subscripts L and R instead of (A,B,C,D) to denote the couplings of the *basic fermions* (that is, u,d,c,s flavors) to the respective gauge mesons. The subgroups $SU(2)^I$ and $SU(2)^{II}$ act on (u,d) and (c,s) flavors, respectively, and $SU(2)^{I+II}$ is their diagonal sum. In other words if the gauge particles of $SU(2)^I$ and $SU(2)^{II}$ are denoted by $\mathbf{W}^I$ and $\mathbf{W}^{II}$, respectively, and the associated currents by $\mathbf{J}^I$ and $\mathbf{J}^{II}$, respectively, then the normalized gauge particles of $SU(2)^{I+II}$ would be denoted by $(\mathbf{W}_I + \mathbf{W}_{II})/\sqrt{2}$, the associated currents would be $\mathbf{J}^I + \mathbf{J}^{II}$. It is easy to see that if the coupling constant associated with the grand unifying symmetry G is denoted by g, then in the limit of exact symmetry those associated with the direct subgroups $SU(2)^I$ and $SU(2)^{II}$ would be g; while that associated with the diagonal subgroup $SU(2)^{I+II}$ would be $g/\sqrt{2}$. Thus the coupling constant of the effective weak $SU(2)$ denoted by g_W is $g/\sqrt{2}$ in the limit of exact symmetry ($g_W = g/\sqrt{2}$). On the other hand, the coupling constant $g^c_{L,R}$ of the effective "strong" symmetries $SU(3)'_{L,R}$, which denote chiral rather than vectorial color, are equal to g in the symmetric limit; this is because $SU(3)'_{L,R}$ are direct subgroups of the parent symmetries $SU(4)'_{L,R}$ (thus $g^c_{L,R} = g$). It then follows that

$$g_W = g_c^{L,R}/\sqrt{2} \quad \text{(in symmetric limit)} . \quad (16.5)$$

Note that the assumption (3) of Georgi-Quinn-Weinberg symbolized by Eq. (16.2) is effectively not respected by the grand unifying symmetry $[SU(4)]^4$ with a hierarchical breaking shown in (16.4). The coupling constant (g_W) associated with the "weak" symmetry group is now lower by a factor $1/\sqrt{2}$ compared to that ($g^c_{L,R}$) associated with the relevant "strong" symmetry group. Compare Eq. (16.2) with Eq. (16.5). (The reader will observe that even though the chiral color symmetry $SU(3)'_L \times SU(3)'_R$ breaks subsequently to the vectorial color $SU(3)'_{L+R}$ through a mass scale $M' < m_W$, the relevant "strong" symmetry for the hierarchy (16.4) for renormalization group purposes is still the chiral symmetry $SU(3)'_L \times SU(3)'_R$, since $M \gg m_W > M'$.)

Now ordinarily a factor of $\sqrt{2}$ is not that consequential. But this coupling constant ratio squared appears as a coefficient in front of the log term, which determines M (see Eq. (16.3)). Thus M is highly sensitive to the hierarchical constraints such as (16.2) or (16.5). With some algebra the factor 2 in fact becomes more like $7/3 = 2.3$ and for some cases 2.7, hence the enormous reduction in the value of M. To be precise, using the renormalization group equations for $[SU(4)]^4$ subject to the hierarchical constraint (16.5), one obtains:

$$\frac{\alpha_s}{\alpha} = 2(8/3) \left[1 - (77\alpha/3\pi) \ln (M/\mu)\right]^{-1} . \tag{16.6}$$

Compare (16.6) with (16.3). It is clear that for a given value of (α_s/α) at a fixed low value of $\mu \approx 3$ GeV, (16.6) will yield a drastically lower value for the grand unifying mass scale M than (16.3). One obtains for this case,

$$M \approx 10^5\text{--}10^6 \text{ GeV} , \tag{16.7}$$

$$M_X \approx 10^4\text{--}10^5 \text{ GeV} . \tag{16.8}$$

For this case $\sin^2\theta_W \approx 0.28$ without radiative corrections. Present experimental value for $\sin^2\theta_W \approx 0.22 \pm 0.02$. (Radiative corrections can in general lower the value 0.28. See also remarks later for cases permitting low-mass unification with values of $\sin^2\theta_W$ lower than 0.28.)

Now if M_X is as light as $10^4\text{--}10^5$ GeV, signals for exotic quark-lepton interactions can appear in processes such as $e^-e^+ \to q\bar{q}$ (or equivalently in $p\bar{p} \to \mu\bar{\mu}$) at center of mass energies as low as $M_X/10 \approx 10^3\text{--}10^4$ GeV through graphs exhibited in Fig. 16.4. Thus for symmetry structures such as $[SU(4)]^4$ signals of grand unification can manifest themselves in high-energy accelerators in planning (for example, at Isabelle or its immediate successors) and in cosmic ray studies.

It is worth noting that if the hierarchy exhibited in (16.4) leading to low-mass unification is realized in nature, chiral color $SU(3)'_L \times SU(3)'_R$ ought to be a relatively good low-energy symmetry. Now chiral color cannot remain an exact symmetry; it must break to vectorial color at least through quark-mass terms. This would generate a relatively light massive octet of axial color gluons with a Lagrangian mass $m_A = M' < m_W$ accompanying the familiar octet of vector color gluons.

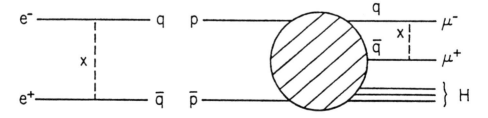

Fig. 16.4 Exotic quark-lepton interactions.

If the effective mass m_A of the axial color gluons is either much lighter than 1 GeV or is as large as about m_W, then the vector QCD predictions for deep inelastic scattering would not be altered to any noticeable degree.[21] However if the effective constituent mass of the axial gluons is as light as 1 to few GeV, exchange of such axial gluons would generate in contrast to vector exchange an appreciable spin-spin term in the $q\bar{q}$-potential, which would survive in the nonrelativistic limit. Such a spin-spin force can be noticeable through a careful study of the charmonium together with the still heavier bottomonium systems.[22] It remains an interesting open question as to whether QCD is truly purely vectorial or whether it possesses an axial component. Note however that low-mass unification can be realized (for the example at hand) even if the Lagrangian mass of the axial gluons is as heavy as $m_W \sim 100$ GeV. Such heavy axial gluons would have no noticeable effect either in deep inelastic phenomena (for example, $eN \to eX$) or in the binding of the quarkonium systems.

It is worth remarking further that the essential mechanism that permitted low-mass unification is the *dichotomy* between flavor and color descents,[23] one that led to $g_W < g_c$ in the symmetric limit. Such a dichotomy neither requires the unifying symmetry to be simple, nor does it require chiral color to be necessarily an effective low-energy symmetry in the sense of the flavor electroweak symmetry $SU(2) \times U(1)$. For example, for $[SU(6)]^4$, if the flavor $SU(6)$ descends to the diagonal symmetry $SU(2)^{I+II+III}$, where $SU(2)^{I+II+III}$ acts on (u,d), (c,s), and (t,b) doublets, and if chiral $SU(6)$ color descends to vector $SU(3)$ color (at the primary stage of spontaneous symmetry-breaking), then such a dichotomy is generated, leading to flavor $SU(2)$ coupling constant, which is smaller than the color $SU(3)$ coupling constant in the symmetric limit. This generates a relatively low unifying mass scale $\sim 10^6$ GeV. This trick can be generalized further to $[SU(2n)]^4$. Recently Elias and Rajpoot have shown that such descents with $n > 3$ lead to values of $\sin^2\theta_W$ lying between 0.25 and 0.21, which spans the present experimental range.[24]

In summary we see that depending upon the nature of the symmetry group and its hierarchical breakdown, the unification mass scale M can vary easily by as much as ten orders of magnitude: from the "never never" land of 10^{16} GeV to the more approachable region of 10^5–10^6 GeV. The latter low-mass unification can be realized consistent with all low-energy data, including deep inelastic scattering, quarkonium physics, and values of $\sin^2\theta_W$. It remains to be seen whether nature permits us to see the intriguing signals of grand unification at such low mass. It is to be hoped that cosmic ray studies and in particular high-energy accelerators in planning can shed light on the question in the near future.

3.2 *The Manner of Violations of B, L, and F*

The second question of detail involves the manner of violations of the fundamental quantum numbers B, L, and F. This is important for the origin of instability of the lightest baryon — the proton — and for the origin of excess baryons during cosmic evolution.

As stressed in the introduction of this chapter, with quarks and leptons in the same multiplet, and within the extended symmetry requiring fermions and antifermions in the same multiplet, we expect that the quantum numbers B, L, and F would be violated at least spontaneously even if they are conserved in the basic Lagrangian and therefore the lightest baryon — the proton — would be unstable against decay into leptons. However, there are two distinct ways in which such a violation may occur.

For symmetry structures such as $[SU(4)]^4$, the extended symmetry $SU(16)$, or in general $[SU(n)]^4$ and its suitable extensions, the quantum numbers B, L, and F are all conserved in the basic Lagrangian. But they are all violated spontaneously, for example through the unavoidable mixing of gauge particles and/or through the mixing of the Higgs scalars and so on (see Fig. 16.5). For example, within a symmetry scheme such as $SU(n)_{\text{flavor}} \times SU(n)_{\text{color}}$ with quarks and leptons in the same multiplet, the very act of generation of photon for the case of integer-charge quarks inevitably forces through SSB a mixing of the flavor W and the leptoquark X gauge particles, which generates B and L violations.[25] (See Fig. 16.5.) Likewise fermion number is violated within the more extended symmetry due to a spontaneous mixing of gauge particles $Y(F = +2)$ with $Y'(F = -2)$ carrying different fermion numbers as well as baryon and lepton numbers (see Fig. 16.5).

This *manner* of violations of baryon and lepton numbers (and also fermion number within the more extended symmetry) has been an integral part of the approach to unification, which Salam and I have followed over the years. It appears somewhat natural to us that *basic quantum numbers such B, L, and F are violated not by a choice and thus not by the basic Lagrangian*. They are violated, if at all, because they have to be. Even though they are conserved in the basic Lagrangian, the minimum of the potential by way of determining the true ground state of the system *forces* a (spontaneous) violation of these quantum numbers. For instance in the process of generation of the photon, W and X gauge particles must mix spontaneously (for the case of

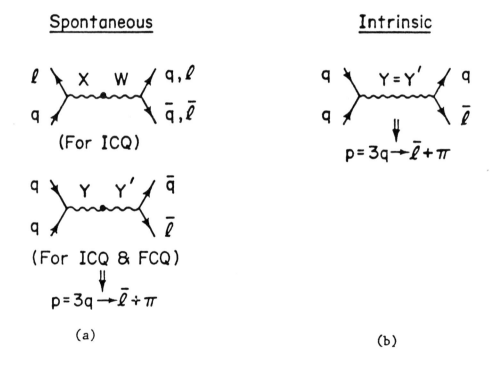

Fig. 16.5 Spontaneous versus Intrinsic Violations of B, L, F. By convention B denotes quark number, which is $+1$ for quarks, $+3$ for protons, and 0 for leptons.

integer-charge quarks), as mentioned previously; this induces B and L violations. Likewise, within the more extended symmetry involving fermion-number gauging, $F = +2$ gauge particles would inevitably mix spontaneously with $F = -2$ and $F = 0$ gauge particles leading to B, L, and F violations of varying complexions; *such a violation would be operative not only for integer but also for fractionally charged quarks.*

This manner of violation of quantum numbers appears less arbitrary than a violation introduced explicitly into the basic Lagrangian. It is in line with the more universal hypothesis that perhaps *all* symmetries (for example, $SU(3)$, isospin as well as parity, charge conjugation, and CP), which are not exact, are broken spontaneously.

By contrast within unifying symmetries such as $SU(5)$, $SO(10)$, E_6, and E_7, the quantum numbers B, L, and F are violated intrinsically in the basic Lagrangian. For these symmetries, the gauges are so "squeezed" in the process of choosing these subgroups that gauge particles Y and Y' coupling respectively to $\bar{q}q$ and $\bar{l}q$ currents within the maximal symmetry become one and the same within these subgroups (see the right-hand side of Fig. 16.5), which induces B, L, and F violations intrinsically in the basic Lagrangian. In other words, B, L, and F violations occur intrinsically for these subgroups not just because one puts quarks and leptons in one multiplet, but because one *chooses* to gauge these specific subgroups of the maximal groups permissible by the fermion content.

3.2.1 Practical similarities and differences between the two schemes: Proton decay. Regardless of the origin of the violations of B, L, and F — be it spontaneous or intrinsic — proton is predicted to be unstable in both cases; allowing for fermion number violation this should hold for either integer or fractionally charged quarks. This is the most important common prediction of both schemes. There are some differences in the lifetime predictions. The case of decaying integer-charge quarks within for example $[SU(4)]^4$ leads to a proton lifetime that can be as short as about 10^{29} years, but as long as about 10^{33}–10^{34} years, while for the $SU(5)$ model with its superheavy gauge mass determined from coupling constant renormalization arguments to be around 10^{15} GeV, the recent prediction of proton lifetime is about $10^{32\pm1}$ years.[26] In other words, given the uncertainty in the predicted range, the predictions of the two schemes do coincide in the longer end (10^{31}–10^{34} years).

For the case of spontaneous violations of B, L and F, especially for maximal unifying symmetries such as $SU(16)$ or its extensions, proton may in general decay via *four* alternative modes satisfying a change in fermion number by 0, -2, -4 and -6. The modes are: (1) $p \to 3\nu + \pi^+$, $e^- + 2\nu + 2\pi^+$ ($\Delta F = 0$); (2) $p \to e^- + 2\pi^+$ ($\Delta F = 0$); (3) $p \to e^+\pi^0$ ($\Delta F = -4$) and (4) $p \to 3\bar{\nu} + \pi^+$, $e^+ + 2\bar{\nu} + \pi^0$ ($\Delta F = -6$). Some of these modes can even co-exist with competing rates. For the cases of $SU(5)$ and $SO(10)$, on the other hand, only the $\Delta F = -4$ mode is permissible.

3.2.2 Difference between the two schemes at high energies and high temperatures. The two schemes of B, L, F violations — spontaneous versus intrinsic — differ radically from each other in one respect. The violations of spontaneous origin are operative only at "low" temperatures. At high temperatures sufficiently above the masses of the superheavy X-like particles, the masses of X and W as well as X-W mixing mass and therefore B, L, F violations are likely to vanish due to phase transitions.

The violations of intrinsic origin on the other hand are damped at low temperatures by the mass of the superheavy gauge particles. They become comparatively as strong as all other interactions at sufficiently high temperatures at which Xs are massless.

This difference would have its important consequence in the development of the early universe (in particular for the generation of excess baryons as mentioned later).

3.2.3 *The possibility of a stable proton?*

Even within the unification hypothesis, it is possible to ensure proton stability for instance by introducing new fermionic components and cleverly choosing their assignments, which prevent proton decay on energetic grounds. There exist additional mechanisms that permit the conservation of a global quantum number and thereby ensure proton stability.[27] But let me at least say that the simplest and the most straightforward attempts at grand unification end up in making the proton unstable (unavoidably especially for the case of spontaneous violations of B, L, and F).[28]

To summarize, regardless of whether the violations of B, L, and F are intrinsic or spontaneous, we expect within the hypothesis of grand unification that the proton would be unstable. The detailed mechanism leading to proton decay can vary from one model to another, as also the decay modes. However its predicted lifetime for a number of alternative models happens to lie in the range of 10^{29}–10^{34} years.

3.3 The Question of Quark Charges

The third question of detail, which concerns quark charges, is important for determining what signals one ought to look for to see the fundamental constitutents: the quarks and gluons.

It is clear that photon (and therefore electric charge) gets defined within the gauge approach only after SSB. Depending upon the choice of the symmetry group and the nature of SSB there can be two alternative complexions for the photon and correspondingly two alternative patterns for quark charges: integral and fractional.

To see briefly how this comes about, consider a unifying symmetry of the form $G_{\text{flavor}} \times G_{\text{color}}$. To be specific, choose this symmetry to be $[SU(4)]^4$. Scalar multiplets of the type $C = (4,1,\bar{4},1)$ together with three additional multiplets related to C by left $\leftrightarrow$ right and color $\leftrightarrow$ flavor discrete symmetries are needed to break $[SU(4)]^4$ spontaneously so as to generate the photon.[29] The vacuum expectation value of C, depending upon the choice of Higgs–Kibble parameters, can in general take the form:

$$<C> = \begin{bmatrix} c_1 & & & \\ & c_1 & & \\ & & c_1 & \\ & & & c_4 \end{bmatrix} \qquad (16.9)$$

c_4 gives mass to W_L^+. Thus

$$gc_4/2 \lesssim m_{W_L^+} ,$$

where g is the effective flavor gauge coupling constant.

The parameter c_1 can in general be zero or nonzero. If it remains zero in spite of radiative corrections, then $SU(3)$ color as a local symmetry remains unbroken; the octet of color gluons remains massless and quarks acquire fractional charges. It is conceivable that the forces generated by these massless gluons confine quarks and gluons permanently, which is commonly assumed.

If, on the other hand, c_1 is nonzero, however small, $SU(3)$ color as a local symmetry is broken, even though a good global $SU(3)$ color symmetry is still preserved. The octet of gluons acquire a light mass $= \bar{f}_s c_1/\sqrt{2}$, where $\bar{f}_s$ is the appropriate effective $SU(3)$ color gauge coupling constant (see below). Electric charge now becomes a *symmetrical combination* of $SU(4)$ flavor and $SU(4)$ color generators. This gives rise to the integer-charge quark. Simultaneously and *unavoidably* there is induced a spontaneous mixing between Ws and the leptoquark gauge particles Xs with a strength $= \bar{f}_s g c_1 c_4$. The mixing term thus generated is given by:

$$\mathcal{L}_{\text{mixing}} = \bar{f}_s g c_1 c_4 (XW + \text{h.c.}) \ . \tag{16.10}$$

Such a mixing induces violations of baryon and lepton numbers satisfying $\Delta B_q = -\Delta L = \pm 1$ with $\Delta F = 0$, which in turn makes the quarks unstable against decay into leptons.[30] The case of unconfined unstable integer-charge quark with liberated physical masses $\sim$ few to 10 GeV and lifetime $\sim 10^{-6}$–10^{-9} sec does not contradict observation (see subsequent discussions). This in turn removes the necessity for the assumption of "absolute" confinement of quarks.

The two fundamental alternatives of confined fractionally charged quarks and liberated integer-charge quarks thus arise as allowed logical possibilities within the same unifying symmetry $[SU(4)]^4$ or extensions thereof possessing the structure $G_{\text{flavor}} \times G_{\text{color}}$. Which of these two cases corresponds to reality depends entirely on whether SSB assigns vanishing or nonvanishing value to c_1. The origin of these two alternatives is outlined schematically in Table 16.1.

Now for unifying symmetries (which incidentally are necessarily non-Abelian), there are in general restrictions on quark charges arising from the constraint that $SU(3)$ color must be preserved as a good global symmetry even if its local form is broken spontaneously. It turns out that this constraint is easily satisfied within semisimple unifying symmetries of the form $G_{\text{flavor}} \times G_{\text{color}}$ (for example $[SU(4)]^4$) subject to the introduction of scalar multiplets of the type C. However, it does not seem feasible to satisfy this constraint for example within $SU(5)$ at least with the presently available mechanism for generating SSB. Thus for $SU(5)$, because of this a posteriori constraint, quark charges need to be fractional. But for $[SU(4)]^4$, depending upon the nature of SSB (that is, depending upon whether $c_1 = 0$ or $c_1 \neq 0$) quark charges can be either fractional or integral.

The two alternative patterns for quark charges can be distinguished experimentally in the near future in a number of ways:

1. First, they can be distinguished by a study of structure functions for deep inelastic μN and neutral current νN scatterings at high energies. Such a study should exhibit rises in $F_2 \sim 10$–30 percent depending upon x for μN (and analogous rises for n.c. νN scatterings)[31] once $W = [M_N^2 + Q^2 (1-x)/x]^{1/2}$ exceeds its threshold value for production of liberated color octet states, if quarks acquire integer charges spontaneously.[32] No such rises should be seen for the case of *fcq*. Recent studies of μN scattering by the Michigan-Fermilab collaboration[33] show a threshold-rise in F_2 just of the above sort corresponding to a threshold ≈ 9 to 10 GeV. (The relevant calculation is being carried out by A. Janah and M. Özer.) If this rise is confirmed by further high statistics experiments, it will be a clear indication of liberated color and integer nature for quark charges.

Table 16.1

The Two Alternatives

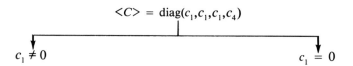

$\langle C \rangle = \text{diag}(c_1, c_1, c_1, c_4)$

$c_1 \neq 0$ | $c_1 = 0$

1. $Q_{em} = Q_{flavor} + Q_{color}$
2. Quark-lepton charges

$$\begin{bmatrix} 0 & 1 & 1 & 0 \\ -1 & 0 & 0 & -1 \\ -1 & 0 & 0 & -1 \\ 0 & 1 & 1 & 0 \end{bmatrix}$$

3. Massive octet of gluons $(m_{gluon})_{out} \sim$ few to 10 GeV

4. Spontaneous violations of B and L through X-W mixing
   UNCONFINED UNSTABLE INTEGER-CHARGE QUARKS

$$q \to l\pi$$

$$\tau_q \sim 10^{-6}\text{-}10^{-9} \text{ sec}$$

for $(m_q)_{out} \sim 5\text{-}10$ GeV

5. QUARKS, GLUONS, AND ALL COLOR CAN BE LIBERATED SUBJECT TO PARTIAL CONFINEMENT

---

1. $Q_{em} = Q_{flavor} + Q_0$
2. Quark-lepton charges

$$\begin{bmatrix} 2/3 & 2/3 & 2/3 & 0 \\ -1/3 & -1/3 & -1/3 & -1 \\ -1/3 & -1/3 & -1/3 & -1 \\ 2/3 & 2/3 & 2/3 & 0 \end{bmatrix}$$

3. Massless octet of gluons

4. STABLE FRACTIONALLY CHARGED OBJECTS

5. QUARKS, GLUONS, AND ALL COLOR MAY BE CONFINED PERMANENTLY

---

2. The two alternatives may also be distinguished by a study of two-photon processes, for example $e^-e^+ \to e^-e^+ +$ hadrons at PETRA, PEP, and LEP energies. *The cases of fractional versus gauge icq lead to different predictions for the cross-sections of such two-photon processes even below threshold for color production.*[34]

3. A startling distinction between the two cases would be the discovery of the fundamental constituents in their liberated form: the decaying integer-charge quarks and the decaying gluons. The distinctions based on leptoproduction of color on the one hand and those based on signatures for decaying *icq* and gluons on the other have been discussed in detail in a number of recent papers and in review work. For purposes of the present paper I shall limit myself to discussing only the

signatures for decaying integer-charge quarks and the implications of quark decays for cosmological evolution at the early stage of the universe on the one hand and for proton decay on the other. This is considered in the next section.

A comparative summary of the details pertaining to the questions of the unifying mass scale M, $\sin^2\theta_W$, manner of violations of B, L, and F, and quark charges for alternative unifying symmetries is presented in Table 16.2.

4. Decaying Integer-Charge Quarks and the Unstable Proton

4.1 The Mechanisms for B,L-Violating Decays of Integer-Charge Quarks

The mechanisms for quark decays have been discussed in some detail elsewhere.[37] Here I shall present the gist of these mechanisms emphasizing the results realized recently[38] that the quark lifetimes are expected to be a few orders of magnitude larger than the estimates given previously.

First we see that the mass of the gluon as well as the baryon-lepton number violating W-X mixing (see Eq. (16.10)) are proportional to the VEV parameter c_1. The parameter c_1 is defined by

Table 16.2

Detailed Features of Models

Unifying Symmetries	Unif. Mass Scale[a]M	$\sin^2\theta_W$[b]	Quark Charge[c]	Baryon Number Violation	τ_{proton}
$[SU(4)]^4$	Chiral color 10^5–10^6 GeV	.28–.30	Either icq or fcq	Spontaneous	10^{29}–10^{34} years
$SU(16)$	Vector color 10^{16} GeV	.20	icq or fcq	Spontaneous	10^{29}–10^{33} years
$SU(5)$	10^{16} GeV	.20	fcq	Intrinsic	$10^{32\pm1}$ years[d]
$SO(10)$	10^{16} GeV	.20	fcq	Intrinsic	$10^{32\pm1}$ years[d]
E_7	$> 10^{16}$ GeV	.67	fcq	Intrinsic	$> 10^{34}$ years
E_6	$> 10^{16}$ GeV		fcq	Intrinsic	$> 10^{34}$ years

[a]The unifying mass scale M, at which the effective coupling constants merge, is typically one order of magnitude larger than M_X. For $[SU(6)]^4$, one can obtain low unifying mass scale without requiring low-energy symmetry to include chiral color. (See Elias and Rajpoot, op. cit. in n. 24.) In this case $\sin^2\theta_W$ will be lower than 0.28.
[b]The values quoted correspond to the low-energy electroweak symmetry G_W being $SU(2)_L \times U(1)$.
[c]See the reference cited in note 35 for a restriction on quark charges within unifying symmetries.
[d]For $SU(5)$, the predicted lifetime is around $10^{32\pm1}$ years. See however note 36.

choosing to renormalize the Lagrangian coupling constants and masses arising through SSB at a momentum scale $\gtrsim 1$ GeV, such that the effective strong interaction coupling constant $\alpha_s = \bar{f}_s^2/4\pi$ at this momentum is small enough to permit a perturbative treatment of strong interactions. With this prescription, experimental constraints such as "precocious" scaling suggest[39] that the gluon masses are light $\lesssim 300$ MeV at $Q^2 = 1$ GeV^2 analogous to the masses of the up and down quarks. Taking $\alpha_s = \bar{f}_s^2/4\pi \approx 0.3$ at $Q^2 = 1$ GeV^2 and the effective gluon mass[38] $\bar{m}_V(Q^2 = 1$ GeV$^2) = \bar{f}_s(Q^2 = 1$ GeV$^2)c_1/\sqrt{2} \approx (30$ to $300)$ MeV, we obtain $c_1 \approx (220$ to $200)$ MeV. Now there are three basic diagrams that induce baryon-lepton number violating quark decays following the mixing given by (16.10). These are exhibited in Fig. 16.6.

The selection rules based on electric charge and fermion number conservations are such that Fig. 16.6a can lead to emission of only neutral leptons (neutrinos) but not electrons or muons. Figures 16.6b and 16.6c on the other hand can lead to emission of neutral as well as charged leptons. A sample of decay modes following from these diagrams is as follows:

$q_{\text{red,yellow,blue}} \to \nu + $ (pions or other mesons) (Fig. 16.6a) ; (16.11)

$$\left.\begin{array}{l} q^+_{\text{yellow,blue}} \to (\nu + \pi^+) \text{ or } (e^- + \pi^+ + \pi^+) \\ q^0_{\text{yellow,blue}} \to (\nu + \pi^0) \text{ or } (e^- + \pi^+) \\ q^0_{\text{red}} \to (\nu_\mu + \pi^0) \text{ or } (e^- + D^+) \text{ or } (\mu^- + F^+) \\ q^-_{\text{red}} \to \nu_\mu + \pi^- \text{ or } (e^- + D^+ + \pi^-) \ . \end{array}\right\} \quad \text{(Fig. 16.6b) ;} \quad (16.12)$$

$$\left.\begin{array}{l} q^+_{\text{yellow,blue}} \to \nu + (\pi^+ \text{ or } \mu^+\nu \text{ or } e^+\nu) \\ q^0_{\text{yellow, blue}} \to e^- + (\pi^+ \text{ or } \mu^+\nu \text{ or } e^+\nu) \\ q^0_{\text{red}} \to \nu + (D^0 \text{ or } \bar{\nu}_e \nu_\mu) \\ q^-_{\text{red}} \to e^- + (D^0 \text{ or } \bar{\nu}_e \nu_\mu) \end{array}\right\} \quad \text{(Fig. 16.6c) .} \quad (16.13)$$

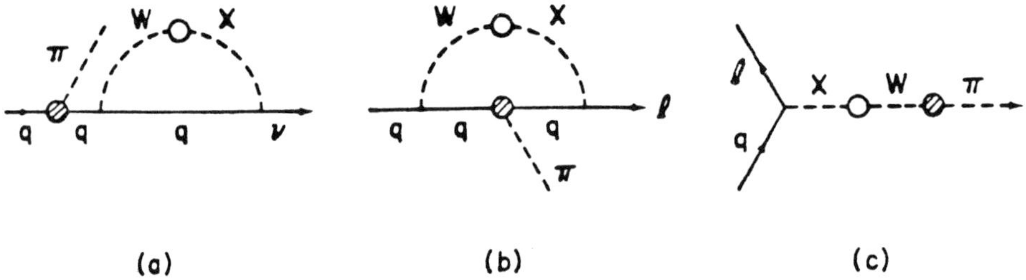

Fig. 16.6 Quark-decay mechanisms. The selection rules are such that diagram a can lead to emission of a neutrino, but not a charged lepton. Diagrams b and c can lead to emission of either neutrino or charged lepton.

In the preceding q stands for u or d flavor as the case may be. In writing these decay modes, I have, for illustration, assumed the constraints following from the underlying gauge symmetry $[SU(4)]^4$, which permits a spontaneous mixing of $X^+{}_{24}$ and $X^+{}_{34}$ with the familiar W^+ mesons as well as the mixing of $X^0{}_{14}$ (coupled to $\bar{q}_{red}\, l$ currents) with $W^0{}_{41}$ (coupled to the $\bar{c}u$ current). Furthermore certain selection rules based on the distinction between e and μ (with μ being assigned as the strange lepton) are also exhibited (16.12) and (16.13).

Note that tree diagram Fig. 16.6c is damped by two large mass denominators ($m_X^{-2}\, m_W^{-2}$), while loop diagram Fig. 16.6a is convergent and is damped by only one such denominator ($m^{-2}{}_X$). Thus one may drop the contribution of Fig. 16.6c to quark decay compared to that of Fig. 16.6a.

$$M \text{ (Fig. 16.6a)} \gg M(\text{Fig. 16.6c}) . \tag{16.14}$$

The relative weights of Fig. 16.6a versus Fig. 16.6b are somewhat more uncertain. If one assumes that the emission of a composite pion by off-shell quarks from inside the loop (Fig. 16.6b) is damped by a form factor behaving like $(1/K^2)$ for large loop momentum K^2, then Fig. 16.6c would be too convergent and therefore also damped by two large mass denominators ($m_X^{-2}\, m_W^{-2}$) like the tree diagram Fig. 16.6c. In this case, only Fig. 16.6a would be relevant. This in turn would imply that quark decays would be dominated by emission of neutrinos rather than charged leptons (see (16.11)). (Charged leptons could still be emitted if $m_q > m_{gluon}$ via a two-step process for charged red quark decays: for example

$$q^-_{red} \xrightarrow{\text{strongly}} V^- + (\text{virtual } n^0_{yellow}) \xrightarrow{\text{Fig. 16.6a}} V^- + \nu_e;$$

the charged gluon V^- could in its turn decay to $e\bar{\nu}$ or $\bar{\mu}\nu$.)

On the other hand, if the pion emission from inside the loop is not damped for large loop momentum (K^2), then the contribution of Fig. 16.6b would be comparable in magnitude to that of Fig. 16.6a and quark decays would yield neutrinos as well as charged leptons (e and μ) in comparable proportions.

With this qualification, quark decay modes may be read off from earlier papers. Basically there are two patterns. Either

$$1) \quad m(q^-_{red}) < m(V^-_{gluon}) , \text{ or } m(q^-_{red}) - m(q_{yellow,blue}) > m_\pi .$$

In this case all quarks decay directly or sequentially into (neutrinos + mesons) and also to (e^- or μ^- + mesons) (if Fig. 16.6b is not damped by $qq\pi$-vertex):

$$q \to \nu + \text{mesons} \tag{16.15}$$

$$q \to e^- \text{ or } \mu^- + \text{mesons (if Fig. 16.6b is not damped by } qq\pi\text{-vertex)} . \tag{16.16}$$

Or, alternatively,

$$2) \quad m(q^-_{red}) > m(V^-_{gluon}) \text{ and } m(q^-_{red}) - m(q_{yellow,blue}) < m_\pi .$$

In this case all except the two charged red quarks (d_{red}^- and s_{red}^-) decay as shown. The two charged red quarks can decay into (charged gluon + neutrino); the charged gluon in turn would decay into a lepton pair or hadrons

$$q_{\text{red}}^- \to V_{\text{gluon}}^- + \nu \qquad (16.17)$$
$$\qquad\qquad\downarrow$$
$$\qquad e\bar{\nu},\, \mu\bar{\nu} \text{ or hadrons} .$$

In addition, they may also decay within $[SU(4)]^4$ directly into leptons + mesons:

$$q_{\text{red}}^- \to \nu + (\pi, \rho, A_1) . \qquad (16.18)$$

In addition to the decay modes listed, if diquarks exist and are lighter than quarks, a quark could convert into a diquark + a virtual antiquark via strong interactions followed by the antiquark converting into an antilepton.

$$q \xrightarrow{\text{strong}} \phi_{qq} + \text{``}\bar{q}\text{''}_{\text{virtual}} . \qquad (16.19)$$
$$\qquad\qquad\qquad\qquad \hookrightarrow \bar{l}$$

In its turn, the diquark (assuming that it is heavier than normal baryons N, Λ, Σ, Ξ) could decay as follows:

$$\phi_{qq} \xrightarrow{\text{strong}} B(qqq) + \text{``}\bar{q}\text{''}_{\text{virtual}} . \qquad (16.20)$$
$$\qquad\qquad\qquad\qquad \hookrightarrow \bar{l}$$

4.2 Baryon and Lepton Number-Conserving Quark Decay Modes

The decay modes listed (16.15)–(16.20) represent baryon and lepton number-violating decays. In addition to these there would, of course, always exist the baryon and lepton number-conserving quark decays, which arise through normal weak interactions as well as through spontaneous mixing of charged color gluons V^+ with the weak gauge particles W^+. (The strength of $V^+ - W^+$ mixing is proportional to c_1^2 and thus to the (mass)^2 of the gluons. Note that $V^+ - W^+$ mixing is unavoidable for the case of integer-charge quarks.) A sample of such baryon and lepton number-conserving weak decays of quarks is as follows:

$$\left.\begin{array}{l} s \to u + \pi^-,\, u + e^- + \bar{\nu}_e \\[4pt] c \to s + \pi^-,\, s + e^- + \bar{\nu}_e,\, u + \pi^0 \end{array}\right\} \text{Amp} \sim O(G_F) . \qquad (16.21)$$

$$\left.\begin{array}{l} u_{\text{red}}^0 \to d_{\text{yellow}}^0 + \gamma \\[4pt] c_{\text{red}}^0 \to s_{\text{yellow}}^0 + \gamma \end{array}\right\} \begin{array}{l} \text{Amp} \sim O(G_F)\, \text{``}m_V^2\text{''} \times [e \ln(m_W^2/m_V^2)] \\[4pt] \text{(through loop diagram involving } V^+\text{-}W^+ \\ \text{mixing)} \end{array} \qquad (16.22)$$

Here m_V determining the perturbation loop diagram denotes gluon and quark Lagrangian masses. As noted before, renormalizing at a momentum scale $\mu \sim 1$ GeV, m_V is of order 300 MeV or smaller; in any case it is much smaller than the physical masses of gluons and quarks. Since $m_c - m_s \sim 1$ GeV and $m_s - m_u \gtrsim 350$ MeV, and since $G_F \gg G_X = f_X^2/m_X^2 \sim 10^{-10} \text{GeV}^{-2}$, we expect that the B-conserving decay modes of the charm and s quarks (exhibited in (16.21) and (16.22)) would dominate over their B-nonconserving decay modes (16.15)–(16.20). In other words, the heavier quarks (strange, charm, and so on) are likely to decay primarily via B-conserving normal $O(G_F)$ weak interactions and thus are expected to possess lifetimes of order 10^{-10} sec or shorter. It is the lighter up and down (u and d) quarks that should decay primarily via the B-nonconserving modes (16.15)–(16.20). (One possible exception could be this: if a modest Q value is available — that is, as long as $m(u^0_{red}) - m(d^0_{yellow}) \gtrsim 50$ MeV — the B-conserving decay mode $u^0_{red} \to d^0_{yellow} + \gamma$ arising through $V^+ - W^+$ mixing (see (16.22)) could compete favorably with or dominate over the B-violating decay modes $u^0_{red} \to \nu_\mu +$ mesons (see (16.11) and (16.12))).

4.3 Lifetimes of Liberated Up and Down Quarks

As discussed previously the B-nonconserving decay modes of quarks are induced dominantly through Fig. 16.6a (and possibly also Fig. 16.6b, which could be comparable to Fig. 16.6a if $qq\pi$ vertex is not damped by a form factor). The amplitude for Fig. 16.6a is given by[40]:

$$M (\text{Fig. 16.1a}) = g_{qq\pi} \left[\frac{f_X^2 g^2 c_1 c_4}{8\pi^2 m_X^2} \ln (m_X^2/m_W^2) \right] \bar{v}(1 - \gamma_5)q$$

$$\equiv g_{qq\pi} [x] \bar{v}(1 - \gamma_5)q \qquad (16.23)$$

As mentioned previously, $c_1 \approx (\frac{1}{4} \text{ to } 3) 100$ MeV. Since X-exchange induces $K_L \to \bar{\mu}e$ decay within $[SU(4)]^4$, from the known upper limit on $K_L \to \bar{\mu}e$ decay (that is, $\Gamma(K_L \to \bar{\mu}e)/\Gamma(K_L \to \text{all}) < 2 \times 10^{-9}$, which in turn implies that $\Gamma(K_L \to \bar{\mu}e)/\Gamma(K^+ \to \mu^+\nu) < 10^{-9}$), one may deduce that (f_X^2/m_X^2) low momentum $< 2 \times 10^{-10}$ GeV^{-2}. As a representative value, I shall take $(f_X^2/m_X^2) \approx 10^{-10}$ GeV^{-2}. (A value for $G_X = (f_X^2/m_X^2)$ of this order is suggested independently as follows: The unifying mass scale M is $\approx (1/2$ to $3) 10^5$ GeV for the parent symmetry being $[SU(4)]^4$ and for chiral color being a good low-energy symmetry; the superheavy gauge mass $m_X \approx M/5$ and that[41] (f_X^2) at low momentum

$$\approx (\tfrac{1}{4} \text{ to } \tfrac{1}{2})g^2 \approx (\tfrac{1}{4} \text{ to } \tfrac{1}{2})(4e^2) \approx \tfrac{1}{2} ;$$

these values imply

$$(f_X^2/m_X^2) \approx (\tfrac{1}{2} \text{ to } 10) \times 10^{-10} \text{ GeV}^{-2} .)$$

Noting that c_4 contributes to the mass of W_L^+, $(gc_4/2) \leq m_{W_L^+}$. I assume in the following that c_4 is at least a major contributor to W_L^+ mass (this is necessarily true for the minimal Higgs system). Thus $gc_4 \equiv (2m_{W_L^+}\xi) \approx (150 \text{ GeV}) \xi$, where $\xi \approx 1/2$ to 1. Finally $g \approx 2e \approx 2/3$. Substituting these values, that is,

$$c_1 \approx 100 \text{ MeV}, \quad gc_4 \approx (150 \text{ GeV})\xi, \quad g \approx 2/3, \quad \text{and} \quad f_X^2/m_X^2 \approx 10^{-10} \text{ GeV}^{-2},$$

the parameter x standing for the expression inside the square bracket in (16.17) is given by[42]:

$$x \approx 2 \times 10^{-10}\xi \qquad (\xi \approx 1/2 \text{ to } 1) . \qquad (16.24)$$

The partial decay rate for $q \to l + \pi$ is given by:

$$\Gamma(q \to l + \pi) = \frac{(m_q)_{\text{out}}}{8\pi} (g_{qq\pi}^2 x^2) \qquad (16.25)$$

$$\approx \left[\frac{(m_q)_{\text{out}}}{1 \text{ GeV}}\right] (g_{qq\pi}^2/4\pi) (10^{+5} \text{ sec}^{-1})\xi^2 . \qquad (16.26)$$

Here $(m_q)_{\text{out}}$ denotes physical "outside" mass of liberated quarks $(m_q)_{\text{out}} \approx 5$ to 10 GeV. The effective coupling parameter $g_{qq\pi}$ is hard to ascertain. Use of PCAC[43] for this case might be grossly misleading since one is dealing with the transition of a massive on *mass-shell* quark ($m_q \approx 5$ to 10 GeV) to an on-shell pion but a highly off *mass-shell* quark with $p^2 = m_v^2 \gtrsim 0$ (see Fig. 16.6a). As a plausible guide, I assume that $(g_{qq\pi}^2/4\pi) \approx 1$ to 10 (however, one may not be too surprised if $(g_{qq\pi}^2/4\pi)$ relevant to physical quark decay (Fig. 16.6a) turns out to be as large as even 100). This yields for a 10-GeV quark:

$$\Gamma(q \to l + \pi) \approx (1 \text{ to } 10) (10^{+6} \text{ sec}^{-1}) \xi^2 . \qquad (16.27)$$

The relative weight of multipion emission depends upon for example effective[44] $(g_{qq\pi}^2/4\pi)$. Conservatively it is estimated[45] that the combined partial rate for multimeson emission is *at least one order of magnitude* and perhaps even two orders of magnitude higher than single pion emission for a 10-GeV quark. Thus for the up and down quarks (with $m_q \approx 10$ GeV)

$$\Gamma_q^{\text{Tot}} \approx \Sigma\Gamma (q \to l + \text{mesons}) \approx (10^{+7} - 10^{+9} \text{ sec}^{-1})\xi^2 , \qquad (16.28)$$

where variation by one order of magnitude is due to uncertainty in $(g_{qq\pi}^2/4\pi)$ and the variation by the second order of magnitude is due to uncertainty in the estimate of the relative weight of multimeson modes. Allowing for ξ^2 to be as small as about $(1/10)$ (see, however, note 42), we obtain the following estimate for the lifetime of the lighter up and down quarks decaying predominantly via the B-nonconserving modes:

$$\tau_q \text{ (lighter quarks)} \approx 10^{-6} \text{ to } 10^{-9} \text{ sec } (m_q \approx 10 \text{ GeV}) . \qquad (16.29)$$

While there might still be some uncertainty at the longer end of the quark lifetime, it would appear integrating all the foregoing discussions that the lifetimes of liberated up and down quarks with 10-GeV mass decaying primarily via B-nonconserving modes are *not* much shorter than about 10^{-10} sec. This, it ought to be noted, is almost three to four orders of magnitude *longer* than previous estimates.[46] This big difference[47] between prior and present estimates is in part a consequence of the fact that now we have chosen to do perturbation theory (including perturbation in strong interactions) with respect to a Lagrangian whose parameters (such as c_1) are defined by renormalizing at a momentum scale ~ 1 GeV, where the strong coupling $\alpha_s = f_s^2/4\pi$ is small. (Previously we had chosen c_1 corresponding to the physical mass of the gluons, which would correspond to renormalizing at a low momentum scale ~ 0 at which α_s is large. This would presumably make perturbative treatment with respect to α_s unreliable.) However, admittedly, one's ignorance in handling this delicate matter appears to be sufficiently profound. A satisfactory resolution of the apparent discrepancy between the two estimates would eventually involve an understanding of the long-distance strong-interaction regime of QCD. A second source of the discrepancy is the magnitude of (f_X^2/m_X^2), which is one order of magnitude smaller than the previous estimates (see previous discussions).

I can put the moral of all this discussion, as follows: *Quark lifetimes even for the up and down quarks may quite plausibly be quite a bit longer than one had thought before.* For a 10-GeV quark their lifetimes may be as long as about 10^{-6} or even 10^{-5} sec.

Does this contradict any observation? Such long lifetimes would have contradicted observation if physical quark masses were rather light (≈ 2 to 4 GeV), such that their pair-production cross-section might have been significant enough to be sensitive to existing particle searches. However, for $m_q > (5–6)$ GeV, existing laboratory particle searches at Fermilab and ISR would not have been sensitive enough to detect such quarks, even if they are relatively long-lived, assuming that their pair-production cross-section is smaller than the production cross-section of the 10-GeV Υ-particle. In this regard, recall that one of the most sensitive searches of relatively long-lived particles is the one recently carried out at Fermilab,[48] which reports that charged particles of mass $\lesssim 5$ GeV with pair-production cross-section equal to Υ-production cross-section can not have lifetimes longer than about 5×10^{-8} sec. This search is not, however, sensitive to particle masses exceeding 5 GeV. Cosmic ray searches using emulsions are either sensitive to fractional charges or to particles with lifetimes shorter than about 10^{-10} sec. The one notable exception appears to be the recent search, involving cosmic ray interactions at 10–100 TeV, by Goodman et al. who *do* report observation of three events signifying possible production of new particles with mass exceeding 5 GeV and with lifetimes exceeding about 10^{-7} sec.[49] These events could well signify pair-production of liberated quarks. It would be most desirable to improve such cosmic ray searches with a view to substantiating these events more firmly as well as ascertaining masses, lifetimes, production cross-sections, and eventually decay products of the new particles.

The purpose of this discussion was to note that existing searches do not exclude the existence of liberated relatively long-lived ($\tau \sim 10^{-5}$–10^{-6} sec) massive integer-charge quarks with mass $\gtrsim 5$–6 GeV, and that there even appear to be some candidates for such objects in cosmic ray searches. This is not to say that quarks must necessarily be this long-lived. In accordance with our discussions, they may have a lifetime as short as about 10^{-9}–10^{-10} sec (if $m_q \sim 10$ GeV).

4.4 *Quarkosynthesis with Confined or Liberated Quarks: Decay Modes and Lifetimes of Cosmological Integer-Charge Quarks*

The B-nonconserving decay modes and lifetimes presented refer to decay modes and lifetimes of liberated physical quarks with masses $\sim$ 5 to 10 GeV. Assuming that quarks are integer-charged, do such lifetimes $\sim 10^{-6}$–10^{-9} sec apply to cosmological quarks as well? (By "cosmological quarks" I mean quarks that existed in their quark phase in equilibrium with other particles — for example, photons, electrons, and neutrinos — until the temperature and correspondingly the particle *density* was low enough for quarks to synthesize to form nucleons.) If the answer to the question just raised were in the affirmative, since quarkosynthesis does not take place until a time scale $\sim 10^{-5}$–10^{-4} sec (see below), a major fraction ($\gtrsim (10^5 - 1)/10^5$) of quarks would have decayed out to leptons before they had a chance to synthesize and this would have been disastrous for the alternative of unstable integer-charge quarks decaying into leptons, especially if their lifetimes were shorter than 10^{-6} sec. Okun and Zeldovich[50] in fact noted this ("would be") disaster and concluded that the hypothesis[51] of relatively short-lived integer-charge quarks ($\tau \lesssim 10^{-6}$ sec) decaying into leptons is not compatible with cosmology.

I now remark that this would be disaster is avoided altogether due to the facts that 1) quarks (and gluons) are strongly subject to the so-called Archimedes effect, and that 2) the early universe ($t < 10^{-5}$ sec) was so dense with quarks that average interquark separation $\bar{d}$ was much less than 1 fermi and accordingly they could not possibly acquire their liberated physical mass. They could only possess their light Archimedes mass (< 300 MeV) depending upon and varying with $\bar{d}$. The lifetimes of these Archimedes quarks with their effective light masses (< 300 MeV) are much longer (3 to 4 orders of magnitudes) than those of the physical quarks. As a result only a small fraction ($\lesssim$ 20 percent) of quarks disappear into leptons for the case of integer-charge quarks before quarkosynthesis, which is of no concern. This obviates the Okun–Zeldovich difficulty.

Let me briefly outline the picture of quarkosynthesis viewed along these lines. The basic picture is in fact the same for both alternatives: fractional (confined) or integer-charged (liberated) quarks.

4.4.1 *Quarkosynthesis.* Within a spontaneously broken unified gauge theory involving chiral gauges, quarks are massless in the basic Lagrangian. They acquire their mass spontaneously like the gauge particles Ws through VEV of scalar fields $\langle\phi\rangle \neq 0$. For simplicity, assume that quarks acquire mass through VEV of the same ϕ as the Ws. The transition temperature[52] T_c above which $\langle\phi\rangle = 0$ is around 300 GeV. Somewhere for $T \lesssim T_c \approx 300$ GeV (corresponding to time scale $t_{\text{sec}} \approx 1/T_{\text{MeV}}^2 \approx 10^{-11}$ sec in the process of cosmic evolution), $\langle\phi\rangle$ acquires a nonzero value; the up and down quarks acquire their bare Lagrangian mass $m_q^{(o)} \sim 10$ MeV, the strange and charm quarks acquire appropriately heavier masses. The average interquark separation $\bar{d}$ (which can be estimated[53] from the number density of photons in equilibrium with quarks at a given temperature) is given by $\bar{d} \approx (\frac{1}{2}$ to $1)10^{-10.6}$cm$/T_{\text{MeV}}$ for "m_q" $\ll T$. This implies $\bar{d} \approx 10^{-3}$ fermi for $T \sim 300$ GeV. The quarks at this stage are too close to each other and thus suffer strongly from the Archimedes effect.

As the temperature cools to 1 GeV, correspondingly the time scale t increases to $\approx 10^{-6}$ sec, the interquark separation $\bar{d}$ increases to about ($\frac{1}{4}$ to $\frac{1}{3}$) fermi. The Archimedes mass of up and down quarks also increases correspondingly following renormalization group ideas, perhaps corresponding to a momentum scale $Q \sim 1/\bar{d}$. Thus $\bar{m}_q \sim 25$–100 MeV at temperatures $T \sim 1$ GeV. At this temperature quarks are still too close to each other for three specific quarks to combine to make a specific nucleon. One cannot tell whether a given quark belongs to nucleon$_1$ or nucleon$_2$, which are only 1/5 fermi apart from each other. In other words the "bags" defining the boundaries of different nucleons are not yet too sharp in that they overlap significantly with each other. Quarks at this temperature (~ 1 GeV) are still in quark phase.[54]

As the temperature cools further to about 200–400 MeV (time scale $t \sim 10^{-5}$ sec), the separation $\bar{d}$ increases[55] to about 1 fermi. Somewhere around this temperature a very interesting phenomenon must take place in accordance with the ideas of QCD (confined or liberated). On the one hand, when the temperature cools still further, the quarks would have a tendency to get further apart from each other with $\bar{d}$ exceeding 1 fermi due to the expansion of the universe. This however inevitably brings in the *transition region* where for increasing $\bar{d}$ or decreasing Q the strong interaction suddenly tends to get very strong. The quarks for the liberated case, if they could get sufficiently far from each other ($\bar{d} > 1$ fermi), would tend to acquire their liberated heavy physical mass (~ 5 to 10 GeV). But alas! such a transition is energetically highly disfavored at this stage of the cosmic evolution, since the temperature of the system at this instant is only a few hundred MeV. Now for the confined case, such a transition is absolutely forbidden, in any case, since the liberated masses are effectively infinite. Thus, for both the confined and the liberated case, there is at this stage only one choice left; the quarks can still group themselves into color-singlet combinations (three in each combination forming the "bags" of individual nucleons, each of size ~ 1 fermi). Such a synthesis is strongly favored energetically at this epoch; hence the quarks synthesize. This marks the end of the era of "quark phase" and the beginning of the era of the "nucleonic phase" in cosmic evolution. The quarks remain from now on inside the bag with effective masses ~ 350 MeV. If they are of the liberated type, they can, of course, be liberated provided the nucleon can be shaken hard enough so that the quarks can be pulled out with their *liberated physical masses* ~ 5 to 10 GeV. For the confined case, they remain (unfortunately) confined for ever.

To summarize,

1. Quarkosynthesis temperature T_s, suggested on the basis of evolving quark density, appears to lie somewhere between 200 and 400 MeV for either the confined or the liberated case. Thus quarkosynthesis must have taken place at a time scale $t_s \approx (2 \text{ to } 1/2)10^{-5}$ sec.
2. *For the liberated case, due to the nature of quark density in the early universe, quarks never get a chance to acquire their liberated physical mass at any instant.* This is of crucial importance for the survival of quarks, without which we would not exist (see below).

4.4.2 *Decay modes and lifetimes of cosmological quarks.* We are concerned with the lifetimes of only the lightest (that is, up and down) quarks.[56] These quarks live much longer within the environment of the early universe than the corresponding liberated quarks simply because their mass $\bar{m}_q$ is much lighter ($\lesssim 300$ MeV) than that of the liberated quarks and correspondingly the phase space is much lower. There are two cases depending upon $\bar{m}_q$:

1. $\bar{m}_q < m_\pi$: As discussed previously, for cosmological temperature as low as about 1 GeV (that is, time scale $t \approx 10^{-6}$ sec), up and down cosmological quarks have a mass less than m_π; it approaches m_π from below as temperature cools to about 600 to 800 MeV ($t \sim 2 \times 10^{-6}$ sec). For such low values of $\bar{m}_q$, the primary decay modes of the light quarks ($q \to l + $ pions) are forbidden. The only B-nonconserving decay mode that is allowed[57] energetically and that still possesses the loop-enhancement (as Fig. 16.6a) is "q" $\to \gamma + \nu$. Here photon can be emitted from the quark line

inside the loop without a damping. This specific decay mode is, of course, allowed only for neutral quarks. Charged quarks with mass $< m_\pi$ cannot utilize the $(\gamma + \nu)$-mode and thus must decay to three leptons $q^+ \to l + e^+ + \nu$ via tree diagram (Fig. 16.6c). These would be extraordinarily long-lived. I obtain[58]:

$$\Gamma(``q" \to \gamma + \nu) \approx (\overline{m}_q/100 \text{ MeV}) (3 \times 10^{+2} \text{ sec}^{-1}) \xi^2 \qquad (16.30)$$

$$\Gamma(``q^+" \to l\, e^+\nu) \approx (\overline{m}_q/100 \text{ MeV})^5 (3 \times 10^{-11} \text{ sec}^{-1}) \xi^2 \ . \qquad (16.31)$$

Compare these with the decay rates of physical quarks (Eqs. (16.27) and (16.28)), which are at least about five orders of magnitudes higher.

2. $\overline{m}_q > m_\pi$: For temperatures *below* about (600–800) MeV (that is, $t \lesssim 2 \times 10^{-6}$ sec) up and down cosmological quarks acquire masses exceeding m_π. Now single pion emission ($``q" \to l + \pi$) is permissible with a rate appropriate to the available Q value $(\overline{m}_q - m_\pi)$, which thus increases rapidly starting from zero at $t \sim 2 \times 10^{-6}$ sec to a *maximum* of ≈ 160 MeV, when $\overline{m}_q$ acquires its "maximum" value ≈ 350 MeV at quarkosynthesis, that is, at $t \sim 10^{-5}$ sec.

$$\Gamma(``q" \to l + \pi) \approx [(\overline{m}_q - m_\pi)/(m_q)_{\text{out}}] \left( \frac{g^2_{``q"q\pi}}{g^2_{qq\pi}} \right) \Gamma(q \to l + \pi)_{\text{out}} \ . \qquad (16.32)$$

Here "out" refers to properties of liberated quarks. For the instant at which $\overline{m}_q - m_\pi \approx 100$ MeV, a median value, we obtain using (16.26):

$$\Gamma(``q" \to l + \pi) \approx (g^2_{``q"q\pi}/g^2_{qq\pi}) \, 10^4 \text{ sec}^{-1} \xi^2 \qquad (16.33)$$

Use of PCAC, which may be more justifiable for light quarks than for the heavy liberated quarks, suggests that the first factor on the right is typically less than unity. Thus during the epoch between $t \sim 2 \times 10^{-6}$ sec and $t_s \sim 10^{-5}$ sec, which incidentally is the epoch during which quark lifetimes are the shortest, typical lifetimes for up and down quarks are $\gtrsim 10^{-4}$ sec.

$$\tau(``q")_{\text{up,down}} > 10^{-4} \text{ sec} \ . \qquad (16.34)$$

Since quarkosynthesis time $t_s \sim (\tfrac{1}{2}$ *to* $2) \times 10^{-5}$ *sec, we see that at most only about 20 percent of cosmological quarks could have decayed out prior to quarkosynthesis, even if the transition temperature for B-nonconservation is set to be as high as* ~ 1 GeV.[59]

4.5 The Problem of Baryon Excess

It is worth noting at this point that the rates of baryon number-violating decays and reactions of the type:

$$"q" \to l + \pi, \quad l + \gamma$$

$$l + \gamma \to "q"$$

$$X \to \bar{q}l, \bar{q}q, \bar{l}l \tag{16.35}$$

$$qq \leftrightarrow ql \text{ etc.}$$

are too slow compared to the rate of cosmological expansion at this epoch — that is, the period shortly preceding the period of quarkosynthesis. To be precise, the relevant epoch starts at a time t_0 corresponding to a temperature T_0, which marks the onset of B,L-violation during cosmic evolution. If B,L-violations are induced primarily through W-X mixing, the relevant temperature T_0 may lie somewhere between a few hundred MeV and 1 GeV depending upon c_1 and further details of the Higgs–Kibble mechanism. Correspondingly the starting instant of the epoch $t_0 \approx$ (10 to 1) 10^{-6} sec. The epoch of interest ends with the completion of quarkosynthesis, which corresponds to a temperature $T_s \approx$ 200–400 MeV and a time $t_s \approx$ (2 to 1/2) $\times 10^{-5}$ sec.

During the period spanning from t_0 to t_s the B,L-violating decays and reactions of the type (16.35) are too slow compared to the rate of cosmic evolution at this time; thus they are out of thermal equilibrium. Secondly within symmetry structures such as $[SU(4)]^4$ possessing left-right symmetry, one can introduce the isoconjugate model of CP-violation. Such a model of CP-violation possessing the desirable features:

$$\eta_{+-} = \eta_{00},$$

and

$$|\eta_{+-}| \approx (m_{W_L^+} m_{W_R^+})^2 \sin \delta$$

was proposed some time ago by Mohapatra and myself.[60] Now CP and T-violations in general imply differences between partial rates for the B- and L-violating decays of quarks versus those of antiquarks. Given that B,L-violating processes are out of thermal equilibrium between $t = t_0$ to $t = t_s$ such a difference would give rise to a baryon excess[61] $B - \bar{B} \neq 0$ even though the universe started with $B - \bar{B} = 0$.

A numerical estimate of the ratio (B/γ) arising via the aforementioned mechanism depends somewhat sensitively on the variation of the effective masses of the quarks with the cosmological interquark separation, which in turn varies with the cosmological temperature. An estimate of this ratio remains to be carried out.[62] It is good to note, however, that the hypothesis of decaying integer-charge quarks together with the picture of quarkosynthesis outlined earlier presents a totally novel mechanism for generating B excess, which differs in a number of important ways from the mechanisms considered recently in the literature.[63] The mechanism described here refers to an epoch of time t lying between $\approx 10^{-5}$ and 10^{-6} sec corresponding to a temperature $T \lesssim 1$ GeV. Alternative mechanisms refer to an epoch of time much earlier $t \lesssim 10^{-30}$ sec corresponding to a temperature scale much higher $\gtrsim 10^{12}$ GeV.

The mechanism for generation of baryon excess I have outlined utilizes $\Delta F = 0$, $\Delta B_q = -\Delta L = \pm 1$-transitions. This arises through spontaneous mixing of W and X gauge particles for the case of icq. Within the hypothesis of spontaneous violations of B, L, and F, there is also the possibility that fermion number ± 2 gauge particles Y, Y' and so on (see Fig. 16.1) could mix and induce violations of B, L, and F of a different complexion, which would be relevant for integer- as well as fractionally charged quarks. If Y, Y' as well as the mass parameters denoting their mixing are superheavy $\gtrsim 10^{12}$ GeV, and if in addition there exists nonvanishing T-violation at superhigh temperatures $\gtrsim 10^{12}$ GeV, such violations of B, L, and F would provide an additional mechanism for generation of B excess, which would be operative at an epoch of time much earlier compared to the epoch outlined above. In summary we see that if quarks are fractionally charged, B excess has most likely arisen only at an early epoch of time ($t \lesssim 10^{-30}$ sec), while if quarks are integer-charged, such an excess might have arisen during the early ($t \lesssim 10^{-30}$ sec) as well as the late epoch ($t \sim 10^{-5}$ sec).

4.6 *The Unstable Proton*

With quarks being unstable, the proton becomes unstable as well against decay into leptons. However, assuming that the physical masses of quarks and diquarks are heavier than those of the proton, the proton — a three-quark composite — can decay only provided all three quarks within the proton "simultaneously" convert into leptons (see Fig 16.7a and b for some representative mechanisms for proton decaying into three neutrinos plus π^+). This makes the proton extraordinarily long-lived; yet it must decay. Representing the 4-fermion vertex of the transition of proton to three quarks by a strength $= t_N/m^2$ where $t_N \sim 1$ and m is a characteristic mass for the problem, and substituting the strength (κ "m_q") for the transition of each quark to lepton (see Eqs. (16.23) and (16.24) for definition and magnitude of κ), we obtain[64]:

$$M(p \to 3\nu + \pi^+) \approx \left(\frac{t_N}{m^2}\right) \left[\frac{\sqrt{2}\, g_{NN\pi} \kappa^3 E_{eff}}{(m_N)}\right] (\bar{u}_{\nu_1}(1-\gamma_5)u^c_{\nu_2})(\bar{u}_{\nu_3}(1-\gamma_5)u_p)$$

(16.36)

$$\Gamma(p \to 3\nu + \pi^+) \approx \frac{\left[\dfrac{t_N}{m^2} \times \dfrac{\sqrt{2}\, g_{NN\pi}}{m_N} \kappa^3 E_{eff}\right]^2 (2m_p^7)}{[(2\pi)^5 (768)(60)]}.$$

Here E_{eff} is a factor of order unity representing on the one hand the "average" value of "m_q" $\times$ (quark propagator) (see Fig. 16.7) over the phase space and on the other the inclusion of Fig. 16.7a and 16.7b along with possible important corrections from pion (that is, $q\bar{q}$) and gluon *exchanges* between quark lines in Fig. 16.5b. For an estimate, it appears plausible to permit a range for $E_{eff} \approx$ 1/5 to 5, that is, $E^2_{eff} \approx 25$ (1 to 1/625). Taking $\kappa \approx 2 \times 10^{-10}\xi$ (see Eq. (16.24)), we obtain

$$\Gamma(p \to 3\nu + \pi^+) \approx t_N^2 (m_p/m)^4 \,(1 \text{ to } 1/625)\xi^6 \,(3 \times 10^{-39} \text{ sec}^{-1}) \ .$$

(16.37)

If ($3\nu + \pi^+$) is a dominant mode, the corresponding lifetime is

$$\tau(\text{proton}) \approx t_N^{-2}(m/m_p)^4 \,(625 \text{ to } 1)\xi^{-6} \,(10^{+31} \text{ years}) \ .$$

(16.38)

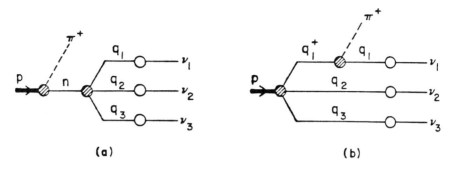

Fig. 16.7 Diagrams for proton decay. Gluons and pions (that is, $q\bar{q}$ pairs) can be exchanged between for example q_1^+ and q_2 or q_3 in diagram b. Such exchanges though not shown are implied and can be important. It is these exchanges which make diagram b distinct from diagram a.

Since $(1/m)$ signifies a correlation length between quarks within the nucleon, which is $\approx 1/2$ to 1 fermi, one might expect that $m \sim (1 \text{ to } 2)m_\pi$ with $t_N \approx 1$. This yields taking $\xi \approx 1/2$ to 1 (see Eq. (16.24))

$$\tau(\text{proton}) \approx 10^{29} \text{ to } 10^{34} \text{ years} \ . \tag{16.39}$$

Note that the variation of lifetime by nearly three orders of magnitude is due to variation in the dynamical entity E_{eff}^2, while that by about 1.5 order of magnitude is due to variation in the kinematical quantity ξ^{-6}.

4.6.1 *Decay modes.*

$\Delta F = 0$ transitions: First consider transitions that conserve fermion number ($\Delta F = 0$) but violate baryon and lepton numbers such that the corresponding effective 4-fermion interaction satisfies $\Delta B_q = -\Delta L = \pm 1$. (By convention baryon number of quarks $B_q = 1$. Fermion number F of quarks and leptons (v_e, e^-, μ^-, v_μ) is $+1$ and $F = B_q + L$.) *Since leptons carry integer charges, such transitions are permissible only for integer-charge quarks.* For such transitions branching ratios of various allowed decay modes of proton depend upon whether the matrix element of Fig. 16.1b is comparable in magnitude to that of Fig. 16.1a or not. (Recall that this in turn depends upon whether $f_{qq\pi}$ is damped at high K^2 or not; see discussion in Sec. 4.1 on quark decays.) The estimated branching ratios corresponding to these two alternatives are as follows:

Branching Ratios

$|M (\text{Fig. 16.1a}) | \gg | M (\text{Fig. 16.1b})|$

$p \rightarrow 3v + \pi^+$ ($\sim$ 80 percent)

$\rightarrow 3v + \pi^+ + \pi^+ + \pi^-$ ($\sim$ 5-8 percent)

$\rightarrow 3v + \pi^+ + \pi^+ + \pi^0$

$\rightarrow 3v + \rho^+$ ($\sim$ 10 percent)   (16.40)

$|M \text{ (Fig. 16.1a)}| \approx |M \text{ (Fig. 16.1b)}|$

$$p \to 3\nu + \pi^+ \qquad (\sim 75\text{-}80 \text{ percent})$$
$$\to \nu\nu + e^- + \pi^+ + \pi^+ \qquad (\sim 5\text{-}8 \text{ percent})$$
$$\to 3\nu + \pi^+ + \pi^+ + \pi^- \qquad (\sim 5\text{-}8 \text{ percent}) \qquad (16.41)$$
$$\to 3\nu + \pi^+ + \pi^0 \qquad (\sim 5\text{-}10 \text{ percent})$$

Note that all the foregoing transitions conserve fermion number, which is 3 initially and finally. As a result there are a *minimum of four particles* in final state.

$\Delta F \neq 0$ transitions. Protons may also decay as mentioned before within the unification hypothesis through effective 4-fermion transitions, which violate fermion number (for example, $\Delta F = \pm 4$, $\Delta B_q = \pm 3$, $\Delta L = \pm 1$). (See Fig. 16.4.) Violation of fermion number can arise spontaneously if we permit a gauging of fermion number,[64] putting fermions ψ_L and anti-fermions ψ_L^c in the same multiplet, and if SSB leads to a mixing of gauge particles carrying $F = +2$ with those carrying $F = -2$. (This would be analogous to the case of $\Delta B_q = -\Delta L = \pm 1$, $\Delta F = 0$ transitions that arise through a spontaneous mixing of X and W gauge particles.) Such a mixing can lead to effective 4-fermion transitions of the type:

$$qq \to \bar{q}\bar{l} \text{ or } qqq \to \bar{l} \quad (\Delta F = -4, \Delta B_q = -3, \Delta L = -1), \qquad (16.42)$$

$$qq \to ql \quad (\Delta F = -2, \Delta B_q = -3, \Delta L = +1). \qquad (16.43)$$

These in turn can lead to proton decays of the type:

$$p \to e^+ \pi^0, \bar{\nu}\pi^+ \quad (\Delta F = -4) \qquad (16.44)$$

$$\to \nu\pi^+, e^-\pi^+\pi^+ \quad (\Delta F = -2). \qquad (16.45)$$

Note that the $\Delta F \neq 0$ transitions (16.42) and (16.43) and therefore *the decay modes (16.44) and (16.45) can be operative for integer as well as fractionally charged quarks.* The rates of such decays would depend, of course, upon the masses of the $F = \pm 2$ gauge particles, which are at least as heavy as the $F = 0$ X gauge particles and the mixing (mass)^2.

Thus we see that the two-body decay modes of proton such as $p \to e^+ + \pi^0$ and $p \to \bar{\nu} + \pi^+$, and so on, which involve change of fermion number by four units are permissible for either integer or fractionally charged quarks.

In other words, observation of the decay modes $p \to e^+ + \pi^0$ or $p \to \bar{\nu} + \pi^+$ would imply that baryon and lepton numbers are violated together with fermion number. *Such an observation by itself cannot, however, distinguish between integer versus fractional charges of quarks; nor can it distinguish between spontaneous versus intrinsic violations of* B, L, *and* F. On the other hand, the decay modes of the type $p \to 3\nu + \pi^+$, if seen with a significant branching ratio, would imply that there exist transitions satisfying $\Delta B = -\Delta L \neq 0$, but $\Delta F = 0$. Such a search should be carried out even if the decay mode $p \to e^+ + \pi^0$ is observed.

5. Summary and Questions

In summary, the hypothesis of grand unification provides the scope for several striking discoveries.

5.1 *The Decay of the Proton.*

As stressed earlier, within the grand unification hypothesis we expect that the proton must decay regardless of whether quarks carry integer or fractional charges. For most models its expected lifetime is around 10^{29}–10^{34} years. The observed baryon excess appears to be already a signal that the quantum numbers B, L, and F are not absolute.

Discovery of proton decay would open a new chapter in the history of particle physics in that the traditional belief in the absoluteness of baryon and lepton numbers would topple. It would boost grand unification ideas; it would sharpen focus on the rather fundamental question: Why does nature choose to conserve just one visible charge Q_{em} absolutely?

5.2 *The Absolute Universality of Quarks and Leptons and Their Forces*

The grand unification hypothesis furthermore provides the scope for the intriguing observation that the distinction between quarks and leptons would disappear at appropriately high energies; simultaneously there would be no distinction between the effective coupling constants of the so-called weak, electromagnetic, and strong forces. This, if seen, would be a true new landmark in physics well beyond the stage of electroweak unification. There exist models within which this new landmark could be visible at an energy scale as "low" as 10^4–10^5 GeV. Optimistically therefore, Isabelle, LEP, or their immediate successors may help us discover this new landmark.

While providing the scope for these discoveries, the unification hypothesis also raises several important questions:

1. Does the concept of absolute confinement (that is, confined QCD) hold only in the limit of small "velocities" like Newtonian concepts to be replaced by the concept of partial confinement and liberated QCD at large velocities? Will quark charges appear integral rather than fractional also at large velocities?
2. Is nature intrinsically left-right symmetric? Are observed parity- and CP-violations consequences of only small "velocities" to disappear at large velocities?
3. Do the signals for grand unification manifest themselves at modest energies (10–100) TeV?
4. Is the elementarity of quarks and leptons true only in the limit of small "velocities"? Will they too exhibit eventually a composite character?

One needs high-energy machines and the helping hand from experiments to find the answers to these basic questions.

> Pure logical thinking cannot yield us any information about the real world. All knowledge of reality starts with experience and ends with it.
>
> —A. Einstein

NOTES

1. A. Einstein, see for example *The Meaning of Relativity,* Appendix II.

2. J. C. Pati and A. Salam, "Lepton-Hadron Unification," reported in the Proceedings of the 15th High Energy Physics Conference, held at Batavia, p. 304, Vol. 2, Sept. (1972) by J. D. Bjorken; J. C. Pati and A. Salam, Phys. Rev. **D8,** 1240 (1973).

   J. C. Pati and A. Salam, Phys. Rev. Lett. **31,** 661 (1973), Phys. Rev. **D10,** 275 (1974), and Phys. Lett. **58B,** 333 (1975). H. Georgi and S. L. Glashow, Phys. Rev. Lett. **32,** 438 (1974).

3. S. Weinberg, Phys. Rev. Lett. **13,** 168 (1967); A. Salam in *Elementary Particle Theory, Nobel Symposium,* edited by N. Svartholm (Almqvist, Stockholm, 1968), p. 367; S. L. Glashow, J. Iliopoulos, and L. Maiani, Phys. Rev. **D2,** 1285 (1970).

   For initial attempts at electroweak-unification see J. Schwinger, Ann. Phys. (N.Y.) **2,** 407 (1957); S. L. Glashow, Nucl. Phys. **22,** 57a (1961); A. Salam and J. C. Ward, Nuovo Cim. **11,** 568 (1959), and Phys. Lett. **13,** 168 (1964).

   Ideas on spontaneous breakdown of local symmetries were developed by P. W. Higgs, Phys. Rev. Lett. **12,** 132 (1964); P. Englert and R. Brout, Phys. Rev. Lett. **13,** 321 (1964); G. S. Guralnik, C. R. Hagen, and T. W. B. Kibble, Phys. Rev. Lett. **13,** 585 (1965); P. W. Higgs, Phys. Rev. **145,** 1156 (1966); T. W. B. Kibble, Phys. Rev. **155,** 1554 (1967).

4. Weinberg, op. cit. in n. 3; Salam, ibid.

5. See the references cited in n. 2.

6. Pati and Salam, op. cit. in n. 2.

7. H. Georgi, H. R. Quinn, and S. Weinberg, Phys. Rev. Lett. **33,** 451 (1974).

8. D. Gross and F. Wilczek, Phys. Rev. Lett. **30,** 1343 (1973); H. D. Politzer, Phys. Rev. Lett. **30,** 1343 (1973); H. D. Politzer, Phys. Rev. Lett. **30,** 1346 (1973).

9. Pati and Salam, op. cit. in paragraph 2 of n. 2.

10. J. C. Pati, A. Salam, and J. Strathdee, Il Nuovo Cim. **26A,** 77 (1975); J. C. Pati, Proceedings of the Second Orbis Scientae, Coral Gables, Florida, Jan. 1975 (p. 253).

11. Georgi, Quinn, and Weinberg, op. cit. in n. 7.

12. At least not with effective coupling bigger than $G_{\text{grav}} \times 10^{-8}$.

13. Electric charge may be conserved absolutely if photon is truly massless.

14. See references cited in paragraph 2 of n. 2.

15. R. N. Mohapatra, J. C. Pati, and P. Vinciarelli, Phys. Rev. Lett. **31,** 494 (1973); S. Weinberg, Phys. Rev. Lett. **31,** 494 (1973); H. Fritzsch, M. Gell-Mann, and H. Leutwyler, Phys. Lett. **B47,** 365 (1973).

16. See A. Davidson and J. C. Pati, preprint COO-3533-125, University of Syracuse (1978).

17. H. Fritzsch and P. Minkowski, Ann. Phys., (N.Y.) **93,** 193 (1975); H. Georgi, in Proceedings of Williamsburg Conference of Division of Particles and Fields, Sept. 1974.

18. F. Gürsey and P. Sikivie, Phys. Rev. Lett. **36,** 775 (1976); P. Ramond, Nucl. Phys. **B110,** 214 (1976).

19. Op. cit. in n. 7.

20. V. Elias, J. C. Pati, and A. Salam, Phys. Rev. Lett. **40,** 920 (1978).

21. Ibid.; J. C. Pati and A. Salam, "Axial Chromodynamics," Nucl. Phys. **B150,** 76 (1979).

22. G. Feinberg and J. Sucher, Phys. Rev. Lett. **35,** 1740 (1975); G. Feinberg, B. Lynn, and J. Sucher, Phys. Rev. **D19,** 2231 (1979).

23. Elias, Pati, and Salam, op. cit. in n. 20.

24. V. Elias and S. Rajpoot, Trieste preprint, 1979.

25. Pati and Salam, op. cit. in paragraph 2 of n. 2.

26. T. Goldman and D. Ross, Cal Tech preprint, 1979; see, however, most recent considerations by N. Chang, A. Das, and J. Perez-Mercades [CCNY preprint, 1979], which yield $\tau_p \sim 10^{35}$ years.

27. See for example P. Langacker, G. Segrè, and H. A. Weldon, Phys. Lett. **73B,** 87 (1978). For a review see M. Gell-Mann, P. Ramond, and R. Slansky, Rev. Mod. Phys. **50,** 721 (1978).

28. Pati and Salam, op. cit. in paragraph 2 of n. 2.
29. Davidson and Pati, op. cit. in n. 16.
30. Pati and Salam, op. cit. in paragraph 2 of n. 2.
31. M. Özer and J. C. Pati, "Liberated Color in Leptoproduction I," preprint 79-067, University of Maryland; A. Janah, M. Özer, and J. C. Pati, preprint 79-068, University of Maryland.
32. See references cited in paragraph 1 of n. 2.
33. R. C. Ball et al., Phys. Rev. Lett. **42**, 866 (1979).
34. See J. C. Pati, "The Unification Puzzle," Proceedings of the Seoul Symposium held at Seoul, Sept. 1978, edited by J. Kim, P. Y. Pac and H. S. Song, published by Seoul National University Press 1979, pages 483-595.
35. R. N. Mohapatra, J. C. Pati, and A. Salam, Phys. Rev. **D13**, 1733 (1976).
36. Chang, Das, and Perez-Mercades, op. cit. in n. 26.
37. J. C. Pati, A. Salam, and S. Sakakibara, Phys. Rev. Lett. **36**, 1229 (1976); J. C. Pati, S. Sakakibara, and A. Salam, "Missing Quark Mystery," Trieste preprint IC/75/93.
38. Pati, op. cit. in n. 34, and "Grand Unification and Proton-Stability," Wisconsin Seminar on Proton Stability, preprint 79-078, University of Maryland, and paper in preparation.
39. Pati, op. cit. in n. 34.
40. This is assuming in a perturbative sense that the vertices, W-X mixing, and quark masses inside loop integration have their constant Lagrangian values. There could be some question as to whether this is legitimate in the low K^2 infrared region, in which quark masses acquire large values (nonperturbatively).
41. V. Elias, preprint 79-005, University of Maryland.
42. To be more accurate ξ should be allowed to represent the uncertainty in c_1 also. Thus ξ may conceivably be as small as 1/5 or even 1/10. In other words one cannot pretend to pin down these parameters too sharply until liberated quarks are seen. I proceed with the estimates taking $\xi \simeq 1/2$ to 1.
43. For curiosity, note that if we use PCAC, taking matrix element of $\partial_\mu A_\mu$ between "physical" quark states of mass "m_q", we obtain "$g_{qq\pi}$" $\approx$ ("m_q"/m_N)$g_{NN\pi}$ $\approx$ (5 to 10)$g_{NN\pi}$ or (1/3)$g_{NN\pi}$ depending upon whether we use physical liberated mass or effective constituent mass for "m_q".
44. Note that for multipion emission, it is possible for one of the intermediate quarks to be nearer to its physical mass shell compared to the case of single pion emission for which ("p_q")^2 of intermediate quark is constrained to equal $p_v^2 \approx 0$ (see Fig. 1).
45. See the references cited in n. 42.
46. See the references cited in n. 37.
47. The sources of the net difference between present and previous estimates of quark decay rates is roughly given by the following bookkeeping:

$(\Gamma(q)_{now}/\Gamma(q)_{before}) \approx [(c_1^2)_{now}/c_1^2)_{before}] \times [(c_4^2)_{now}/(c_4^2)_{before}]$

$\times [(f_x^2/m_x^2)^2_{now}/(f_x^2/m_x^2)^2_{before}] \times [(m_q)_{out}/(m_q)^{before}] \times [(g_{qq\pi}^2)_{now}/(g_{qq\pi}^2)_{before}]$

$\approx (10^{-2}) \times (4) \times (10^{-2}) \times (5) \times (10^{-2}$ to $10^{-1}) \approx \frac{1}{2}(10^{-3}$ to $10^{-4})$.

48. D. Cutts et al., Phys. Rev. Lett. **41**, 363 (1978); R. Vidal et al., Phys. Lett. **77B**, 344 (1978).
49. J. A. Goodman et al., University of Maryland preprint, 1978. I thank R. Ellsworth and G. Yodh for discussions on their cosmic ray search.
50. L. B. Okun and Ya. B. Zeldovich, Comments in Nuclear and Particle Physics, 1975.
51. See references cited in n. 2.
52. D. A Kirzhnits and A. D. Linde, Phys. Lett. **42B**, 471 (1972); Ann. Phys. **101**, 195 (1976); S. Weinberg, Phys. Rev. **D9**, 3357 (1974); L. Dolan and R. Jackiw, Phys. Rev. **D9**, 3320 (1974).

53. I am indebted to Garry Steigman for this estimate and for several enlightening discussions. I am also grateful to Vigdor Teplitz and Duane Dicus for helpful conversations. The estimate referred to goes as follows: n_γ = number density of photons $\approx 10^{31.5} \, T_{\text{MeV}}^3/\text{cm}^3$; n_q = number density of quarks in equilibrium = $(9/2) n_f n_\gamma$ (for $m_q \ll kT$), where n_f = number of relevant quark flavors = 2 to 3. The factor (9/2) includes statistical weights due to Fermi versus Bose factors, spin states of quarks and antiquarks versus those of photon, and the factor of 3 for quark color. Taking $n_q \bar{d}^3 \approx$ (1 to 4), we get $\bar{d} \approx$ (0.6 to 1)$10^{-10.6}$ cm/T_{MeV}. For $m_q \gtrsim kT$, the insertion of Boltzmann factor will reduce n_q and thereby increase $\bar{d}$ for a given temperature T, although this will not change the answer much for $kT \gtrsim$ 300 MeV. A sharper estimate allowing for the variation of the quark masses with temperature is being carried out by Dicus, Teplitz, and myself.

54. Somewhat analogous considerations arising in quark stars have been discussed widely in the literature. See for example J. C. Collins and M. J. Perry, Phys. Rev. Lett. **34**, 1353 (1975); B. Freedman and L. McLerran, Phys. Rev. **D17**, 1109 (1978). However, to my knowledge, the transition from quark to nucleon phase, which did take place in the early universe, does not appear to have been discussed in the literature. This discussion is especially pertinent for the fate of "would be" liberated quarks during cosmic evolution.

55. Correspondingly the quark density n_q in the cosmos (see note 53) begins to be comparable to the quark density within a nucleon or a nucleus at this temperature.

56. The heavier strange and charm quarks decaying via normal weak interactions would have essentially the same lifetime ($\lesssim 10^{-10}$ sec) in the quark phase as in the liberated phase, since the Q values for their decays (for example, $s \to u + \pi^-$, and so forth) are the same in both phases. This incidentally implies that almost all the heavier quarks must have decayed out into the lighter ones long before quarkosynthesis occurred ($t_s \sim 10^{-5}$ sec).

57. Note that baryon nonconservation does not arise until cosmological temperatures go below transition temperatures, at which c_4 and c_1 are nonzero, this was stressed in the paper of J. C. Pati and A. Salam, "Lepton-Hadron Unification," in Proceedings of the 1976 Aachen Neutrino Conference, footnote 55. We expect, since $c_1 \sim 100$ MeV, that the corresponding transition temperature depending upon the details of the Higgs scalar complex would lie somewhere between a few hundred MeV and 1 GeV. In the extreme case of c_1 remaining zero till quarkosynthesis temperature $T_s \sim 200$–400 MeV, baryon-number-violating decays of $\Delta F = 0$ variety would be absent altogether in cosmic evolution. Excess of baryons over antibaryons utilizing B nonconservation must in this case arise only through $\Delta F \neq 0$ transitions; these can be operative at much higher temperatures ($T \gg T_s$) even though they too arise spontaneously, since the relevant mass and mixing parameters are in general associated with heavier gauge particles compared to those for the $\Delta F = 0$ case.

58. Pati, op. cit. in n. 34 and n. 38.

59. See n. 57.

60. R. N. Mohapatra and J. C. Pati, Phys. Rev. **D11**, 566 (1975).

61. A. Yu. Ignatiev et al., Phys. Lett. **76B**, 436 (1978), M. Yoshimura, Phys. Rev. Lett. **41**, 281 (1978); B. Toussaint et al., Princeton preprint, 1978; S. Dimopoulos and L. Susskind, SLAC-PUB-2126, 1978; S. Weinberg, preprint HUTP-78/A040, Harvard Univeristy; J. Ellis, M. K. Gaillard, and D. V. Nanopoulos, CERN preprint TH2596, 1978.

62. Note added in proof: Subsequent to the writing of this manuscript, R. Mohapatra and I have calculated the ratio B/γ for this model (Inst. of Physics, Bhubaneswar, India, preprint (1979), to be published). With reasonable values for the Yukawa coupling parameters $(h^2/4\pi) \sim \alpha$ and maximal phases for CP violation, we obtain $(B/\gamma) \sim 10^{-9 \pm 1}$ in good accord with observation.

63. Ibid.

64. Op. cit. in n. 10.

Discussion

Following the Paper by J. C. PATI

Chairman, ALFRED KASTLER

Kastler: The remark has been made that all this may be science fiction. Remember that one century ago, in 1879, the atomic theory of matter was still considered as science fiction, even by such physicists as Ostwald or Ernst Mach in Germany, Pierre Duhem in France, and that in spite of the fact that the Avogadro number and the size of atoms had already been determined by people like O. Loschmidt and J. Van der Waals. Dr. Lipkin has shown (in an earlier chapter) the reasons that today the atomic structure of matter is considered as a physical reality by all physicists. So we may hope that the same may become a reality for this generation's physics.

H. J. Lipkin (Weizmann Institute): The suggestion that it is possible to measure the difference between fractionally charged and integrally charged quark models below color threshold is not correct. This is explained in detail in my chapter in this volume.

All formulations describing the measurement of quark charges in this way, either with parton models or operator product expansions, implicitly assume that the experiment probes the structure of hadrons at very short distances, and that there is no qualitatively new physics at even shorter distances. This appears either in the assumption that the parton is "point like" and has no structure in its coupling to the electromagnetic field, or in the neglect of "higher-order" terms in the operator product expansion.

The existence of a color threshold above the energy of the experiment implies that these basic assumptions are not valid. There is a shorter distance than the distance probed by the experiment at which new physics occurs. At this point all the justification breaks down for either the pointlike parton model or the neglect of higher-order terms in the operator product expansion. One might argue that below color threshold the energy is simply not high enough for these models to work and that there is no reason for any of their predictions to hold. The conclusions from this argument would be that the experimentally observed scaling and other successful results from the parton model indicate that there is no new physics or color threshold at shorter distances.

An alternative point of view is that an intermediate region exists below color threshold where the effects of the new physics at smaller distances are not observed because they involve operators that are not color singlets whose matrix elements all vanish in the subspace of color singlet states. Thus the results of the parton model and zero-order operator product expansion appear to hold in this energy range but will break down at color threshold. The only measurable observables in this domain are matrix elements of color-singlet operators between color-singlet states, such as the color average of the quark charge.

Which of these two alternatives is correct cannot be decided without a detailed dynamical theory. Either 1) the new physics above color threshold shows itself below threshold in a way that destroys the simple parton model or operator product expansion, and no results hold, or 2) the new physics is hidden and the simple models hold but only for color-singlet observables. The third alternative, which allows both the simple models and the observability of color-octet components of the electromagnetic current is untenable. This approach has internal contradictions that are pointed out in my chapter.

Pati: I see that you are making this comment under the assumption that the two currents are separated by the propagator of a heavy color-octet intermediate state. This is indeed not the case. Although it is true that the physical liberated masses of color-nonsinglet states are heavy $\gg 1$ GeV, following renormalization group approach, the effective running masses of quarks and gluons appropriate to short distance operator product-expansion-approach are extremely light ($\lesssim 10$ to 100 MeV for running momentum $\gtrsim 1$ GeV). Thus for two-photon-processes *as long as the* P_T *transferred to the hadrons is large enough* ($\gtrsim 2$ GeV, say), one needs to use only small effective masses for quarks as well as gluons. This holds regardless of whether the available energy is below or above color threshold, and this is why the two-photon processes, being bilinear in the electromagnetic current, do provide a distinction between integer versus fractional quark charges even below color theshold as long as the constraint of large $P_T \gtrsim 2$ GeV is satisfied. (The relevant calculations are being done by two of my students, Mr. Janah and Mr. Özer, at Maryland.)

H. Pagels (Rockefeller U.): I would like first to make a comment: Many people, including yourself, refer to the possibility that quarks and leptons are composite, and, while this possibility is not ruled out, I am always impressed by the fact that the electron has a standard $g - 2$ and a proton does not. It has an anomalous moment. The electron thus appears to be an elementary structure and the proton is not. That does not allow for the possibility that there are substructures. Now a question: The actual chromodynamics that you present is a theory that seems to me to be ridden with anomalies and is not renormalizable — would you comment on this?

Pati: First on the question of electron not showing any structure on the basis of $g - 2$ and other analogous measurements, these are indeed very striking facts, no question about it, but my attitude at present toward the question of compositeness of leptons as well as of quarks is this. One is seeing already a very revolutionary situation with quarks themselves. The quark-gluon-dynamics with its asymptotically free short-distance behavior and the very strong confining long-distance character is so peculiar that we would have had a hard time imagining it just about a decade ago. Its partial explanation has evolved only over the last few years through the development of QCD, but we still do not fully understand it. Likewise, I believe that there can be another yet completely unfamiliar force, which is perhaps extremely strong and extremely *short range* (range $< 10^{-18}$ $- 10^{-19}$ centimeters, say), which is responsible for binding the so-called preons to make composite quarks and leptons. The quarks and leptons must themselves be "neutral" with respect to the charges associated with this new force, whereas their constituents are not. Such a picture can be compatible with the present "pointlike" behavior of leptons as well as of quarks. One has ideas about this new force, but I am not yet prepared to talk about it.

Now, on the question of anomalies for chiral gauge structures such as $[SU(4)]^4$, which contains chiral chromodynamics $SU(3)'_L \times SU(3)'_R$, it is clear that the group itself is not anomaly-free, nor are the gauge interactions on the space of the basic fermions ($F_{L,R}$), but the full theory is anomaly-free due to the presence of the so-called mirror fermions ($F^m_{L,R}$). The interactions symbolically have the form

$$V_A (\overline{F}_L F_L + \overline{F}_R F_R) + V_B (\overline{F}_R F_R + \overline{F}^m_L F^m_L) ,$$

which as you see is anomaly-free.

17. SUMMARY

Murray Gell-Mann

I am honored to have been chosen to contribute the last of these essays, presented originally as lectures in honor of Albert Einstein's centenary; and I was delighted that the symposium was held in the historic city of Jerusalem, known to all readers of the Bible as the citadel of the Jebusites.

In the early part of this conference much attention was devoted to Einstein's political, philosophical, and musical hobbies. (I was disappointed that his hobby of sailing was neglected; there was no regatta and not even a cocktail party at the Jerusalem Yacht Club.) During the last few days of the symposium we turned to his profession, theoretical physics, of which he was such a magnificent practitioner, the greatest since Newton. It is gratifying that the praise heaped upon Einstein for his scientific work has gone to someone who really deserved it.

Contributors to this volume, in a set of brilliant essays, have clarified many aspects of our present understanding of fundamental physics.

We have heard about the present flowering of research on Einstein's utterly nontrivial theory of gravitation (which he liked to call by the pet name of "general relativity" after its gauge group), and then about the continuing search for unity in the description of physical forces. That search is carried out nowadays within the framework of quantum field theory, and with the inclusion of strong and weak interactions, as well as hitherto unobserved ones, in addition to electromagnetism and gravitation. Such theories have a classical limit of sorts, and can therefore be compared with Einstein's pioneering attempts at classical unification. But his efforts were based on gravitation and electromagnetism alone, and were not successful. His proposed equations are nevertheless represented, I am told, on the reverse of the medal struck for this conference. It is a pity that out of all the equations written down by Einstein, the medal commemorates ones that are probably wrong.

I shall not try to summarize the conference, not even that part devoted to elementary particle physics, but rather just make some observations about the search for unification of forces. This is a remarkable moment in the history of our field, when two of the great problems of our generation in particle physics seem to have been largely solved. QCD may have given us the long-sought description of the strong interaction. The $SU_2 \times U_1$ gauge theory with symmetry violation $\Delta I^{\text{weak}} = \frac{1}{2}$, to which Glashow, Weinberg, and Salam and Ward have made important contributions, seems to have provided an excellent combined description (at least at moderate energies) of the electromagnetic and weak interactions. The quarks, the gluons, the intermediate bosons, and some sort of spontaneous violation of symmetry have all become respectable, which they were not when we first discussed them years ago, and after further experimental and theoretical work the ideas of exact SU_3^{color} and spontaneously broken $SU_2 \times U_1$ may be fully confirmed.

The true nature of symmetry violation is still not understood. A particularly striking case is that of CP violation, a very important experimental discovery now fifteen years old. But the logic of all symmetry violation, even if it is all spontaneous, still escapes us.

The pattern of the quarks and leptons is still mysterious, including the apparent replication of "families." There are many other fundamental puzzles. Nevertheless, the apparent successes have led theoreticians to plunge into a search for a unified description of nearly all the forces, based on broken Yang-Mills theory, or even a synthesis of all the forces, including Einsteinian gravitation.

The attempts at a grand synthesis of everything involve an effective unification around the Planck mass $\sim 2 \times 10^{19}$ GeV. It requires enormous faith, of course, to believe that our present physical ideas can be extrapolated to an energy some 10^{18} higher than energies investigated experimentally today. Nevertheless, it is remarkable that a theory can be exhibited that has many of the features we would expect of a completely unified theory, including the property of being a generalization of Einstein's theory of gravitation. The most impressive example we have is extended $N = 8$ supergravity. The list of particle states that we can read out of the field equations or the representation of the superalgebra has many virtues. Particles of spin > 2 are lacking. There is one and only one graviton of spin 2. There are eight gravitinos of spin 3/2, just right for swallowing the eight Goldstone fermions that arise from the spontaneous violation of $N = 8$ supersymmetry, required in order to avoid the unobserved degeneracy of fermions and bosons. There are $\frac{1}{2}N(N - 1) = 28$ initially massless spin 1 bosons, perfect for gauging the symmetry group SO_8.

Unfortunately SO_8 is too small to contain $SU_3^{color} \times SU_2 \times U_1$; there are too few spin 1 bosons to include the charged intermediate bosons of the weak interaction. Likewise, although there are $(1/6) N(N - 1)(N - 2) = 56$ Majorana spin (1/2) fermions, the list cannot be made to include all the flavors of quark and lepton that we perceive as elementary. There are $(1/24) N(N - 1)(N - 2)(N - 3) = 70$ spinless bosons, but they probably do not enter the Lagrangian in such a way as to behave like Higgs bosons and break the various symmetries spontaneously.

There are some features of the list that are worth remembering. The "spin 3/2 gap" is filled; it was certainly odd that before supergravity we dealt with elementary objects of spin 2, 1, 1/2, and 0, but not 3/2. The spin 1/2 fermions do not lie in a basic representation of the gauge group; rather it is the gravitinos that do so. There are elementary spin 0 bosons present even though they probably do not act as Higgs fields.

If the $N = 8$ supergravity theory is to describe nature, then the list of elementary fields of the theory must have only an indirect relation to the elementary particle spectrum as we perceive it at the very low energies (in comparison to 10^{19} GeV) of today's experiments. But whether or not $N = 8$ supergravity works, we must take into account that there are all sorts of ways in which the list of particles read off from the equations of a quantum field theory may be enlarged. Of course, there are ordinary bound states. There are Goldstone bosons and fermions, which can either be swallowed or receive a mass as a result of perturbations. There may be solitons (such as magnetic monopoles) or other particlelike solutions and bound states involving them. A recent idea is that auxiliary field components may propagate and yield particle states as a result of radiative corrections. In any quantum field theory, we must understand to what extent all these secondary objects can masquerade at present energies as elementary particles.

Now let me turn to the only slightly more modest program of unifying color and flavor interactions in a broken Yang–Mills theory with a simple gauge group G (or perhaps the square, cube, and so forth, of a simple group) in order to have a single coupling parameter g at very high energy. Here the effective unification may take place at a mass somewhat lower than the Planck mass, but still at least of the order of 10^{14} or 10^{15} GeV. It still requires a lot of GUTS to believe that our present ideas are adequate for such an extrapolation.

Besides $SU_3^c \times SU_2 \times U_1$, a decomposition of G may yield other factors, further color or flavor interactions that may be active at energies far below the unification mass. The additional color factors, if any, would be exactly conserved and all known particles would be singlets with respect to them; they would correspond to Yang–Mills theories QCD', QCD", ..., presumably having the confinement property that QCD is thought to have, and their renormalization-group-invariant masses Λ', Λ'', ... would presumably be higher than masses investigated today. The additional flavor factors, if any, could include a group relating the fermion "families" to one another, a "family group."

In the smallest scheme, based on SU_5, there are no such additional flavor or color factors. The left-handed spin 1/2 fermions of each family are assigned to **5̄ + 10**, a lopsided arrangement in which there is no relation between left-handed particles and left-handed antiparticles, i.e., no C operator. The replication of families is, of course, unexplained.

If elementary Higgs bosons are used to give spontaneous symmetry breaking, one usually takes an adjoint **24** of SU_5 to break the group down to $SU_3^c \times SU_2 \times U_1$ without producing fermion masses and then at least a **5** and a **5̄** of SU_5 to break $SU_2 \times U_1$ down to U_1^{e-m}. Actually, **5 + 5̄** gives not only the successful rule $m_b = m_\tau$ (after renormalization) but also the less successful ones $m_d = m_e$ and $m_s = m_\mu$ (after renormalization) and so additional Higgs bosons may be required.

These elementary spinless fields are not only introduced ad hoc in arbitrary representations. They also have arbitrary couplings to one another and further arbitrary couplings to the fermions. Thus the fermion masses are not really calculable (except possibly for the sum rules quoted). The scale of the masses of the intermediate bosons $X^\pm$ and Z^0 for the weak interactions (I hope the temporary aberration of calling them $W^\pm$ and Z^0 does not last much longer) is determined by another arbitrary parameter, which must be dialed to the remarkable value of 10^{-13} or so (10^2 GeV/10^{15} GeV). Furthermore, this small number is completely independent of another small number, the ratio of Λ_{QCD} to the unification mass, around 10^{-15} (1 GeV/10^{15} GeV), which is determined by exp$(- \text{const}/x)$, where x is the common coupling constant near the unification mass ($\sim 1/40$ for SU_5 with three families).

The theory of spontaneous symmetry-breaking with elementary Higgs bosons is indeed renormalizable, as brilliantly proved by 't Hooft, but renormalizability is not enough for fermion and intermediate boson masses; we need calculability. To put it another way, dialing all of these parameters is what is called these days an unnatural act.

I have preached for twenty-five years (true despite my youthful appearance!) that symmetries should become exact in the limit of high energies and involve essentially ratios of unrenormalized quantities. Adjusting renormalized quantities to special small values or to accidentally almost equal values (not reflecting a high-energy symmetry) has never been good form. Today's notion of "naturalness" is not much different. However, in a theory of spontaneous symmetry violation by elementary Higgs bosons there are always some small quantities that should be calculable but are subject to arbitrary renormalization in the theory, and in fact nearly everyone is dialing some parameter to a magic value, invoking a private suspension of the requirement that the operation be "natural" and reserving the epithet "unnatural" for what everyone else is doing.

Now the curious assignment of left-handed fermions of a single family to the reducible representation $\bar{5} + 10$ of SU_5 permits the existence of a conservation law that is exact (except possibly for tiny effects of weak instantons). The trick is one that Ramond, Slansky, and I exploited in our encyclopedia of ways in which the proton could be stabilized. Now, of course, the very nice recent work on baryon creation in the universe gives for the first time (provided CP violation can be made "hard" enough) a good reason why the proton should decay, and one probably does not want to stabilize it. The trick itself, which is an old one, remains very important. If we have a gauged quantum number called X and an ungauged quantum number called Z, both exactly conserved by the Lagrangian, then a simultaneous spontaneous violation of both, with a linear combination exactly conserved in the global sense, gets rid of the massless Goldstone boson that would be associated with the exact conservation of Z by feeding it to the massless gauge boson that would be associated with the exact conservation of X. In the SU_5 scheme the quantum number that is $-3/5$ for $\bar{5}$ and $1/5$ for 10, thus commuting with SU_5, can be used for Z and the generator of SU_5 that is $4/15$ for $\bar{d}$ and $-2/5$ for e^- and $\bar{\nu}$ can be used for X. (In an explicit Higgs boson picture, the Lagrangian must be arranged to conserve Z.) The exact global conservation of $n_{\text{baryons}} - n_{\text{leptons}} = X + Z$ can then result. In the proton decay, $p \to e^+ + \pi^+ + \pi^-$ is allowed, for example, while $p \to e^- + \pi^+ + \pi^+$ is forbidden.

Now although the SU_5 scheme omits the left-handed antineutrino and therefore prevents the neutrino from acquiring a Dirac mass, there is still the possibility of a Majorana mass for the left-handed neutrino, except that the conservation of $n_{\text{baryons}} - n_{\text{leptons}}$ forbids it.

Larger groups than SU_5 have been studied as candidates for G in a unified Yang–Mills theory. Even without explaining the replication of families, one can at least make the theory less lopsided by using SO_{10}, with each family of left-handed fermions assigned to the spinorial representation 16, which gives $10 + \bar{5} + 1$ of SU_5, or by using E_6, with the left-handed fermions of each family in a 27, which gives $10 + \bar{5} + 1 + 5 + \bar{5} + 1$ of SU_5. Suppose the symmetry is broken down to SU_5, with a huge SU_5 singlet mass for the fermions wherever that is possible. (Here that is the same as requiring that any fermion mass invariant under $SU_3^c \times SU_2 \times U_1$ be huge.) Then for the 16 of SO_{10} the SU_5 singlet is carried away to high mass, and in the 27 of E_6 the two SU_5 singlets mate with each other and so do the extra 5 and $\bar{5}$. In each case just the $\bar{5}$ and 10 are left at low mass.

Ramond, Slansky, and I (and no doubt others) have looked at these cases with an eye to what happens to the neutrino mass. Let us take the SO_{10} example. The SU_5 singlet in the 16 representation is the missing left-handed antineutrino, here elevated to a very high mass by a Majorana mass term. For a single family, the fermion masses belong to $(16)_{\text{symm}}^2 = 10 + 126$. The SU_5 singlet component in 126 is precisely that Majorana mass term. Meanwhile, the low-lying fermions would acquire their masses, in violation of SU_5, in fact in violation of $(SU_2 \times U_1)^{\text{weak}}$ by $\Delta I^{\text{weak}} = \frac{1}{2}$. To the extent that these masses come from the 10 term alone, we would have not only relations $m_b = m_\tau$, etc., but also $m_{\text{Dirac}}(\nu_\tau) = m_t$, etc.; if there are contributions from 126 as well (and from 120, which can enter when more than one family is included) we would still expect the Dirac mass of each neutrino to be of the same order of magnitude as the mass, adjusted for renormalization, of the corresponding quark of charge $+2/3$. The observed left-handed neutrino would then acquire an effective Majorana mass of roughly $m_{\text{Dirac}}^2/m_{\text{Majorana}}((\bar{\nu})_L)$. Say we guess a Dirac mass of 30 GeV for ν_τ; then a Majorana mass of $\sim 10^{12}$ GeV for $(\bar{\nu}_\tau)_L$ would give about 1 eV for the effective mass of $(\nu_\tau)_L$, providing a significant contribution to the gravitational closure of clusters of galaxies and pushing the universe a bit closer to asymptotic flatness or closure. A much lower Majorana mass for $(\bar{\nu}_\tau)_L$ would cause difficulties with cosmology; a higher mass is possible but would lead to negligible cosmic effects.

It is not known whether the masslessness or near-masslessness of each left-handed neutrino is caused by symmetry principles (exact or nearly exact) that forbid both a Majorana mass for it and a Dirac mass connecting it to a left-handed antineutrino or by the forbiddenness merely of the Majorana mass for ν_L coupled with a guarantee of a huge Majorana mass for $(\bar{\nu})_L$, with the Dirac mass of normal size. With the second point of view, certainly, neutrino masses near the cosmological limit of a few eV are not unlikely; experimentally, neutrino mass matrix elements would show up most easily in the form of neutrino oscillations from one species to another.

Of course it is not only the smallness or vanishing of the neutrino masses that is mysterious. As I mentioned before, it is crucial to understand why the intermediate boson masses are only about 10^2 times larger than Λ_{QCD} when both are so tiny compared to the scale of unification. Likewise, we need to understand why the fermion masses are in the same general range, and why, for each nonzero value of fermion electric charge, the fermion mass matrix seems to be dominated by one diagonal matrix element, with the other matrix elements looking like first and second corrections.

Since in an elementary Higgs boson scheme all these questions are not only unanswered but probably unanswerable, we should look for some other insights into symmetry violation. Of course if there were symmetry principles governing the existence, the self-coupling, and the Yukawa couplings of elementary spinless fields, as in supersymmetric theories, then most of the arbitrariness would be removed, but unfortunately supersymmetry does not seem so far to lend itself very well to spontaneous symmetry violation of the kind required. (There is a great deal of interesting work on this subject by Fayet.)

We are tempted to look at hypothetical dynamical symmetry-breaking as a way of explaining all or part of the spontaneous violation in a unified Yang–Mills theory. Of course spinless elementary fields may be present, playing a different role from that of the arbitrary elementary Higgs bosons, but for simplicity let us consider just spin 1/2 fermions and spin 1 gauge bosons, with the spin 0 Higgs states arising dynamically.

In such a theory the replication of families poses difficult problems. If a representation of G were repeated two or three or more times, then there would be an ungauged non-Abelian symmetry group SU_2, SU_3, ... commuting with G and with the Lagrangian. Spontaneous violation of this unitary symmetry would result either in the appearance of true Nambu–Goldstone bosons or else in the global conservation of an isomorphic symmetry by means of the trick discussed earlier in the Abelian case of "$X + Z$." Since neither of these situations fits the facts, we conclude that in the type of theory under discussion the fermions must lie in an irreducible representation of G or possibly in a sum of distinct irreducible representations (leading to global conservation of Abelian quantities only). Let us assume for simplicity that we have an irreducible representation (which tends, of course, to be huge).

Steven Weinberg and others have shown how under these conditions dynamical symmetry-breaking would work. There would be nonzero vacuum expectation values of operators violating symmetry under G. If $H \supset G$ is the symmetry group of the kinetic energy, then we can define H_1 and G_1 to be the subgroups of H and G, respectively, that are left invariant by all the vacuum expected values. Then the generators of G_1 would correspond to exact conservation laws and massless gauge bosons, those of G/G_1 to massive gauge bosons that have swallowed dynamical Higgs bosons, those of H_1/G_1 to approximate conservation laws, and those of $(H/H_1)/(G/G_1)$ to modified Nambu–Goldstone bosons that are permitted to have nonzero mass because of the gauge interaction. The pion would belong to this last category.

Let us think of QCD in the context of a unified Yang–Mills theory with $G \supset SU_3^c \times SU_2 \times U_1$. We expect a nonzero vacuum expected value for all sorts of QCD operators, including

$$\Sigma_{A=1}^{8} G_{A\mu\nu} G_{A\mu\nu}$$

and also $\bar{q}q$ for each flavor. The latter is the basis of PCAC and we have known for many years that $\langle \bar{u}u \rangle \approx \langle \bar{d}d \rangle \approx \langle \bar{s}s \rangle \approx \Lambda^3$; for heavier quark flavors, the approximate equality may be violated. The gluon operator violates G, of course, while preserving $SU_3^c \times SU_2 \times U_1$, and the quark operators would break $SU_2 \times U_1$ down to U_1^{e-m}.

Now Susskind and collaborators point out that if we stick to just QCD, then the pions would be swallowed by $X^\pm$ and Z^0 in the case of dynamical symmetry-breaking and the resulting intermediate boson masses would be of the order of $e\Lambda$, more than 1,000 times too small. They suggest therefore that one make use of the possibility that we mentioned earlier, of at least one additional factor for the color group, leading to a QCD´, with a renormalization-group-invariant mass Λ', which is supposed to be $\gg \Lambda$. Then nonzero vacuum expected values for $\bar{f}'f'$, where the fermion fields f' carry primed color, would lift the $X^\pm$ and Z^0 masses to around $e\Lambda'$, so that Λ' should be around 10^3 GeV. Just as Λ divided by the unification mass is around $\exp(-\text{const}/x)$, where x is the coupling strength around the unification mass and the constant is calculable from group theory, so a different group-theoretical situation is supposed to yield a different constant and the needed value of Λ'. The intermediate boson mass scale would then be calculable and not the result of adjusting an arbitrary parameter.

The pions (with a small admixture of π') would now be observable modified Nambu–Goldstone bosons while some π' (with a small admixture of π) would be eaten by the intermediate bosons.

"Ultraviolet" fermion masses would not exist in such a theory, but instead the very high "soft" masses of primed fermions ($\sim \Lambda'$ and soft on the scale of Λ') would be shared, through radiative corrections, with the observed fermions, giving them calculable masses that would act at usual energies like "ultraviolet" masses. Large representations for fermion masses would no longer frighten us because we would not be introducing elementary Higgs bosons to produce them. (We recall the **126** representation of SO_{10}; the analog for E_6 is **351**´; for larger groups the relevant representations are even bigger.)

These ideas of hypothetical dynamical symmetry-breaking are not free of difficulties. We must see whether the very large symmetry violations such as those that break off color (and primed color!) from G can be explained by the same kind of mechanism that we have just discussed for violating $SU_2 \times U_1$ once color and primed color are separated. Can the pattern of symmetry violation be made self-consistent? We must also see whether there is any combination of group G and fermion representation for which the evolution of the coupling constants between moderate energies and the unification mass comes out right. Do the coupling constants become large again at very high energies?

The two kinds of mechanism we have discussed for fermion mass generation have important similarities and important differences. In each case, vacuum expected values reduce the symmetry of G and of H and it may be possible to consider the reduction of symmetry in stages. Consider for simplicity the first stage, with a vacuum expectation value reducing a symmetry group F to a subgroup F_1. The operator in question belongs to a particular component of a irreducible representation of F since we are barring accidentally comparable contributions of different representations as "unnatural." By definition F_1 leaves the component invariant, but under the action of the operators of the quotient space F/F_1 it is turned into a set of essentially equivalent components called an orbit. The irreducible representation, the subgroup, and the orbit describe the symmetry violation.

We may recall that the spinless fields of a nonlinear σ-model span a quotient space F/F_1 and that the nonlinear σ-model may be obtained by applying constraint conditions to the fields of a linear σ-model based on any representation of F that contains a singlet of F_1. To digress further, I may point out that space-time is a quotient space, given by the Poincare group divided by the Lorentz group, and superspace is another. Quotient spaces and orbits are of transcendent importance.

Now when fermion masses are generated by the elementary Higgs boson mechanism the vacuum expected value of a Higgs boson field belongs to an orbit (of an irreducible representation) invariant under F_1; and the Yukawa coupling, supposed to be rather weak, gives the fermion mass term in lowest order so that it too belongs to that orbit.

In the hypothetical dynamical scheme for symmetry violation, the vacuum expected value of some $\bar{\psi}\psi$, for example, leads to a "soft" mass for certain fermions and then indirectly, through radiative corrections, to a simulated "ultraviolet" mass for lighter fermions, which need not transform in the same way.

In the elementary Higgs boson scheme, then, orbit theory is to be applied directly to the fermion masses, which can be described by successive reductions of symmetry; while in the hypothetical dynamical symmetry-breaking scheme, it is the "condensations" (nonzero vacuum expected values) and not the simulated "ultraviolet" fermion masses that are described by orbit theory and a succession of symmetry violations.

In the attempts to construct a unified Yang-Mills theory, each scheme seems to offer some advantages. While the idea of dynamical symmetry violation permits, in principle, calculation of physical quantities and relieves us from the necessity of introducing vast numbers of spinless fields and couplings, the explicit Higgs boson theory, with its various small parameters, has the virtue that the selection rule $\Delta I^{\text{weak}} = 1/2$ can be utilized *in lowest order* to give the mass ratio of charged and neutral intermediate bosons. Also, it is easier, in the explicit scheme, to obtain very high masses ($\Delta I^{\text{weak}} = 0$) for unwanted fermion states in zeroth order, ordinary fermion masses ($\Delta I^{\text{weak}} = 1/2$) in first order, and possibly left-handed neutrino masses ($\Delta I^{\text{weak}} = 1$) in second order.

I am not sure that we have today a real understanding of how symmetry violation takes place, and perhaps a future successful theory will have some of the features of each of the schemes we have discussed. Nor do I believe that we know what is the right gauge group, if any, for a unified Yang-Mills theory.

I conjecture, though, that quantum field theory will permit us to extrapolate our fundamental ideas to energies much higher than our experiments have reached today, and that our notions of unification, while daring, are not useless or silly. I think we will have to know a great deal more about the nature of symmetry violation and about the relation between fields in the Lagrangian and states in the spectrum before we can really understand or predict the phenomena that occur when the energy is increased by many powers of 10 over today's values.

In closing, I should like to point out two lessons we can learn from Einstein's work and from observation of difficulties that arise in our research in theoretical physics. One is that while cultivating successful ideas in physical theory we must be careful to prune away any unnecessary intellectual foliage that accompanies them, assumptions that we accept out of laziness or vested interest but that we do not require for success. (It may in fact be the tendency to hang on to such assumptions that makes it harder for us to do good theoretical work as we grow older.) When Einstein showed that absolute simultaneity was an unnecessary concept and later when he did the same for flat space, he was providing dramatic examples of what I mean; a somewhat more prosaic example was his correct choice of a tensor theory for gravitation when other physicists thought themselves restricted to a scalar or vector theory. The second lesson is to take very seriously ideas that work and see if they can usefully be carried much further than the original proponent suggested. We recall that in 1905, in a single volume of the *Annalen der Physik*, Einstein published three great articles. In one of them he took the quantum seriously and explained the photoelectric effect. In another he took seriously the physical existence of molecules and showed how to measure the number of them in a unit volume. In the third he took seriously the symmetry group of Maxwell's equations and discovered special relativity.

Acknowledgment

This review was partly contributed within the framework of the Caltech-Tel Aviv collaboration sponsored by the Binational U.S. Israel Science Foundation, contract 2033/79.

Chairman's Concluding Remarks

ALFRED KASTLER

Let me on behalf of all the foreign guests invited to the symposium express our deepest thanks to the organizing committee. The Einstein centennial symposium will remain for us an unforgetable experience. It succeeded in establishing a fine equilibrium between the personal tribute to Einstein, to the physics of his time, and to the physics of our time. In one of the recent issues of *Physics Today*, I found this statement made by Millikan: "Physicists are more interesting than physics."

I believe Einstein in his modesty would not have approved it. He lived for his work. In an interview to Bernard Cohen a few weeks before he died, he said that the birth of his own ideas in his mind was for him a mystery. It will remain probably for his historians. One of the highlights of our meeting was the session dedicated to the personal reminiscences of those who had approached Einstein and worked with him. We had the impression that he was among us with his fascinating personality. What impressed me profoundly was that this man who had such an acute sense of logical analysis kept, in front of the wonders of nature, the admirative astonishment of his childhood. This astonishment at looking on the magnetic needle, this feeling which became for the adult Einstein the religious feeling in front of the mystery of the cosmos. This feeling expressed also in the music of Mozart (which he liked so much) and in the painting of Chagall.

A week before he died, Albert Einstein signed what is known as the Russell-Einstein manifesto: "As human beings we appeal to human beings. . . ." It became his spiritual testament. Alas, the nuclear arms race goes on today on a monstrous scale and threatens mankind more and more. It is our duty as scientists and as citizens to renew this appeal until it is taken seriously by responsible governments.

For Albert Einstein there was a principle of equivalence between his feeling of belonging to the Jewish people and his feeling of being a world citizen and pacifist. Let us take it as a symbol that during the week of our symposium here in Jerusalem, a decisive step towards peace has been made and let us hope that this step will be followed by other steps for the benefit of mankind.

List of Participants

Alexander, Gideon, Tel-Aviv University, Tel-Aviv, Israel
Ashery, Daniel, Tel-Aviv University, Tel-Aviv, Israel
Auerbach, Naftali, Tel-Aviv University, Tel-Aviv, Israel
Avni, Yoram, Weizmann Institute, Rehovot, Israel
Bacher, Robert, California Institute of Technology, Pasadena, California, USA
Bacry, Henry, University of Aix, Marseilles, France
Banks, Thomas, Tel-Aviv University, Tel-Aviv, Israel
Bekenstein, Jacob D., Ben-Gurion University, Beer-Sheba, Israel
Bell, Daniel, Harvard University, Cambridge, Massachusetts, USA
Benary, Odette, Tel-Aviv University, Tel-Aviv, Israel
Bergmann, Peter, Syracuse University, Syracuse, New York, USA
Berlin, Isaiah, Oxford University, Oxford, England
Berlad, Gideon, Technion, Haifa, Israel
Bialynicki-Birula, Iwo, University of Warsaw, Warsaw, Poland
Biedenharn, Larry C., Duke University, Durham, North Carolina, USA
Bishari, Mordechai, Weizmann Institute, Rehovot, Israel
Blaugrund, Abraham, Weizmann Institute, Rehovot, Israel
Bleuler, Konrad, University of Bonn, Bonn, Federal Republic of Germany
Broude, Cyril, Weizmann Institute, Rehovot, Israel
Carmeli, Moshe, Ben-Gurion University, Beer-Sheba, Israel
Casher, Aharon, Tel-Aviv University, Tel-Aviv, Israel
Cohen, Robert S., Boston University, Boston, Massachusetts, USA
Dar, Arnon, Technion, Haifa, Israel

Dashen, Roger, The Institute for Advanced Studies, Princeton, New Jersey, USA
d'Espagnat, Bernard, Université de Paris-Sud, Paris, France
De Groot, Sybren R., Royal Netherlands Academy of Arts & Sciences, The Netherlands
Dirac, Paul A. M., Florida State University, Tallahassee, Florida, USA
Dothan, Yossef, Tel-Aviv University, Tel-Aviv, Israel
Doty, Paul, Harvard University, Cambridge, Massachusetts, USA
Dresden, Max, Institute for Theoretical Physics, State University of New York, New York, USA
Dvoretzky, Aryeh, Israel Academy of Sciences & Humanities, Israel
Eckstein, Shulamit, Technion, Haifa, Israel
Eckstein, Yakov, Technion, Haifa, Israel
Eilam, Gad, Technion, Haifa, Israel
Eisenberg, Yehuda, Tel-Aviv University, Tel-Aviv, Israel
Elkana, Yehuda, Hebrew University, Jerusalem, Israel
Erikson, Erik H., Harvard University, Cambridge, Massachusetts, USA
Ezrahi, Yaron, Hebrew University, Jerusalem, Israel
Feld, Bernard T., Massachusetts Institute of Technology, Cambridge, Massachusetts, USA
Forman, Paul, Smithsonian Institution, Washington, D.C., USA
Fraenkel, Zeev, Weizmann Institute, Rehovot, Israel
Freedman, Daniel Z., State University of New York, New York, USA
Freund, Peter G. O., Enrico Fermi Institute, University of Chicago, Illinois, USA
Gatto, Renaldo, University of Geneva, Geneva, Switzerland
Geheniau, Jules, Académie Royale des Sciences, des Lettres et des Beaux Arts de Belgique, Bruxelles, Belgium
Gell-Mann, Murray, California Institute of

Technology, Pasadena, California, USA
Glashow, Sheldon L., Harvard University, Cambridge, Massachusetts, USA
Goldhaber, Gershon, University of California, Berkeley, California, USA
Goldhaber, Maurice, Brookhaven National Laboratory, Upton, New York, USA
Goldman, Yizhak, Weizmann Institute, Rehovot, Israel
Goldring, Gvirol, Weizmann Institute, Rehovot, Israel
Gomberoff, Luis, Tel-Aviv University, Tel-Aviv, Israel
Gotsman, Asher, Tel-Aviv University, Tel-Aviv, Israel
Gowing, Margaret, Oxford University, Oxford, England
Graham, Loren R., Columbia University, New York, New York, USA
Gronau, Michael, Technion, Haifa, Israel
Gürsey, Feza, Yale University, New Haven, Connecticut, USA
Harari, Haim, Weizmann Institute, Rehovot, Israel
Harkabi, Yehoshafat, Hebrew University, Jerusalem, Israel
Harman, Abraham, Hebrew University, Jerusalem, Israel
Hass, Michael, Weizmann Institute, Rehovot, Israel
Hoffmann, Banesh, Queens College, CUNY, Flushing, New York, USA
Holton, Gerald, Harvard University, Cambridge, Massachusetts, USA
Horn, David, Tel-Aviv University, Tel-Aviv, Israel
Horowitz, Larry, Tel-Aviv University, Tel-Aviv, Israel
Jakobson, Roman, Harvard University, Cambridge, Massachusetts, USA
Jammer, Max, Bar-Ilan University, Ramat-Gan, Israel
Janner, Aloysio, Katholieke Universiteit Nijmegen, The Netherlands
Joffe, Anatole, University of Montreal, Montreal, Canada
Karshon, Uri, Weizmann Institute, Rehovot, Israel
Kastler, Alfred, Académie des Sciences, Institut de France
Kelson, Itzhak, Tel-Aviv University, Tel-Aviv, Israel
Kirson, Michael, Weizmann Institute, Rehovot, Israel

Klein, Martin J., Yale University, New Haven, Connecticut, USA
Kovetz, Attay, Tel-Aviv University, Tel-Aviv, Israel
Kugler, Moshe, Weizmann Institute, Rehovot, Israel
Kuper, Charles, Technion, Haifa, Israel
Lacoste, Paul, University of Montreal, Montreal, Canada
Leibowitz, Elhanan, Ben-Gurion University, Beer-Sheba, Israel
Lindenbaum, J., City College of New York, New York, New York, USA
Lipkin, Harry, Weizmann Institute, Rehovot, Israel
Maier-Leibnitz, Heinz, Deutsche Forschungsgemeinschaft, Bonn, Federal Republic of Germany
Malin, Shimon, Ben-Gurion University, Beer-Sheba, Israel
Mandula, Jeffry, Massachusetts Institute of Technology, Cambridge, Massachusetts, USA
Maor, Uri, Tel-Aviv University, Tel-Aviv, Israel
Mayer, Meinhard, University of California at Irvine, Irvine, California, USA
Mazeh, Zvi, Tel-Aviv University, Tel-Aviv, Israel
Merton, Robert K., Columbia University, New York, New York, USA
Michel, Louis, Institut des Hautes Études Scientifiques, France
Miller, Arthur I., University of Lowell, Lowell, Massachusetts, USA; Harvard University, Cambridge, Massachusetts, USA
Milgrom, Mordechai, Weizmann Institute, Rehovot, Israel
Møller, Christian, Royal Danish Academy of Sciences & Letters, Denmark
Moshe, Moshe, Tel-Aviv University, Tel-Aviv, Israel
Nambu, Yoichiro, University of Chicago, Chicago, Illinois, USA
Ne'eman, Yuval, Tel-Aviv University, Tel-Aviv, Israel
Nilsson, Jan S., Institute of Theoretical Physics, Goteborg, Sweden
Novozhilov, Yuri, Leningrad & UNESCO
Nussinov, Shmuel, Tel-Aviv University, Tel-Aviv, Israel
Owen, David, Ben-Gurion University, Beer-Sheba, Israel
Pagels, Heinz, New York Academy of Sciences, New York, New York, USA
Pais, Abraham, Rockefeller University, New York, New York, USA

List of Participants

Patera, Jiri, University of Montreal, Montreal, Canada
Pati, Jogesh, University of Maryland, College Park, Maryland, USA
Peres, Asher, Technion, Haifa, Israel
Revel, Daniel, Weizmann Institute, Rehovot, Israel
Ron, Amiram, Technion, Haifa, Israel
Ronat, Elchanan, Weizmann Institute, Rehovot, Israel
Rosen, Joseph, Tel-Aviv University, Tel-Aviv, Israel
Rosen, Nathan, Technion, Haifa, Israel
Rotenstreich, Nathan, Hebrew University, Jerusalem, Israel
Rubinstein, Hector, Weizmann Institute, Rehovot, Israel
Salaman, Esther, Highgate, London, England
Sambursky, Shmuel, Hebrew University, Jerusalem, Israel
Schapiro, Meyer, Columbia University, New York, New York, USA
Schulman, Larry, Technion, Haifa, Israel
Scholem, Gershon, Hebrew University, Jerusalem, Israel
Schwarz, Boris, Queens College, CUNY, Flushing, New York, USA
Schwimmer, Adam, Weizmann Institute, Rehovot, Israel
Sciama, Dennis, Oxford University, Oxford, England
Sen, Ratindra Nath, Ben-Gurion University, Beer-Sheba, Israel
Shaham, Jacob, Hebrew University, Jerusalem, Israel
Shapira, Arieh, Weizmann Institute, Rehovot, Israel
Shor, Amir, Tel-Aviv University, Tel-Aviv, Israel
Sijacki, Djordje, University of Koln, Federal Republic of Germany
Slater, Joseph E., Aspen Institute for Humanistic Studies, New York, New York, USA
Smilansky, Uzy, Weizmann Institute, Rehovot, Israel
Sosnowski, Leonard, International Union of Pure and Applied Physics
Stachel, John, The Institute for Advanced Study, Princeton, New Jersey, USA
Stern, Fritz, Columbia University, New York, New York, USA
Straus, Ernst G., University of California, Los Angeles, California, USA
Tal, Uriel, Tel-Aviv University, Tel-Aviv, Israel
Talmi, Igal, Weizmann Institute, Rehovot, Israel
Tauber, Gerald, Tel-Aviv University, Tel-Aviv, Israel
't Hooft, Gerald, State University of Utrecht, Utrecht, The Netherlands
Tserruya, Yizhak, Weizmann Institute, Rehovot, Israel
Wain, Ralph L., The Royal Society, London, England
Weinberg, Steven, Harvard University, Cambridge, Massachusetts, USA
Wheeler, John A., The University of Texas at Austin, Austin, Texas, USA
Woolf, Harry, The Institute for Advanced Study, Princeton, New Jersey, USA
Yang, Chen N., State University of New York, Stony Brook, New York, USA
Yavin, Avivi, Tel-Aviv University, Tel-Aviv, Israel
Yankielovitz, Shimon, Tel-Aviv University, Tel-Aviv, Israel
Yekutieli, Gideon, Weizmann Institute, Rehovot, Israel
Wager, Zeev, Weizmann Institute, Rehovot, Israel
Zak, Joshua, University of Michigan, Ann Arbor, Michigan, USA; Technion, Haifa, Israel

Indexes

Author Index

Numbers in parentheses indicate the numbers of the references when these are cited in the text without the names of the authors.

Numbers set in *italics* designate the page numbers on which the complete literature citation is given.

Abbott, L. F., 171(2), *183*
Achiman, Y., 32(35), *36*, 200(1), 209(14), *217*
Akulov, V. P., 83, 86(7), *96*, 100(4), *112*
Al'tshuler, B. L., 74(16), *76*
Arafune, J., 122(22), *126*
Arnowitt, R., 42(2), *58*, 88, 95, *96, 97*
Atiyah, M. F., 31(27), *35*
Aubert, J. J., xxvIII(26), *xxxi*, 118(5), *126*
Augustine, J. E., xxviii(26), *xxxi*, 118(5), *126*

Ball, R. C., 235(33), *253*
Barbieri, R., 179(18), *183*, 187(8), *195*
Bardeen, J., 46(28), *58*
Barnett, R. M., 171(2), *183*
Barr, S., 216(26), *218*
Bars, I., 27(9), *35*
Behrends, F. A., 92(15), *96*
Bekenstein, J. D., 43(10–11), 44(13–15), 44(17–18), 45(19–21), 46(27), 47(31), 48(34), 48(36), 49(37), 51(42), 52(45), 52(47), 53(48), 55(51), 56(52–53), *58, 59*
Belavin, A. A., 10, *11*, 33, *36*, 131(6), *133*, 137(2), *137*
Beltran-Lopes, V., 72(7), *75*
Benvenuti, A., xxvii(21), *xxxi*
Bergman, P. G., 42(2), *58*
Bjorken, J. D., 221(2), 222(5), 235(32), 244(51), *252, 253, 254*
Boulware, D. G., 47(29), *58*, 209(11), *217*
Branco, G. C., 191(13) *195*
Brans, C., 69, *75*
Brezin, E., 130(4), 131(5), *133*
Brill, D. R., 75(19), *76*
Brink, L., 88, *96*
Brout, R., xxvii, *xxxi*, 222(3), *252*
Buras, A. J., 201(2), 211(17), 212, *217, 218*
Burhop, E. H. S., xxix(27), *xxxi*
Burlankov, D. E., 33(46), *36*

Cabibbo, N., 189(9), 191(15), *195, 196*
Callan, C., 120(15), *126*, 137(2–3), 137(5), *137*
Candelas, P., 47(30), *58*

Carter, B., 46(28), *58*
Chang, N., 233(26), 237(36), *253*
Chern, S. S., 29(23), *35*
Chodos, A., 137(4), *137*
Christodoulou, D., 43, 43(12), *58*
Churilov, S. M., 43(10), *58*
Clark, R. W., xvi(4), xxv(12), *xxx, xxxi*
Coleman, S., 185(3), *195*, 206, *217*
Collins, J. C., 245(54), *254*
Cowan, C. L., Jr., 213(20), *218*
Cremmer, E., 25, 33(44), 34(47–48), *35, 36*, 93(17), 94, 94(19), *96*
Crombrugghe de, M. A., 179(17), 180(19), 181(22), 182(24), *183*, 191(13), *195*
Crouch, M. F., 213(20), *218*
Crutchfield, W. Y., 131(9), *133*
Cutts, D., 243(48), *254*

Das, A., 94(19), *96*, 233(26), 237(36), *253*
Dashen, R., 137(2–3), 137(5), *137*
Davidson, W., 64, 75, 226(16), 234(29), *252, 253*
de Crombrugghe, M. A., see Crombrugghe de, M. A.
Dehnen, H., 74, *76*
Derman, E., 179(18), *183*
Deser, S., 28(16), *35*, 42(2), *58*, 89(11), 95(22), 95(25), 95(27), 95(30), *96, 97*, 100(2), 106(11), *112*, 209(11), *217*
DeSitter, W., 27, *35*, 74, *75*
DeWitt, B. S., xxvii, *xxxi*, 42(2), *58*, 100(2), 106(11), *112*
Diff, J., 29(20), *35*
Dimopoulos, S., 33(38), 33(42), *36*, 213(22), 214, *218*, 247(61–62), *254*
Dirac, P. A. M., 10, *11*, 42(2), *58*
Dolan, L., 244(52) *254*
Drever, R. W. P., 72(7), *75*
Drinfeld, V. G., 31(27), *35*
Dyson, F., 115(3), *117*

Ebrahim, A., 179(18), *183*, 191(13), *195*
Einstein, A., xv(1), xvi(2–3), *xxx*, 3(1), 9, *11*, 27, 28, *35*, 42(1), *58*, 63(1), 64, 74(9), *75*, 101(7), *122*, 221(1), *252*

Eisenhart, L. P., 27(14), *35*
Elias, V., 174(6), 175(11), 177(13), *183*, 228, 231(21), 231(23-24), 241(41), *252, 253*
Ellis, J., 190(11), *195*, 213(22), *218*, 247(61-62), *254*
Englert, F., xxvii, *xxxi*, 222(3), *252*

Fairchild, E. E., 100(2), 106(11), *112*
Fang, J., 92(15), *96*
Federman, P., 144(11), *153*
Feinberg, G., 213(20), *218*, 231(22), *253*
Ferber, A., 88, *96*
Ferrara, S., 88, 95(20), 95(27-28), *96, 97*
Feynman, R. P., xxvi, xxvii, *xxxi*, 118(4), *126*, 209(11), *217*
Fischler, M., 94(19), *96*
Flaherty, E. J., 29(23), *35*
Floyd, M., 43(8), 43(12), 45(24), *58*
Fock, V., 8, *11*
Fradkin, E. S., 89(12), *96*
Freedman, B., 245(54), *254*
Freedman, D. Z., 88(10), *96*, 185(1-2), *195*
Frere, J. M., 191(13), *195*
Freudenthal, H., 26(7), *35*
Freund, P. G. O., 88, *96*, 118(3), 119(10), *126*
Fritzsch, H., 31, *35*, 129(1-2), *133*, 179(17), 180(19), 181(22), 182(24), *183*, 191, 192(17), 193(18), *195, 196*, 200(1), 209(14), *217*, 225(15), 227(17), *256*
Fronsdal, C., 92(15), *96*

Gaillard, M. K., 190(11), *195*, 213(22), *218*, 247(61-62), *254*
Gates, J., 88, *96*
Gatto, R., 179(18), *183*, 187(8), 191(15), *195, 196*
Gell-Mann, M., xxvi, xxvii, *xxxi*, 25(1), *35*, 33(43), *36*, 92(14), *96*, 115(1-2), *116*, 118, *126*, 129(1-2), *133*, 185(1-2), 186(5), *195*, 225(15), 234(27), *252, 253*
Georgi, H., 31, 33(39-41), *35, 36*, 142(7), 143(10), 144, *153*, 175(8), 175(10), 176(12), 179(18), *183*, 191, 195, *195, 196*, 200(1), 201, 201(3), 202(4), 205, 209(14), 211, *217*, *218*, 221(2), 222(5), 223(7), 224(11), 225(14), 227(17), 227(19), 244(51), *252, 254*
Gerlach, U., 47(29), 53, *58, 59*
Gervais, J. L., 120(20), *126*
Gildener, E., 206(7), 206, *217*
Gilman, R. C., 74, *76*
Glashow, S. L., xxvii, *xxxi*, 31, *35*, 142(7), 143(10), 144, *153*, 175(8), 175(10), 179(18), *183*, 185(1-2), 186(5), 186(7), 191, 193, *195, 196*, 200(1), 209(14), *217*, 221(2), 222(3), 222(5), 225(14), 244(51), *252, 254*
Gödel, K., 74, *75*
Goldberg, H., xxvii, *xxxi*, 115(2), *116*
Goldhaber, G., xxix, *xxxi*
Goldhaber, M., 213(20), *218*
Goldman, T. J., 205, 212(19), *217, 218*, 233(26), *253*
Golfand, Y. A., 83, *96*
Gomberoff, L., 115(4), *116*
Goodman, J. A., 243(40), *254*
Greenberg, O. W., xxviii, *xxxi*, 115(5), *117*, 118(2), *126*, 143(9), *153*

Grisaru, M. T., 86(5), 95(24), 95(26), *96, 97*
Gross, D., xxviii, *xxxi*, 116(7), *116*, 118(6), *126*, 130(3), *133*, 137(2), *137*, 137(3), 137(5), 223(8), *252*
Günaydin, M., 26(6), 27(9), 29(24-25), *35*
Guralnik, G. S., xxvii, *xxxi*, 222(3), *252*
Gürsey, F., 26(7), 28(21), 29(24-25), 30(26), 31(28), 32(34-36), *35, 36*, 115(3), *116*, 185(1-2), *195*, 200(1), *217*, 209(14), 227(18), *252*

Haag, R., 93(16), *96*
Hagen, C. R., xxvii, *xxxi*, 222(3), *252*
Hagiwara, T., 179(18), *183*, 191(13), *195*
Han, M. Y., xxviii, *xxxi*, 115(5), *116*, 118(3), 119(10), *126*, 129(1-2), *133*
Harari, H., 158(1), *158*, 171(3), 174(5), 175(9-10), 179(16), 179(18), 180(20), 182(23), *183*, 190(11), 191(13), *195*
Hartle, J. B., 47(29), 49, *58, 59*, 75(19), *76*
Hasert, F. J., xxvii, *xxxi*
Haut, H., 179(18), *183*, 191(13), *195*
Hawking, S. W., 43(5), 43(8), 43(12), 45, 46(26), 46(28), 47(29), 48(35), 49, *58, 59*, 91, *96*
Hayashi, K., 101(7), *112*
Hehl, F. W., 104(10), 107(12), 108(13-14), 109(16-17), *112*
Helgason, S., 25(3), 32(32-33), *35*
Herb, S. W., 118(5), *126*, 142(4), *153*
Heyde von der, P., 104(10), 107(12), 108(13), 109(16-17), *112*
Higbie, J. H., 74, *76*
Higgs, P. W., xxvii, *xxxi*, 222(3), *252*
Hoffmann, B., xxiv(9), *xxx*, 64, *75*
Hogan, C. E., 119(11), *126*
Hönl, H., 74, *76*
Hooft 't, G., *see* 't Hooft, G.
Howe, P. S., 88, *96*
Hughes, V. W., 72, *75*

Ignatiev, A. Y., 213(22), *218*, 247(61-62), *254*
Iliopoulos, J., xxvii, *xxxi*, 86(6), *96*, 222(3), *252*
Infeld, L., 64, *75*
Innes, W. R., 142(4), *153*
Isaacson, R. A., 75(19), *76*
Isham, C. J., 101(6), *112*
Ishihara, S., 28(20), *35*

Jackiw, R., 137(2), *137*, 191(15), *196*, 244(52), *254*
Janah, A., 235(31), *253*
Jarlskog, C., 212(18), *218*
Jaynes, E. T., 47(32), 48(33), *59*
Johnson, K., 101(5), *112*
Julia, B., 25, 33(44), 34(47), *35, 36*, 93(17), 94, *96*

Kac, V. G., 26(8), *35*
Kähler, E., 28, *35*
Kay, J., 95(25), 95(27), 95(30), *97*
Kerlick, G. D., 109(17), *112*
Kibble, T. W. B., xxvii, *xxxi*, 222(3), *252*
Kim, J. K., 147(16), *154*

Author Index

Kirzhnits, D. A., 244(52), *254*
Kitazoe, T., 179(17), 180(19), 181(22), 182(24), *183*, 191(13), *195*
Kobayashi, M., 178(15), *183*, 190, *195*
Kobayashi, S., 29(22), *35*
Kodaira, K., 28(19), *35*
Koestler, A., 4(2), *11*
Kogut, J., 129(1-2), *133*
Kokkedee, J. J. J., 115(3), *117*

Lanczos, C., xxv(12), *xxxi*
Langacker, P., 234(27), *253*
La Rue, G. S., 119(8), *126*
Lederman, L. M., xxx, *xxxi*
Lee, T. D., 209(12), *217*
Lee, H. K., 147(16), *154*
LeGuillou, J. C., 130(4), 131(5), *133*
Leutwyler, H., 129(1-2), *133*, 225(15), *252*
Likhtman, E. P., 83, *96*
Linde, A. D., 244(52), *254*
Lipatov, L. N., 130(4), 131(5), *133*
Lipkin, H. J., 129(1-2), *133*, 140(1), 141(2-3), 143(8), 144(13), 145(14), 147(15), 147(17), 148(18), *153, 154*
London, F., 8, *11*
Lopuszanski, J. T., 93(16), *96*
Lord, E. A., 109(17), *112*
Lynden-Bell, D., 74(16), *76*
Lynn, B., 231(22), *253*

Machacek, M., 212(18), *218*
MacLerran, L., *see* McLerran, L.
Maheu, R., xxv(12), *xxxi*
Maiani, L., xxvii, *xxxi*, 191(15), *196*, 222(3), *252*
Mandelstam, S., 120(16), *126*
Manin, Yu. I., 31(27), *35*
Marchio, G., 187(8), *195*
Marciano, W., 136(1), *137*, 205, 212(19), *217, 218*
Marshak, R., xxvi, *xxxi*
Maskawa, K., 178(15), *183*, 190, *195*
McLerran, L., 245(54), *254*
Meisels, A., 49(37), 52(44-45), 52(47), *59*
Mills, R. L., xxvi(14), *xxxi*, 129(1-2), *133*
Minkowski, P., 31, *35*, 200(1), 209(14), *217*, 227(17), *252*
Misner, C. W., 42(2), 42(4), 43(9), 45(23), *58*
Miyamoto, Y., 118(3), 119(10), *126*
Mohapatra, R. N., 173(4), *183*, 191(13), 192(15), *195, 196*, 225(15), 247, *252, 253, 254*
Morpurgo, G., 119, *126*
Morrow, J., 28(19), *35*
Moufang, R., 26, *35*

Nambu, Y., xxviii, *xxxi*, 10(14), *11*, 115(5), *116*, 118(3), 119(10), 120(17-18), 120(20), *126*, 129(1-2), *133*
Nanopoulos, D. V., 179(18), *183*, 190(11), 195, *195, 196*, 200(1), 209(14), 213(22), 216(26-27), *217, 218*, 247(61-62), *254*
Nath, P., 88, 95, *96*, 97
Ne'eman, Y., xvi(5-6), xxvii, *xxx, xxxi*, 1(1-2), 2, 88, 96, 100(3-4), 102(8-9), 109(17), 110(18-20), 111(21-22), *112*, 115(1-2), *116*, 158(2-3), *158*, 185(1-2), *195*
Neveu, A., 120(20), *126*
Nieuwenhuizen, P. van, 80(1), *80*, 88, 89(12), 95(22-24), 95(27-29), *96*, 97, 100(2), 106(11), *112*
Nitsch, J., 108(14), 109(16), *112*

Oakes, R. J., 186(5), *195*
Ogievetsky, V., 88, *96*
Okubo, S., 174(5), *183*
Okun, L. B., 244(50), *254*
Oszvath, I., 74, 74(13), *75*
Özer, M., 235(31), *253*

Page, D., 57(55), *59*
Pagels, H., 136(1), *137*, 192(16), *196*
Pakvasa, S., 179(18), *183*, 191(13), *195*
Panangaden, P., 49, *59*
Papastamatiou, N. J., 213(22), *218*
Parker, L., 48(35), *59*, 213(22), *218*
Parisi, G., 120(17), *126*, 130(4), 131(5), 131(11), *133*
Pati, J. C., 33(39), *36*, 119, *126*, 173(4), 174(4), 175(7-8), 175(10-11), 177(13), 185(1-2), *183, 195*, 200(1), 209, 210, *217, 218*, 222(2), 222(5), 223(6), 223(9-10), 225(14-15), 226(16), 228, 231(21), 231(23), 232(25), 234(28-29), 235(30), 235(31-32), 236(34), 237(37-38), 238(38-39), 243(46), 244(51), 245(57), 246(58-59), 247, 250(63), *252, 253, 254*
Pendelton, H., 86(5), *96*
Penrose, R., 43, 43(8), 43(12), 44(16), *58*
Perelomov, A. M., 28(21), *35*
Perez-Mercades, J., 233(26), 237(36), *253*
Perl, M., xxx, *xxxi*
Perry, M. J., 245(54), *254*
Piron, C., 26(6), *35*
Politzer, H. D., xxviii, *xxxi*, 116(7), *116*, 118(7), *126*, 130(3), *133*, 223(8), *252*
Polonyi, J., 94(19), *96*
Polyakov, A. M., 120(15-16), *126*

Quinn, H. R., 33(39-41), *36*, 176(12), *183*, 201, 201(3), 202(4), 205, 211, *217, 218*, 223(7), 224(11), 227(19), *252*
Quigg, C. 142, *153*

Radicati, L., 115(3), *116*
Rajpoot, S., 231(24), *253*
Ramond, P., 32(34), 32(36), *35, 36*, 200(1), 209(14), *217*, 227(18), 234(27), *252, 253*
Rebbi, C., 121(21), *126*, 137(2), *137*
Regge, T., 88, *96*, 102(9), *112*
Reines, F., 213, *218*
Reinhardt, M., 73, *75*
Renner, B., 186(5), *195*
Robinson, H. G., 72(7), *75*
Roček, M., 94(19), *96*
Roman, J. C., 88, *96*
Rosen, N., xxv(11), *xxxi*, 71(6), *75*

Rosner, J. L., 142, *153*
Ross, D. A., 205, *217*, 233(26), *253*
Rubinstein, H. R., 144(11), 147(15), 148(18), *153, 154*
Ruegg, H., 26(6), *35*
Ruffini, R., 43(7), 43(12), *58*
Rujula de, A., 142(7), 143(10), 144, *153*, 179(18), *183*, 191, *195*
Rumpf, H., 109(15), *112*

Sakakibara, S., 237(37), 243(46), 250(63), *253, 254*
Salam, A., xxvi, *xxxi*, 9(10), 9, *11*, 33(39), *36*, 85, 85(4), 86(7), *96*, 101(6), *112*, 118(6), 119, *126*, 171(1), 173(4), 174(6), 175(7-8), 175(10-11), 177(13), *183*, 185(1-2), *195*, 200(1), 209, 209(14), 210, *217, 218*, 221(2), 222(3), 222, 222(5), 223(6), 223(9-10), 225(14), 228, 231(21), 231(23), 232(25), 234(28), 235(30), 235(32), 237(37), 243(46), 244(51), 245(57), 246(59), 250(63), *252, 253, 254*
Sartori, G., 187(8), 191(15), *195, 196*
Schachinger, L., 142(6-7), *153*
Schafer, R. D., 26(5), *35*
Scherk, J., 34(48), *36*, 93(17), 94(20), *96, 97*
Scherrer, W., 101(7), *112*
Schnitzer, H. J., 191(15), *196*
Schrock, R. E., 190, *195*
Schücking, E. L., 74, 74(13), *75*
Schwarz, J. H., 34(48), *36*, 94(20), *97*
Schweizer, M., 108(14), *112*
Schwinger, 222(3), *252*
Sciama, D. W., 47(30), *58*, 74, *76*
Segrè, G., 179(18), *183*, 191(15), *196*, 216(26), 234(27), *253*
Senjanovic, G., 191(13), *195*
Sezgin, E., 80(1), *80*
Shafi, Q., 192(16), *196*, 200(1), 209(14), *217*
Shapiro, J., 88, *96*
Shaw, R., xxvi(14), *xxxi*, 129(1-2), *133*
Sherry, T. N., 111(21-22), *112*
Shirafuji, T., 101(7), *112*
Shupe, M. A., 158(1), *158*
Sidhu, D. P., 192(16), *196*
Siegel, W., 88, *96*
Šijački, D., 102(8), 110(18), 110(20), 111(21-22), *112*
Sikivie, P., 32(34), 32(36), *35, 36*, 200(1), 209(14), *217*, 227(18), *252*
Sitter, W. de, *see* DeSitter, W.
Slansky, R., 234(27), *253*
Sohnius, M., 93(16), *96*
Sokachev, E., 88, *96*
Soroka, V. A., 88, *96*, 100(4), *112*
Starobinski, A. A., 43(10), *58*
Stech, B., 32(35), *36*, 200(1), 209(14), *217*
Steigman, G., 213(21), *218*
Stelle, K., 95(25), 95(27), 95(29-30), *97*, 100(2), 106(11), *112*
Strathdee, J., 85, 85(4), 86(7), *96*, 101(6), *112*, 223(10), *252*
Straumann, N., 108(14), *112*
Strocchi, F., 179(18), *183*, 187(8), *195*
Sucher, J., 231(22), *253*

Sudarshan, E. C. G., xxvi, *xxxi*
Sugawara, H., 179(18), 181(13), *183, 195*
Susskind, L., 33, 33(42), *36*, 129(1-2), *133*, 208(10), 213(22), 214, *217, 218*, 247(61-62), *254*

Talmi, I., 144(11), *153*
Tanaka, K., 179(17), 180(19), 181(22), 182(24), *183*, 191(13), *195*
Taub, A. H., 75(20), *76*
Tavkhelidze, A., 118(3), 119(10), *126*, 140(1), *153*
Taylor, J. G., 88, *96*
Thierry-Mieg, J., 1(1-2), *2*, 100(4), 102(9), *112*
Thirring, H., 69, *75*
Thompson, S. P., 5(3), *11*
't Hooft, G., xxviii(20), *xxxi*, 95(21-22), *97*, 116, *117*, 120, 122, *126*, 131(7-8), 131(10), 132(12-13), *133*, 137(2), *137*
Thorne, K., 42(4), *58*
Tolman, R. C., 42(3), *58*
Tomboulis, E., 95(26), *97*
Tonin, M., 191(15), *196*
Toussaint, B., 213(22), 214, 214(23), *218*, 247(61-62), *254*
Townsend, P. 89(12), *96*
Treiman, S. B., 190, *195*
Tsao, H. S., 100(2), 106(11), *112*
Tucker, R. W., 88, *96*
Tze, C., 28(21), 31(28), *35*

Unruh, W., 45(24), *58*
Utiyama, R., 100(2), 106(11), *112*

Van Nieuwenhuizen, P., *see* Nieuwenhuizen, P. van
Vasiliev, M. A., 89(12), *96*
Veltman, M., 95(21-22), *97*
Vermaseren, J., 95(24), *97*
Vinciarelli, P., 225(15), *252*
Vidal, R., 243(48), *254*
Volkov, D. V., 83, 86(7), 88, *96*, 100(4), *112*
Von der Heyde, P., *see* Heyde, P. von der

Wald, R. M., 47(29), 48(35), 49, 50(41), *58, 59*
Wang, L., 190, *195*
Ward, J. C., 222(3), *252*
Waylen, P. C., 74, *76*
Weil, A., 10(16), *11*
Weinberg, S., xxvi, *xxxi*, 9, 9(12), *11*, 33(39-42), *36*, 118(6), 119(13), *126*, 129(1-2), *133*, 171(1), 173(4), 176(12), 179(17), 180(19), 181(22), 182(24), *183*, 185(1-4), 186, 186(5), 186(7), 191, 191(12), 193, *195*, *196*, 201(2-3), 202(4), 205, 206, 206(7), 208(10), 209(11), 209(15), 211, 213(22), 214, 214(24), 216(26-27), *217, 218*, 222, 222(3), 223(7), 224(11), 225(15), 227(19), 244(52), 247(61-62), *252, 254*
Weitzenböck, R., 101(7), *112*
Weldon, H. A., 179(18), *183*, 191(15), *196*, 216(26), *218*, 234(27), *253*
Wess, J., 83, 88, *96*, 100(4), *112*
West, P. C., 95(29), *97*
Wetterich, C., 192(16), *196*

Author Index

Weyers, J., 179(18), *183*, 191(13), *195*
Weyl, H., xxvi(13), *xxxi*
Wheeler, J. A., 42(2), 42(4), 54(50), *58, 59*, 74, 75(21), *76*, 91, *96*
Wilczek, F., xxviii, *xxxi*, 116(7), *116*, 118(6), *126*, 130(3), *133*, 177(14), 179(17), 180(19), 181(22), 182(24), *183*, 191, *195*, 223(8), *252*
Wilson, K. G., 120, *126*, 129(1-2), *133*
Witt, B. S., de, *see* DeWitt, B. S.
Witten, E., 147(16), *154*
Wolf, J. A., 27(12), 32, *35, 36*
Wu, C. C., 95(23), *97*
Wu, T. T., 10(18), *11*
Wyler, D., 191(13), *195*

Yang, C. N., xxvi(14), *xxxi*, 8(6-8), 10(14-15), 10(18), *11*, 33, *36*, 100(2), 106(11), *112*, 129(1-2), *133*, 209(12), *217*
Yndurain, F. J., 212(18), *218*
Yoshimura, M., 33(42), *36*, 213(22), *218*, 247(61-62), *254*

Zaycoff, R., 101(7), *112*
Zee, A., 179(17), 180(19), 181(22), 182(24), *183*, 191, *195*
Zel'dovich, Y. B., 43(9), 45(23), 51(43), *58, 59*, 244(50), *254*
Zinn-Justin, J., 130(4), 131(5), *133*
Zumino, B., 28(16), *35*, 83, 86(6), 88, 89(11), *96*, 100(4), *112*
Zweig, G., xxvii(23), *xxxi*, 115(2), *116*, 118, *126*

Subject Index

Abrikosov flux tube, 120
affine geometries, 27, 80
affine gravity, 111
affine group, 101, 109, 113
Ampere's law, 3
asthenodynamics, 2, 99, 100, 157
asymptotic freedom, 200
Arnowitt Nath superRiemannian supergravity, 95

background radiation, 39, 40, 113, 213
bag, 101, 138, 244, 245
bandors, 110, 111
baryon-entropy ratio, 216, 217
baryon excess (surplus), 167, 216, 246, 260
baryon number (nonconservation), 39, 164, 167, 209, 210, 211, 213, 214, 222, 232, 234, 235, 237, 241, 244, 245, 246, 250
binary pulsars, 108
b-quark, 171
black hole, 39, 42, 43, 46, 47, 48, 49, 50, 52, 53, 54, 55, 57, 60
black hole entropy, 43, 44, 61
black hole-maximum mass, 57

Cabibbo angles, 2, 166, 178, 179, 182, 190, 191
charm, xxvii, xxix, xxx, 155, 156, 164, 224
charmonium, xxix, 142
Clifford algebras, 85
color, chiral, 176, 229, 230, 256
conformal group, 37, 38, 80
confinement, 118, 119, 120, 123, 124, 132, 140, 182, 251
constituent quarks, 140, 141
Coriolis field, 64, 69
cosmological constant Λ, 34, 74, 75, 80, 99
Coulomb's law, 3
$CP_n{}'$, 28
CP violation, 2, 33, 39, 166, 167, 178, 190, 216, 247, 251, 258

diffeomorphisms—*see* general coordinate transformations
distant parallelism—*see* teleparallelism
dual models, 101

E_6 grand unification group, 31, 32, 33, 170, 209, 227, 233
E_7 grand unification group, 209, 227, 233
E_7 of extended $N = 8$ supergravity, 25, 27, 33

Einstein's asymmetric theory, 16
Einstein equation, 73
Einstein spaces, 27
Einstein-Cartan theory, 89, 106, 109
Einstein-Kähler geometries, 28, 29, 31, 34
electroweak group, 171, 172, 173, 176, 179, 200, 227, 228
electroweak interaction, xxi, 173, 174, 176, 178, 210, 222
equivalence principle, 6, 7
ether, 5
exceptional groups, 25, 26, 27, 29, 30, 37
see also E_6, E_7
exceptional supergroups, 26, 27
expanding universe, 71, 214
extended supergravity, 92, 98, 100
see also extended $N = 8$ supergravity
extended supersymmetry, 90
extended $N = 8$ supergravity, 25, 27, 33, 93, 94, 258

Faddeev-Popov ghost fields, 1, 98, 158
Faraday's law, 3
fiber bundle, 1, 10, 23, 24, 37
flavors, number of, 130

$\overline{GA}$ (4,R) of affine gravity, 109, 113
GL (4R), 113
$\overline{GL}$ (4,R) of spinning bags, 110
$\overline{GL}$ (4,R) of affine gravity, 109
Gauss's law, 3
general coordinate transformations, 6, 24, 80, 100, 110
general covariance group—*see* general coordinate transformations
generations, 157, 168, 171, 172, 175, 177, 180, 181, 182, 202, 204
geometrization, xxii, 16, 20, 22, 27
gluons, xxviii, xxix, 132, 140, 151, 152, 155, 162, 163, 174, 182, 185, 201, 230, 231, 234, 235, 236, 243
gluons, charged color, 240
gluon jet, 174
Godel's universe, xxiii, 74
Goldstone fermions, 258
Goldstone-Nambu fields, 1
graded Lie algebra, 83, 98, 111, 158
grand unification, xxi, 157, 175, 176, 200, 201, 202, 204, 221, 222, 225, 228, 234, 251
grand unification scale, 205, 207, 209, 227, 228, 230
Grassmann algebras, 18, 85, 86
gravitational constant, 71

gravitational radiation, 108
gravitational red shift, 42
gravitational waves, 75
group manifold, 102

Han-Nambu model, 141, 148
HP_n, 28
high temperature, 10
horizontal symmetry, 168, 177
hypercharge, 224
hypercomplex numbers—*see* quaternions
hyperweak interaction, 208

instantons, 1, 24, 25, 28, 31, 34, 120, 131, 137
integrally charged quarks, 119, 133, 236, 237, 243, 244, 249, 250, 255
isospin, 224
isotopic spin, 8, 9

Kaluza-Klein theory, 17, 34, 93
Kerr black hole, 45
Kerr solution, 2, 53

lattice gauge theory, 120
lepton number, 224
lepton number nonconservation, 209, 210, 222, 232, 234, 235, 237, 250
logarithmic potential, 145
low-temperature, 9, 233

magnetic charge, 123
magnetic moments, 140, 142, 143, 144, 156
magnetic superconductivity, 132
Majorana mass, 260, 261
Malcev algebra, 29
manifield, 110
Maxwell's equations, 3, 5, 6, 264
Meissner effect, 132
merons, 1
monopoles, magnetic, 1, 10, 25, 37, 116, 120, 122, 123, 125
Moufang plane, 26
muon, xxvi, 119, 164

Nambu-Goldstone boson, 258, 260, 261, 262
neutral current, xxvii, 164, 174, 177, 186, 223, 235
nonintegrable phase factor, 8
nuclear disarmament, xvii, 265

octonions, 22, 26, 27, 29
$OSp(4/1)$, 25

parastatistics, 118
partons, 118, 146, 149, 151, 153, 163, 255
phase, 8, 179, 180
Planck mass, 2, 91, 209, 214, 217
positive definiteness of the gravitational field energy, 80
projective spaces, 25
proton decay, 167, 212, 213, 233, 241, 248, 249, 251

pseudo-Goldstone bosons, 208
pseudoparticle, 10
Psi meson, xxvii, xxviii, xxix, xxx

QCD, xxi, xxviii, 100, 115, 118, 119, 133, 136, 163, 164, 171, 210, 231, 251, 257, 259
QED, xxvi
quadratic gravitational lagrangians, 33, 106
quarkosynthesis, 243, 244, 245, 246, 247
quark search, 119
quantum electrodynamics, xxv, 24, 99
quantum gravity, 95
quaternions, 22, 25, 27, 28, 29, 30, 31

renormalization, xxv, 1, 25, 34, 47, 80, 86, 94, 99, 132, 165, 166
renormalization group, 203, 205, 206, 207, 230
Regge trajectories, xxvi, 101, 121, 135
$SA(4,R)$ of bags, 101
$SA(2,R)$ of strings, 101, 111
$\overline{SA}(4,R)$ of "spinning bags", 102, 103
$\sin^2 \Theta_w$, 33, 129, 158, 166, 171, 172, 173, 174, 176, 177, 182, 186, 205, 217, 222, 230
$SL(2,C)$, 102
$SL(2,C)$ of Lorentz transformations, 25
$SL(2,R)$, 101
$SL(2,R)$ of strings, 110
$\overline{SL}(3,R)$ of "spinning" bags, 111
$\overline{SL}(4,R)$, 101, 102
$\overline{SL}(4,R)$ of "spinning" bags, 102, 110, 111
$SO(N)$ of extended supergravity, 25, 90, 91, 93
$SO(8)$ of extended $N = 8$ supergravity, 25, 27, 33, 92, 93
$SO(10)$ grand unification groups, 31, 32, 33, 165, 175, 176, 209, 216, 226, 227, 228, 233, 260
$SO(n)$, 29, 80
$Sp(n)$, 26, 29
spin 3/2 fields, 20, 80, 82, 88, 89, 91, 92, 100, 110, 258
$SU(2)$ of isospin, xxvi, 9
$SU(2) \times U(1)$, xxvi, xxvii, 25, 165, 172, 185, 193, 200, 201, 206, 207, 210, 225, 227, 257, 259
$SU(2)_L \times SU(2)_R \times U(1)$, 174, 185, 192
$SU(2/1)$ of gauge asthenodynamics, 158
$SU(3)$ of color, xxviii, 25, 28, 29, 33, 80, 92, 115, 127, 157, 163, 165, 171, 174, 200, 201, 210, 224, 225, 228, 231, 234, 235, 259
$SU(3)$ Ne'eman—Gell-Mann classification group, xvii, xxvi, xxvii, 2, 115, 139, 143, 144, 150, 157, 190
$SU(5)$ grand unification group, 31, 32, 33, 157, 165, 167, 175, 176, 205, 209, 216, 227, 228, 233, 235, 257, 259, 260
$SU(6)$ grand unification group, 165
$SU(6)$ spin-unitary spin independence, 100, 115
$SU(8)$ of extended $N = 8$ supergravity, 25, 27, 33, 80, 93
$SU(11)$ grand unification group, 168
$SU(16)$ grand unification group, 226
$SU(n)$, 29

scale invariance, 7
scaling, 101, 115, 238
Schwarzschild solution, 2
seriality, 157
  see also generations
shear, 103, 105
sigma model, 30, 31
simple unification, 173, 174
spontaneous emission, 48, 49
stimulated emission, 43, 50, 51, 52, 54
strangeness, 164
string, xxvi, 80, 101, 110, 116, 120, 121, 125, 132
strong gravity, 101, 113
superaffine group, 80
superfields, 87
supergravity, 2, 18, 20, 25, 26, 27, 33, 86, 88, 90, 93, 94, 95, 96, 100, 113
supergroup, 158
superheavy mass, 33, 175, 241
superheavy particles, 208, 215, 217
superradiance, 43, 45, 50, 51
super-Higgs mechanism, 94
superspace, 26, 86, 88
supersymmetry, 18, 20, 80, 82, 83, 84, 85, 86, 87, 88, 90, 92, 94, 98, 100, 111
superunification, 200
symmetric group, 194
symmetric spaces, 25, 32, 37

tachyons, 80
tau lepton, xxx, 164
tau-neutrino, 171
Taub universe, 75
technicolor, 168
teleparallelism, 27, 80, 101
tensor-scalar theory, 17, 99
tetrads—*see* vierbeins
time-reversal invariance, 214, 247
t-quark, 32, 166, 168, 171
tunneling, 39, 131, 137
twistor, 18

unavoidability of singularities, 17
$U(1)$ of electromagnetism, 25, 92
upsilon meson, xxx, 142, 156

vierbeins, 28, 34, 89

Wilson loops, 120, 121
W mesons, xxvi, xxvii, 25, 171, 172, 185, 201, 208, 221, 223, 233, 239, 259

X mesons, 211, 215, 216, 232, 233, 241

Zionism, xvii, xxiv, xxv
Zweig rule, 138